I0406978

MANUAL OF LINEAR INTEGRATED CIRCUITS

MANUAL OF LINEAR INTEGRATED CIRCUITS

Operational Amplifiers and Analog ICs

Sol D. Prensky

Professional Writer

Associate Professor, Emeritus
Fairleigh Dickinson University

Reston Publishing Company, Inc.
Reston, Virginia 22090
A Prentice-Hall Company

Library of Congress Cataloging in Publication Data

Prensky, Sol D

Manual of linear integrated circuits.

Bibliography: p.

1. Linear integrated circuits. I. Title.

TK7874.P73 621.381'73 73–15979

ISBN 0–87909–466–4

Reston Publishing Company, Inc.

A Prentice-Hall Company

Box 547

Reston, Virginia 22090

10 9 8 7 6 5 4 3 2 1

Printed in the United States of America.

to my sons

Jimmy, Paul and Hank

and my grandchildren

Ellen Aviva, Kalon and Reuben

ACKNOWLEDGMENTS

A great deal of credit for supplying valuable data for newly-developed LIC models must go to the highly-informative Data Sheets and Applications Notes that are freely supplied by the manufacturers and authors, who are cited in the text footnotes and in the illustrative diagrams.

Considerable help should also be credited to Mr. Robert A. Hirschfeld (Lithic Systems) for topics in Chapters 2 (IC Design and LIC Arrays), 8 (Communication), 9 (Regulators), and 10 (Digital-Interface Circuits). My special thanks for splendid part-time secretarial help to Misses Barbara Shapiro, Hilary Sherry Kerman, Cindy Brooks and Sandy Syr. Finally, I wish to freely assure my wife Dinah that a thankful sigh of relief on her part, at the completion of this latest writing, is a richly deserved one, and her gracious forbearance is gratefully acknowledged.

CONTENTS

PREFACE

Integrated circuits have arrived and matured and their development has been a remarkably fast one. Over a period of only about ten years, both the digital and linear forms have won wide acceptance, and each branch has grown large and important enough to warrant major treatment in a separate book.

With regard to the linear branch of integrated circuits (LICs), their enthusiastic acceptance is due not only to the microminiaturization in size (mostly a minor factor), but, more importantly, to their superior performance. Following the original adoption of general-purpose and improved types of operational amplifiers (OP AMPS), there have been progressively successful introductions of a wide spectrum of important functions; they include *regulators*, *comparators*, *audio power amplifiers*, and a host of *consumer/communication applications*, in addition to *digital-interface circuits*. Hence, in light of the continuing trend for these LICs to displace more and more of the discrete transistor circuits, it becomes increasingly vital for the progressive engineers and technologists to become familiar with the practical as well as theoretical aspects of the LICs in their increasingly varied applications.

The manual supplements courses in basic transistor circuits and proceeds from there to supply the basic and applied device-information needed for upgrading electronic circuitry for professional engineers and technicians. The main concentration is aimed at the last two years of the four-year engineering-technologist level; but, with selective emphasis, it can easily be adapted for use in the first two years of technical institutes.

While the manual contains the essential information needed for a basic understanding of the linear (or analog) segment of the integrated-circuit field, the main objective is directed toward practical, ready-to-use information. The practical-manual aspect of the text has been enhanced by the generous use of *tables for performance-comparison*. These tables highlight the *outstanding features of specific IC types offered by a variety of manufacturers*. This approach contrasts with that of most other texts, where, on one hand, a basically general approach is used, or, on the other hand, a specific approach is employed which concentrates heavily on examples from a single manufacturer. Thus, the use here of numerous independent tables of salient performance information serves to alert the reader to specific innovative improvements from many different manufacturers.

Going beyond the information gathered together from varied manufacturers, the complex picture of diverse LIC specifications and models has been clarified by emphasizing *representative examples for each basic type and specific application circuits* for using them. The fairly "*standardized*" OP AMPS and other general-purpose types are handled in this way in the early chapters (5 through 10) to clearly identify them (along with a recognition of similar types that are "second-sources" of that type, even though bearing different type numbers from other manufacturers). The more special types are discussed later as Precision/Instrumentation types (in Chapter 11) and Specialized Applications (in Chapter 12). In between the circuit information, the practical means for *Breadboarding and Testing* the LIC devices are given in Chapter 6.

Further strengthening the practical approach, comprehensive listings are given in the Appendices: *Selection Guides for* OP AMPS in Appendix II; a *Cross-Reference List*, along with an identifying description, of hundreds of currently-available LICs in Appendix III; and Manufacturers' Addresses in Appendix IV. Also, sources for additional information from manufacturers' data and the literature are summarized in the bibliographical summary at the end of Chapter 12, to aid in keeping up-to-date in this active field.

For both the student and the practicing technical personnel the presentation of basic material, along with information on practical selection and application of the linear ICs, should be helpful in cutting through the maze of details and multitude of types that are characteristic of a developing art. It is most beneficial for the reader to arrive at making usable sense of the IC field, which is essentially a profound simplification in modern electronic circuitry.

SOL D. PRENSKY

1

INTRODUCTION TO LINEAR INTEGRATED CIRCUITS

1-1. GENERAL SIGNIFICANCE OF LINEAR INTEGRATED CIRCUITS

Integrated circuits (ICs) are designated as *linear* to distinguish them from the other category of *digital* integrated circuits. The linear integrated circuit (LIC) is regarded as belonging to the *analog type of circuit*, as opposed to the digital type. In a broad sense this classification separates the digital types, in which the active elements are primarily concerned with switching functions between on and off states, from the linear types, in which the circuits are mainly concerned with *smoothly varying and reasonably faithful amplification* of various analog-signal voltages. Although some circuits are a cross between the two, there is generally sufficient difference between the essential functions of each type to readily identify the two categories.

This text concentrates on linear ICs and on the wide variety of their analog functions. Included are all sorts of amplifiers, oscillators, detectors, active filters, and similar electronic functions. Between linear and digital circuits are comparators and regulators, and also the group of digital-interface circuits (such as sense amplifiers, line-drivers, and converters); these will be considered as belonging to the linear IC type.

1-2. DEVELOPMENT OF LINEAR INTEGRATED CIRCUITS

The linear group of ICs attained its importance and prominence as a group when it eventually became possible to provide, at a reasonably low cost, amplifiers of the *operational-amplifier type.* The development can be traced from the time (approximately 1964) when pairs of transistors, forming *integrated differential amplifiers,* were first fabricated on a single silicon chip. This step represented a significant advance over the use of discrete transistors in the basic circuit; it demonstrated the ability of the IC to greatly reduce the troublesome temperature dependence of discrete transistors, even when the separate transistors were laboriously matched (as discussed later). Since both transistors of the differential pair could be fabricated simultaneously on the same IC chip, it was possible, by this step alone, to make an improvement by whole orders of magnitude in alleviating the temperature-drift problem—from millivolts for the discrete transistor to just microvolts per degree Celsius for the integrated pair.

About a year later, another enormous step forward was taken with the introduction of the *integrated general-purpose operational amplifier.* This IC was able to take good advantage of the integrated differential-amplifier stage for the first stage of a multistage amplifier and was capable of open-loop gains well beyond 10,000—all in a very compact package. By the use of an external feedback resistor, the IC operational amplifier (of the 709[1] type, for example) became a low-cost and highly versatile amplifier that incorporated 9 transistors and 12 associated resistors in a conveniently small package. This relatively simple operational amplifier could easily perform many of the functions of discrete amplifiers and could do so at a lower cost and in a much more reliable and versatile form. The IC operational amplifier[2] offered designers a highly flexible tool that approached a *basic building block* around which a great number of desired circuits could be devised quite simply.

1-3. USING A SIMPLE LINEAR INTEGRATED CIRCUIT

The ease of use and the great utility of LICs can be illustrated by selecting a simple example in the form of a general-purpose operational amplifier (commonly called simply an OP AMP). Contrasted with the use of discrete components (such as transistors, resistors, and small capacitors) in a multistage amplifier, the use of an OP AMP can often make the design of a desired amplification function (either direct or alternating current) a straight-

[1] A popular IC type, characterized in Section 1-7.

[2] J. Gifford, "The Linear Integrated Circuit," Chapter 1 of *Designing with Linear Integrated Circuits,* J. Eimbinder, ed., John Wiley & Sons, Inc., New York, 1969.

forward operation; in many cases it would simply mean choosing values for two external resistors (R_f and R_i) to use with a general-purpose op amp, such as the very popular 741 type[1] illustrated in Fig. 1.1.

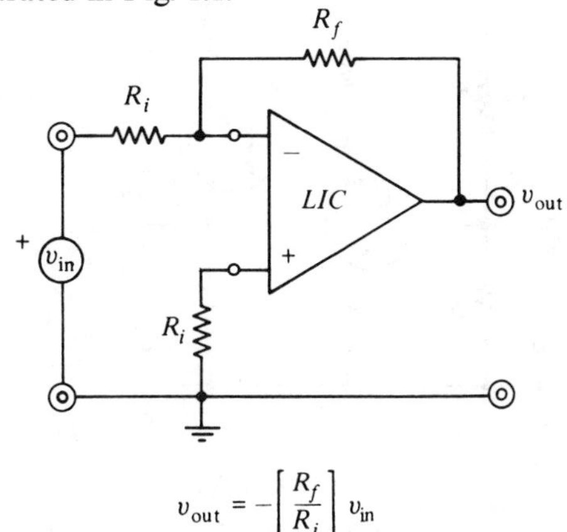

FIG. 1.1 Functional symbol diagram of a linear integrated circuit (LIC); the simple gain formula applies to a typical operational amplifier (OP AMP).

To gain a clear appreciation of the superior convenience and flexibility offered by the IC OP AMP compared to a traditional amplifier made up of discrete components, it is instructive to trace the steps needed in the case of both amplifiers to accomplish a typical amplification function.

Let us assume a project requiring a direct-coupled amplifier with a stable gain of, say, 500, where the input comes from a light sensor that provides a slowly varying dc signal ranging from 1 to 10 mV. (This would call for an output from the amplifier of $\frac{1}{2}$ to 5 V, as displayed on a dc voltmeter). In the next two paragraphs we can follow the necessary procedures to accomplish the same purpose in each case.

In the case of the *discrete amplifier*, we might choose the circuit of Fig. 1.2, calling for an NPN transistor (Q_1) followed by a PNP type (Q_2), arranged in the "compound" type of circuit, where the bias voltages can be handled more conveniently in this type of complementary-transistor connection. Making use of simplified design relations,[1] we would first establish the Q-point for the output stage to ensure that we stay within the linear operating range. With a supply voltage of around 20 V, this would call for a collector voltage of about half the supply, or approximately 10 V, for V_{CE2} at the Q-point, ensuring no distortion. Then, as suggested by Lenk,[2] we would choose proper values for the resistors; for example, we might choose R_{L_2} as 1 kΩ, requiring collector current of Q_2 to be about 10 mA. Proceeding from this, values of R_{E_2} and then of R_{L_1} and R_{E_1} are selected for the proper gain relationship of R_L/R_E for each

[1] Another very popular IC type with internal compensation, characterized in Section 1-7.

[2] J. D. Lenk, *Handbook of Simplified Solid-State Circuit Design*, Prentice-Hall, Inc., Englewood Cliffs, N.J., 1971.

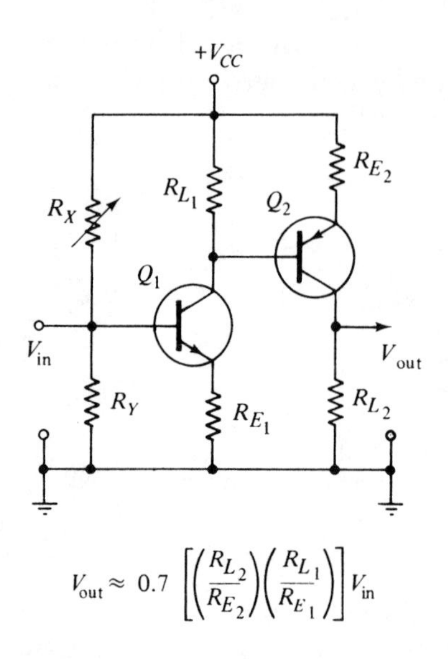

FIG. 1.2 Discrete form of direct-coupled amplifier: it requires calculation of proper values of at least 6 resistors to accomplish desired gain; compare this to a simple ratio of 2 resistors for the LIC in Fig. 1.1.

stage; then the selection of approximate values of R_x and R_y is made for obtaining proper bias currents. As an additional simplification, R_x is shown to be variable, so that it can be easily set to obtain the correct meter reading for collector current of Q_2; finally, a final adjustment of R_x is again made, if required, so that the full output range is obtained without distortion. It will be noted that even when a bit of cut-and-try process is used here to reduce unnecessarily precise calculations, the simplified procedure still involves working with six resistor values in order to accomplish the proper operation of the amplifier in producing a direct-coupled gain of about 500, together with a satisfactory output swing between $\frac{1}{2}$ and 5 V, as desired.

By contrast, when using the general-purpose IC OP AMP (shown in Fig. 1.1), the procedure involves only the selection of two resistors to obtain the proper ratio of $R_f/R_i = 500$. Thus R_f might be 500 kΩ, with R_i equal to 1 kΩ (the third resistor is generally made roughly equal to R_i). With these three resistors and the dual power supply (such as ± 15 V) connected, the amplifier circuit is ready for the signal, as the proper bias and stability conditions have already been taken care of by the internal design of the IC. Additionally, since the IC is capable of at least a ± 10 V output swing, the required output indication here (up to 5 V) is easily handled, without concern for any distortion arising from overdriving. A further advantage lies in the ease of changing the amount of gain, and also, when required, in easily obtaining higher input impedances, as will be shown later.

If we now add to the preceding evidence the pertinent fact that the IC, incorporating more than a dozen transistors, practically always costs less than an equivalent discrete circuit, we can clearly see the great potential of the linear IC in *reducing both complexity and expense* in a multitude of applications. *A note of caution* is in order here for those who might jump to the conclusion that the welcome simplicity in the use of the LIC now makes it unnecessary to study and understand the design considerations of discrete transistor circuits. It should be kept in mind that to make effective use of the linear IC in its many forms it is still necessary to thoroughly understand the basic transistor principles—knowledge needed to determine the gain, stability, frequency-response, and power-capability characteristics of any transistor circuit, be it in the discrete or in the integrated form.

Although our example is admittedly a limited case of simple amplifier function (since it has been selected here particularly for its simplicity), it is still a valid one for a great many practical applications. When more specific applications are needed, there will naturally be other amplifier elements to be considered regarding bandwidth, power output, and the like; each of these desired amplifier elements will be discussed later in the light of the great variety of LICs available, including not only various types of op amps, but many other linear IC types as well.

1-4. TYPES OF LINEAR-INTEGRATED-CIRCUIT FUNCTIONS

Linear ICs are designed to perform an extremely wide variety of analog functions. If we start with a very rough division of analog functions into the two main areas of *amplification* and *oscillation* (tentatively leaving the *switching* function to the digital ICs), we find that the analog functions call for many subdivisions of amplifier types. Citing just a few, for example, we have *low-level* direct-coupled amplifiers for instrumentation, *wide-band* amplifiers for general use, and *tuned amplifiers* (with associated *power amplifiers* for use in communication-type circuits).

In each of these cases, the circuit design can take different forms, as follows:

1. Using *discrete components*, in the traditional method.
2. Use of a *monolithic IC*, when both active and passive components are fashioned on a single chip. (See Fig. 1.3.)
3. Use of a *hybrid IC*, involving a combination of chips or processes in a single package. (See Fig. 1.4.)
4. A *combination* of any of these forms (generally called a *hybrid circuit*, as distinguished from a hybrid IC).

Although the most effective form, of course, depends upon the particular circuit application, the great convenience and potential versatility of the

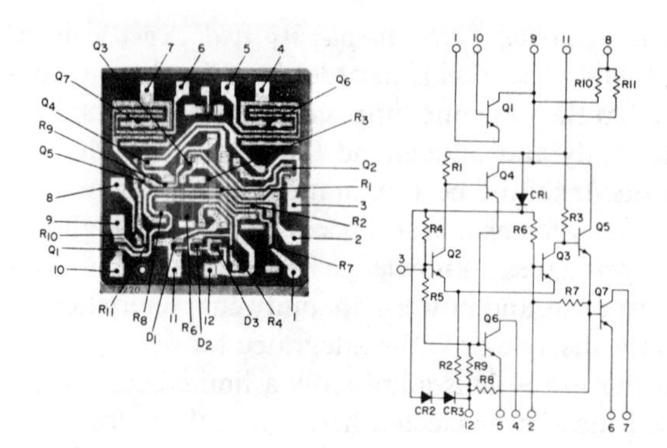

FIG. 1.3 Enlarged internal view of monolithic integrated circuits showing component location and circuit diagram (KD2115); actual size of IC chip is less than 1/16" square. (*RCA "Solid-State Hobby Circuits Manual"*)

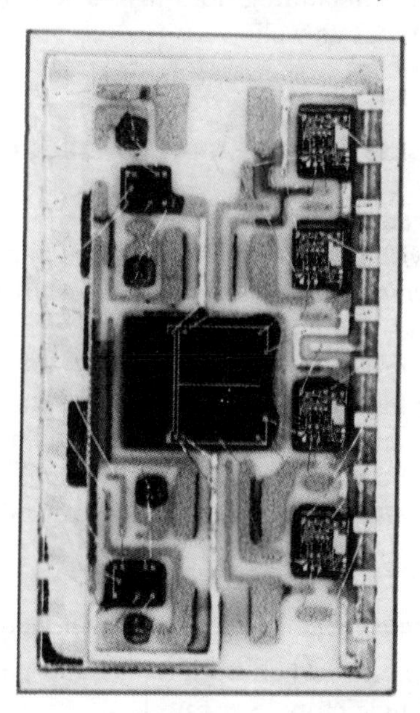

FIG. 1.4 Enlarged internal view of hybrid (modular) IC. (Courtesy of *Hybrid Systems Corp.*)

monolithic linear IC package have made it a preferred choice in those cases where it is possible to develop a circuit of general utility in a practical and economical form. As a result, a host of generally useful monolithic linear ICs has been made available in the variety of types discussed in Section 1-5. Hybrid ICs and combinations with discrete components are also exemplified later in those instances where more specific applications identify them as the most effective choice.

1-5. TYPES OF LINEAR INTEGRATED CIRCUITS

A first glance at any long list of presently available linear ICs reveals that the greatest number of them fall into the group of operational amplifiers (OP AMPS). There are also at least five other sufficiently numerous groups that warrant consideration in an initial classification. These six main groups may be classified as follows:

1. Operational amplifiers.
2. Comparators.
3. Audio amplifiers.
4. Wide-band and radio-frequency circuits.
5. Voltage regulators.
6. Digital-interface circuits.

The characteristics and applications of each of these groups (including their subdivisions) will be discussed, in turn, in separate chapters. It is enough, at this point, to appreciate the wide versatility of linear IC types available in the six groups, plus some others that are covered in Chapter 12. This text does not attempt to cover the equally numerous field of digital ICs widely used in computer and other pulse-circuit applications. However, the linear ICs that are employed to interface with digital circuits are covered in a separate chapter, as the sixth of the aforementioned groups.

1-6. MEDIUM-SCALE AND LARGE-SCALE INTEGRATION

The emergence of newer (and more complicated) IC types is directly related to the progressively larger number of elements that can be successfully integrated on the single and multiple chips that make up the IC packages. The previously mentioned 741 type, for example, involves the integration of 22 transistors and over 12 associated resistors and is called *medium-scale integration (MSI)*. Moreover, the tendency to include larger numbers continues as a development of *large-scale integration (LSI)*, especially in the digital field (as shown in Fig. 1.5). The designation of LSI has come to mean the integration of, at least, over 50 (and often over 100) active components in the resulting IC package. Presently, most linear ICs fall into the MSI class of integration.

FIG. 1.5 Application of large-scale integration (LSI): calculator-on-a-chip uses a single LSI chip for all arithmetic; the LSI package with its numerous transistor functions is less than 1″ square. *(Monroe Calculator Co., model 20)*

1-7. STANDARDIZATION TENDENCIES FOR LINEAR INTEGRATED CIRCUITS

Standardization of type numbers for semiconductor devices, in general, although highly desirable in theory, has always been a difficult goal to achieve in the highly competitive electronics industry—as evidenced by the fact that over 6,000 registered types of 2N transistors alone are presently available, to say nothing of the thousands of registered numbers of the diode–1N type. However, there is a more hopeful indication of a developing tendency in the linear IC field, in that a growing number of manufacturers are producing one another's popular types (or *second-sourcing* such ICs). This becomes apparent from the growing practice by manufacturers of retaining the type number of a particular IC type in their own type number, usually preceded by their own identifying initials. For example, the aforementioned 741 type of *internally-compensated* OP AMP can be obtained from various manufacturers under such numbers as μA**741** (originally Fairchild) and second-sourced as shown in part in Table 1-1.

TABLE 1-1

Similar (Second-Sourced) Type Numbers of Various Manufacturers*

μA**741** (Fairchild)	CA3**741** (RCA)
MC1**741** (Motorola)	S5**741** (Signetics); now μA**741**
LM**741** (National)	SN52**741** (Texas Instruments)

* A selected but more complete list of linear IC manufacturers (and addresses) is given in Appendix IV.

Many other examples of this second-sourcing process may be cited; for example, the dual OP AMP MC1**458**/**1558** (Motorola) is second-sourced as CA3**458**, CA3**558**; and super-beta OP AMP LM**108** (National) is second-sourced as SN55**108**. Other examples of types being standardized are summarized later.

It will be noted that these examples of this tendency toward standardization have included only six of the dozens of IC manufacturers. It would not be fair to conclude from this that many other firms do not produce important and innovative IC designs. A list of the more generally known *IC firms and their addresses is given in Appendix IV*, and many useful and significant *data sheets* and *application notes* can be obtained from them; such freely given information will be found to be of great help in rounding out the picture.

Within the limitations of this text, it is naturally necessary to limit our discussion to the typical examples that represent a given point, irrespective of a particular manufacturer. There are, of course, many other special and desirable types. Such information must be obtained from other sources, such as those given at the end of Sec. 12-4; a comprehensive compilation of LICs from practically all manufacturers, with periodically up-dated issues, is the *D.A.T.A.* reference cited there.

1-8. MAKING A PRACTICAL CHOICE OF AN INDUSTRY-STANDARD OP-AMP TYPE

A most practical way of establishing a firm base for working with the mass of presently available LIC models is given in the Appendices at the end of the book. Appendix II gives a *highly selective list for choosing a fairly standardized* OP-AMP type for a particular purpose, while a *comprehensive listing* for identifying *over 400 current LIC models* is given in the cross-reference list in Appendix III.

1-9. SEQUENCE OF PRESENTATION

In presenting the diversity of LICs, a logical progression will be followed, going from the basic conditions underlying the simpler transistor configurations in *arrays*, followed by the more complex functional IC devices.

Starting with general *physical conditions governing the integration of LIC arrays* (containing transistors and differential-amplifier stages, relatively uncommitted to a particular function), we proceed to the versatile *amplification functions of the op amps*, and then to the more *complex subsystems*, which may combine a number of special functions in the single LIC package.

2

GENERAL CONDITIONS GOVERNING LINEAR-INTEGRATED-CIRCUIT DESIGN

2-1. MONOLITHIC VERSUS DISCRETE COMPONENTS

Mass-produced monolithic integrated circuits can provide large cost savings when replacing conventional "discrete" electronic components such as transistors, diodes, resistors, and capacitors; and this cost advantage increases as the monolithic circuit is made more complex, since added components on an integrated circuit do not increase its cost of manufacture proportionately. Mass production of linear ICs relies upon the ability of LIC designers to use a basic chemical and photographic process originally intended for the making of silicon discrete transistors. Advances in this technique make it possible now to simulate the characteristics of the other active and passive components needed for complete circuit operation, to reproduce more than a single silicon transistor on the same "chip," and to provide an interconnection pattern—all forming the marvelously compact IC device.

Because the process is optimized for best transistor characteristics, other needed components are usually a practical compromise. In some respects, such as tolerance, breakdown voltage, and range of available values, monolithic components are inferior to inexpensive discrete components. For this reason, the earliest ICs to achieve wide usage were the less critical *digital-computer types*, as their on–off switching functions did not depend heavily upon accurately controlled components. Early attempts to build LICs simply as copies of existing discrete designs resulted in some still available older LIC types that were marginally producible with great care in controlling the monolithic process but that could never be manufactured so easily as to allow the necessary low pricing of today's LICs.

Modern LICs are the result of new circuit design techniques, developed over several years, that are less dependent upon the uncontrollable characteristics of the IC process and capitalize on the unique advantages of microscopic, photographically reproduced elements. Among these advantages are the *inherent precise matching* of identical adjacent elements; the improved frequency response allowed by microscopic, *low-capacitance interconnections* between elements; and the availability of *multiple-transistor structures*, not limited by discrete-transistor cost considerations. Moreover, unique elements have been developed, using simple photomask-drawing techniques as well as slight variations in the standard IC process, which have no parallel in discrete components, and, in some cases, which can only become useful when the inherent matching of a monolithic circuit is available. These elements include the *super-beta transistor* and *multiple structures*, such as the multiple-emitter NPN transistor and multiple emitter–multiple collector PNP transistors. Additionally, *compatible field-effect transistors* and other new structures are continually evolving from new LIC requirements.

2-2. MONOLITHIC NPN TRANSISTORS

The technology of monolithic LICs has evolved from the development of modern epitaxial, diffused, photochemically fabricated discrete transistors. Such devices are produced by the successive localized placement in a single-crystal silicon "wafer" of P-type and N-type material to form collector, base, and emitter regions, with PN junctions between them. A comparison of a discrete and monolithic NPN transistor structure, Fig. 2.1, shows that the structures are essentially the same, having an insulating layer of glass material above through which contact holes are etched and aluminum connections are made to external circuitry. The monolithic transistor, however, is surrounded on the sides and bottom by a P-type region, as is its neighbor transistor. Discrete transistors are physically separated, after being produced thousands at a time, on a single wafer. To achieve electrical separation in the IC, a potential is applied to the surrounding P-type "isolation" region, which reverse-biases the PN diode junction between it and each transistor's collector region, giving an effective open circuit between adjacent transistors. Similar isolation is provided between all other elements on the IC.

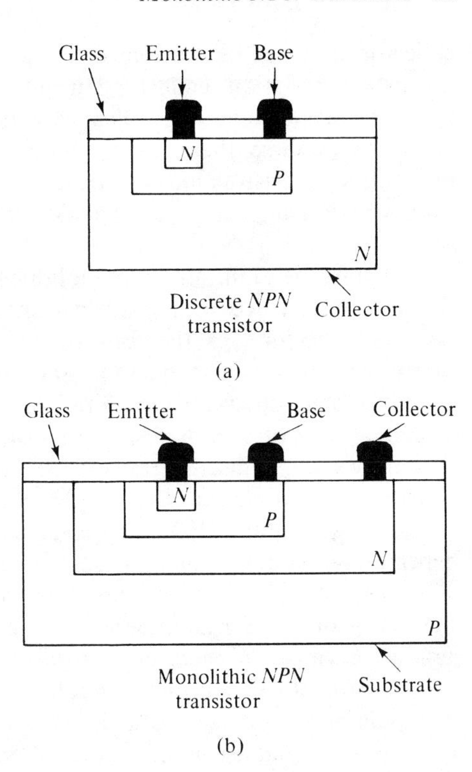

FIG. 2.1 Discrete vs. monolithic transistor (NPN).

Thus, for discrete and monolithic NPN transistors having the same collector, base, and emitter shapes (as viewed from above), electrical characteristics are remarkably similar—but not the same. Where as the discrete transistor has only three terminals, the monolithic transistor now has four; emitter, base, collector, and isolation. While the isolation is reverse-biased under normal operating conditions, it still has an added parallel capacitance to the collector, which can degrade high-frequency response. Furthermore, the four-layer NPNP monolithic transistor structure is the same as the four-layer structure of a silicon-controlled rectifier, which is known to be a *latching mechanism*. Normally, the fourth, isolation layer remains reverse-biased, and the "latch" mechanism is inoperative. But in some practical applications users of LIC's may inadvertently provide external bias voltages that permit "latch-up," sometimes causing internally destructive currents to flow. (The input stage of the "first-generation" popular type 709 OP AMP is notoriously subject to this failure mechanism, which is never shown on circuit schematics by its various manufacturers; it occurs when either input is raised above the positive supply voltage.)

Another difference visible in Fig. 2.1 is that collector current flows vertically downward from the emitter of the discrete transistor, through the collector, and out the bottom of the structure. Because of the monolithic transistor's isolation, collector current must flow down from the emitter, then sideways through the collector region, then back up to a top collector contact. Of course, there is more series resistance in this longer path; thus monolithic transistors almost invariably have higher saturation resistances than identical discrete counterparts and consequently are *less efficient in handling power.*

Despite their limitations, monolithic NPN transistors are quite competitive in performance with discrete transistors. With appropriately precise photomask techniques, they can achieve microwave frequency response or, with larger geometries, can handle amperes of current. High-voltage processes allow breakdown into the hundreds of volts, usually at the expense of current gain or saturation resistance, just as with discrete transistors. Unlike randomly selected discrete transistors, monolithic NPNs can be relied upon to have very *closely matched characteristics* for current gain (h_{fe}), forward emitter–base voltage (V_{be}) and other characteristics, allowing designs that depend for their operation upon such precise matching. After all, if the masks used to photoreproduce the transistors are geometrically matched, and if the transistors on a wafer all undergo identical chemical exposures, matching is assured. Because the monolithic NPN is the closest structure to its discrete counterpart of the various monolithic elements, and because inherent matching allows effective LIC design, the NPN transistor dominates most LIC mask layouts in number. Indeed, some existing LIC designs consist entirely of transistors, with no resistors and other elements—a design technique that would be costly with discrete designs.

2-3. MONOLITHIC DIODES

Since a diode is formed from any PN junction, there are several regions in the monolithic structure that may be used. The key to feasibility and economy of the LIC is that all elements are formed from the same process steps as the NPN transistors, the only difference being the photomask geometry and the electrical function to which each region is subjected. Available diodes, seen in Fig. 2.2, include the (a) collector–isolation, (b) collector–base, (c) emitter–base, and (d) aluminum–collector diodes.

The *collector–isolation diode* offers high breakdown voltage, but is of limited use, since the isolation side is common to the LIC's most negative supply potential in most applications.

The *collector–base diode* offers breakdown voltage nearly as high as that of the collector isolation, but it has the added advantage of being isolated from other diodes or elements in the IC. Its use is unfortunately restricted by the existence of a *parasitic* unwanted PNP transistor structure formed by the

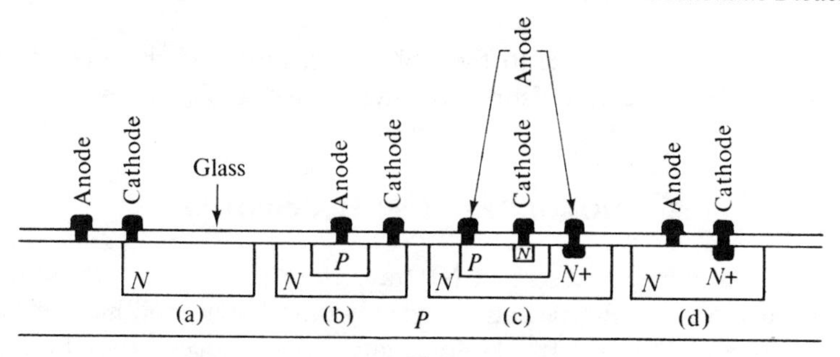

FIG. 2.2 Four types of diode: (a) collector–isolation; (b) collector–base; (c) emitter–base; (d) aluminum–collector. *(Schottky)*

P-type diode anode, N-type cathode, and the P-type isolation. Any time three such regions exist in mutual contact, with one junction forward-biased and the other reverse-biased, transistor action will take place. Thus, when the collector–base (CB) diode is forward-biased, the isolation's reverse biasing does no good, as parasitic current will flow into the isolation. Thus the CB diode is of limited usefulness also.

The *emitter–base (EB) diode*, operated within an isolated N-type collector region, is also capable of parasitic current flow, by the turning on of the "SCR" four-layer latching mechanism. If, however, the P-base material is deliberately shorted to the surrounding N-collector region, such parasitics are prevented. Thus the most widely used diode structure in LICs is the *emitter-base, which is really a transistor, with collector and base shorted together*. Unfortunately, optimum NPN transistor processing requires an EB reverse breakdown between 5 and 7 V, but most LIC diodes operate in portions of the circuit where this is all that is needed. Moreover, the low reverse voltage is reliably determined by the process, allowing the EB diode to be used as a monolithic Zener Diode.

Newest of the diode structures to be used is the *aluminum collector*, or *Schottky diode*. Aluminum is a natural P-type doping material. When an aluminum contact is normally made in an LIC to a collector region, it is preceded by an extra-heavy dose of N-type doping, so that the interconnect contact will be *ohmic*, that is, a low-resistance contact, rather than forming a PN junction. By eliminating the heavy N contact region and maintaining very clean chemical processes, a deliberate *Schottky or hot-carrier diode* is formed. While having a low reverse breakdown, like the EB diode, the Schottky diode has a much lower forward drop (about 0.2 rather than 0.7 for a silicon PN junction), and has very fast recovery time. This newer structure, still completely within the normal LIC process, has found use in nonsaturating logic gates, fast-recovery comparators, low-power FM-IF amplifiers, and high-frequency mixers and limiters.

It should be remembered that monolithic diodes, like monolithic transistors, have extra capacitance to the isolation region, and this may degrade their performance in comparison to two-terminal discrete diodes.

2-4. MONOLITHIC PNP TRANSISTORS

While it is easy to make a discrete PNP transistor (simply reverse the sequence of chemicals to make P-type and N-type regions), the monolithic PNP must share its process with the NPN. Unfortunately, if the process is best for NPNs, it makes rather poor PNPs. LIC designers have learned to live with and design around the monolithic PNP's deficiencies, and, in some cases, have used it for functions that discrete PNPs cannot perform.

The *vertical PNP*, Fig. 2.3, is formed from the base, collector, and isola-

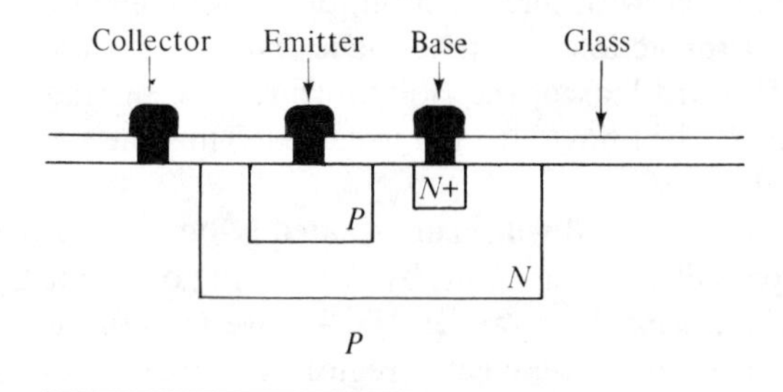

FIG. 2.3 Vertical PNP transistor.

tion regions, and is the same as the unwanted parasitic transistor encountered with CB diodes. Because the "doping" levels are optimized for NPNs, it usually has a low current gain of less than 50, and may be used only in grounded-collector (or emitter-follower) configurations, since its collector is also the common isolation region. Thus it is useful only as an emitter follower. Because of the thickness of its base region (ordinarily the NPN collector), its frequency response is also poor compared to the NPN.

The *lateral PNP*, Fig. 2.4, has both emitter and collector formed from

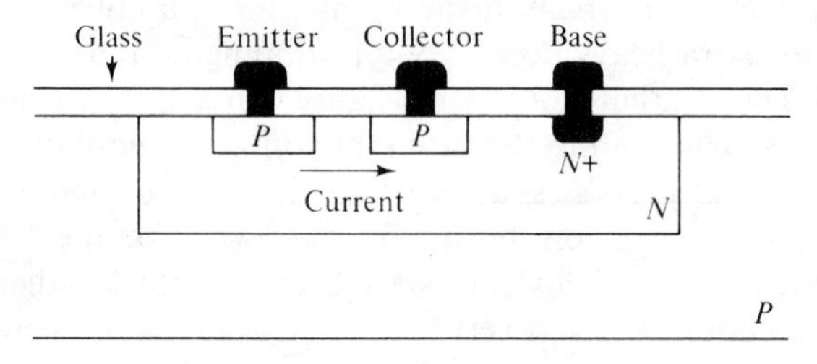

FIG. 2.4 Lateral PNP transistor.

the same material as the NPN's base, and its base from the NPN's collector N-type region. As it is an isolated transistor, it can be used in any configuration, but because its emitter–collector spacing, dictated by photomask resolution, is wide, it has poor current gain and frequency response. LIC processing and mask layout techniques have improved upon early lateral PNPs to the point where they are universally used in dc and audio LIC devices, and in the dc biasing sections of high-frequency LICs.

While special processes have from time·to time been announced to allow high-quality PNPs to coexist with high-quality NPNs, the additional, critical chemical processes have prevented low-cost mass production thus far.

2-5. MONOLITHIC RESISTORS

If one were to cut out a chunk of any material, such as the silicon from which LIC's are made, and measure between two probes at different spots, some resistance could be determined, depending upon the resistivity of the basic material, the impurities in the material, and the size and spacing of the probe contacts. In monolithic circuits the objective is to use the available resistivity of regions to optimize the NPN transistor performance, without having to add or adjust processes to achieve desired resistor values. In fact, resistivities can vary considerably in the LIC process, so that it is not in the interest of mass-production to expect resistors made at the same time as transistors to have accurate values.

A monolithic resistor is made by contacting two points in a region and isolating that region from everything around it, typically with a reverse-biased junction. Most popular is the base resistor; other types, such as emitter resistors, collector resistors, and *pinch* resistors, are used less frequently (see Fig. 2.5). The value of the resistor increases with increased

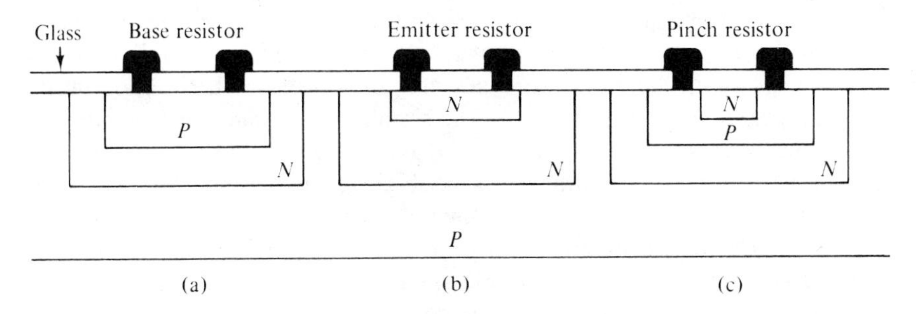

FIG. 2.5 Three types of monolithic resistors.

length or with decreased width. Photochemical resolution limits minimum width, so that, for a given resistivity in a region, very large resistances take up very large lengths and are uneconomical. Similarly, very low resistances require spacing of contacts too close together; thus, there are limitations to both the maximum and the minimum resistance that is available. Base resistors are generally between 100 Ω and 20 kΩ. Emitter resistors allow lower resistance, from 2 to 100 Ω, while collector resistors can reach several hundred thousand ohms. Even higher values are achieved with pinch resistors, in which a base resistor is partially pinched off by an intervening emitter region, just as a rubber water hose is pinched off.

The base resistor offers the best predicted accuracy, but even this is only ±20 or ±25 percent from a nominal value. The other types can vary ±50 or ±95 percent in some situations, so circuit biasing that is established by their value is not recommended. The higher-valued collector or pinch types are often used where a noncritical "bleeder" resistance is needed, while the very low valued emitter type is sometimes used for degeneration, parasitic-oscillation suppression, or to allow mask layout topologies where two metal strips must cross. In the latter case the emitter resistor is used as a "cross-under" and is usually not shown in the circuit schematic by the LIC manufacturer.

While circuit designs relying on absolute resistor values are not advisable in LICs, extensive use is made of the matching of two adjacent resistors. Two resistors might be off by 25 percent, but they will be within 1 or 2 percent of each other. This is especially useful in the widespread use of *differential amplifiers*, where resistor matching, not absolute value, is essential.

2-6. MONOLITHIC CAPACITORS

While junctions and resistors can be made as small as the photographic processes allow and still compete with larger discrete components, capacitor values depend primarily on their area, given available dielectric constants. Because of this, *only very small capacitors are possible in monolithic form*, usually less than 200 pF. Two types are compatible with the monolithic process and are shown in Fig. 2.6. The simplest is the *junction capacitor*, in which the natural capacitance between P and N regions is used. Any junction can be used, but the EB junction gives highest capacitance per unit area, although at the 5- to 7-V EB breakdown voltage.

The capacitance of any semiconductor junction varies with applied reverse voltage; thus the monolithic capacitor may be used as a voltage-variable element in oscillators and tuned circuits.

The second type, the *oxide capacitor*, uses the silicon as its lower plate; the natural insulator, silicon dioxide (ordinary sand component), which covers the IC, as the dielectric; and part of the aluminum interconnection as the upper plate. It has a high breakdown voltage, but may be hard to produce in volume unless the silicon-dioxide part of the LIC process is

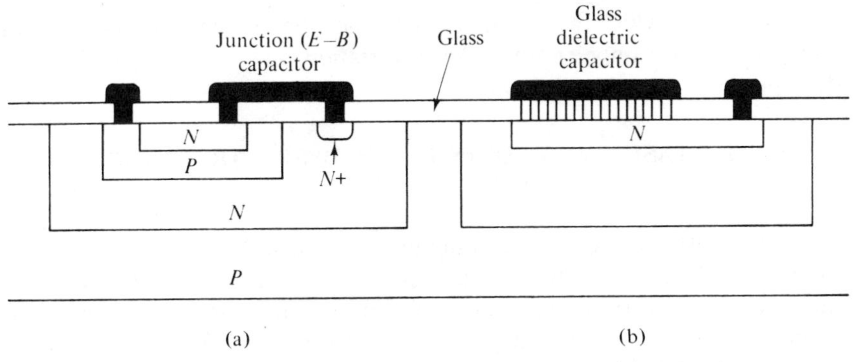

Fig. 2.6 Monolithic capacitors: (a) junction type; (b) dielectric type.

unusually free of localized *pinhole* defects, any one of which can short out the capacitor.

At least one "plate" of either type of capacitor is always in close contact with the isolation or some other region on the device; thus such capacitors have an unwanted, or "parasitic," capacitance to these regions, which sometimes degrades performance when compared with discrete capacitors. Monolithic capacitors are used primarily in frequency shaping of operational amplifiers and as RF bypasses, both of which are relatively noncritical of the uncontrollably wide tolerances, typically ± 50 percent.

2-7. MONOLITHIC COMPONENTS WITH NO DISCRETE COUNTERPARTS

Newer LICs contain unique structures that have resulted from the constant search to circumvent the weaknesses of existing monolithic components. Since creation of a new transistor or other structure is often no more than a mask-layout exercise using existing processes, it is easy to see that an NPN transistor with a single emitter region may be redrawn to contain *two or more emitters* (the basis for the popular TTL logic family); or, again, if a lateral PNP contains two P-type regions designated emitter and collector by the biasing connected to them, it is a simple matter to add one or more P-regions for use either as *multiple emitters or multiple collectors. Super-beta NPN transistors* are ordinary NPNs in which the emitter diffusion has been extended to create a very shallow base region, giving h_{fe} from 2,000 to 10,000, but at the price of a very low collector–emitter breakdown voltage (often less than 1 V) and very small collector currents.

The versatility of the basic monolithic LIC process has by no means been fully exploited. Components will continue to appear in mass-produced new products, which will seem to defy the obvious conventional circuit analysis, yet are only another clever variation on the available standard LIC process.

2-8. BASIC TECHNIQUES IN THE DESIGN OF LINEAR INTEGRATED CIRCUITS

Although both conventional and unique components are used in the design of the actual LIC, some basic rules can be summarized here. Older designs, in which fewer of these rules were understood, are consequently harder to manufacture, cannot give large eventual price reductions, and often are restricted in performance capability. Briefly, a well-designed LIC uses the following techniques:

1. Design relies on *matching* rather than absolute value of the components.
2. *Transistors are substituted* for resistors, diodes, and other components whenever possible. Transistors are the most controllable element available; they take less space than most elements replaced, and are inherently matched.
3. *Use differential structures.* If the biasing and other characteristics of a single-stage amplifier vary considerably because of uncontrolled components, it may be balanced against an identical amplifier that varies in the same way. (This is discussed in detail in Chapter 3, showing that an output sensitive to the difference between the two halves of the differential structure will tend to cancel out the effect of variations.)
4. If a more complex structure, such as a *compound transistor*, can give better results than a simple structure, use the more complex. It probably will cost less, all things considered.
5. Since component cost is not a consideration, the *combination of two components*, each with a weakness, can give performance superior to that possible with one premium component. For example, when a very high current gain is needed in conjunction with high breakdown voltage, as in the input of an operational amplifier, the combination of a "Super-Beta" low-breakdown transistor with an ordinary low-beta, but high-voltage, transistor in a cascade connection gives the best of both characteristics.
6. At some point, the performance of a monolithic component, such as a large capacitor, is uneconomical compared to a discrete component. Properly designed LICs have left the monolithically impractical components to be added *external to the IC*. As design and process techniques improve, more and more of these compromised elements will appear on the IC chip itself.

2-9. MONOLITHIC VERSUS HYBRID INTEGRATED-CIRCUIT DESIGNS

Most of the statements in this book apply to monolithic LICs, that is, to circuits mass-produced on a single silicon chip [mono, single; lithos (Greek), stone or crystal]. Another family of micro-circuit devices, sometimes imprecisely called integrated circuit, is the *hybrid type* in which individual, microscopic components are placed on a single insulating substrate and connected by fine wires. While the external package seems the same size as the package in which monolithic LICs are housed, the internal size is quite different, as is the cost of construction.

A hybrid circuit can be constructed with exactly the same design techniques as an ordinary discrete-component circuit and with the same availability of precise, controlled components. It need not have any of the peculiar parasitic problems of the monolithic circuit. In general, however, it will be more expensive than a larger-sized discrete circuit because of the precise microscopic assembly labor required, and will not be as practical to repair, for the same reason. The monolithic IC, however, cannot be repaired at all. Most likely, the monolithic IC will be cheap enough to throw away if defective, and its probability of failure is much lower than that of the hybrid.

Why consider hybrids at all, then? Simply because they offer a size-reduction comparable to monolithic circuits but with performance comparable to discretes. In some cases, as for *very high power* or for *high frequencies*, the hybrid can perform better than a monolithic circuit. But the performance advantage depends on the state of the monolithic art. One-Watt Audio Power Amplifiers, for example, were only feasible as hybrids in the mid 1960s. Today, 1- to 5-W monolithic devices are readily available as LICs. What is today only available in hybrid form is likely to be tomorrow's monolithic.

2-10. LINEAR-INTEGRATED-CIRCUIT ARRAYS

Because the *monolithic process* is governed by photo-reproduction techniques and is inherently one of *duplication*, LIC manufacturers can easily build dual, triple, or higher combinations of identical or similar functions into a package. Manufacturing economics have reached the point where the cost of a single LIC "die" or "chip" is much less than the price it is sold for in packaged form, and, in many instances, the material and labor cost of packaging into a hermetic or plastic container exceed the cost of manufacturing even a complex "die."

Simplest of these combinations is the *Transistor Array*, in which several transistors on one die are enclosed in a multilead container costing less than the same number of transistors individually packaged and, in certain applications, offering superior performance. More complex circuit functions, such as Operational Amplifiers, Voltage Regulators, Comparators, and Power Amplifiers, are now also found in pairs, triples, or sometimes quadruples; each of these functional types is discussed later in separate chapters. This chapter will limit itself to the structurally simpler transistor arrays, which offer circuit designers not only cost reduction, but enable "monolithic"-type discrete circuits to be constructed, using their inherent matched characteristics.

2-11. TRANSISTOR ARRAYS

Consisting of groups of monolithic transistors, with some or all terminals available to the user, *transistor arrays* have become popular in constructing special-purpose linear circuits that may not be available in completely monolithic form, but in which monolithic design techniques are useful, at costs competitive with the use of discrete transistors. Transistor arrays can replace hand-matched dual transistors in many critical applications, such as differential-amplifier input stages, at much lower cost. While some arrays, such as the *CA3046* and *CA3018/18A*,[1] make nearly all transistor terminals available externally, more efficient pin usage with a large number of array transistors usually requires internal commitment of several terminals, giving less universal arrays (such as the *CA3026*), which are nevertheless extremely versatile (see Fig. 2.7, 2.8, and 2.9).

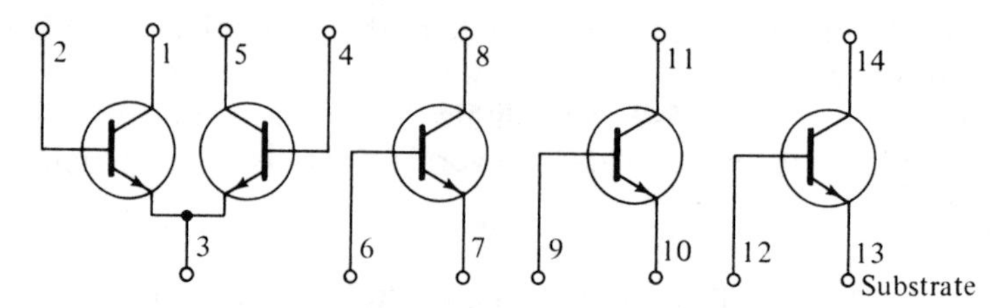

FIG. 2.7 CA3046 five–transistor array. *(RCA)*

Newer transistor array types are aimed at specific extremes of performance. For example, types *CA3081*, *CA3082*, and *CA3083*[2] are high-current large-geometry transistors for display-driver applications, with correspondingly poor high frequency response. Type *CA3084* is made of "lateral PNP" transistors where such devices are needed, and offers excellent matching but with relatively poor frequency response and dc current gain (see Fig. 2.10, 2.11, and 2.12).

[1] High-voltage versions are CA3146E/AE, CA3118T/AT.
[2] High-voltage versions are CA3181, CA3182, and CA3183E/AE, respectively.

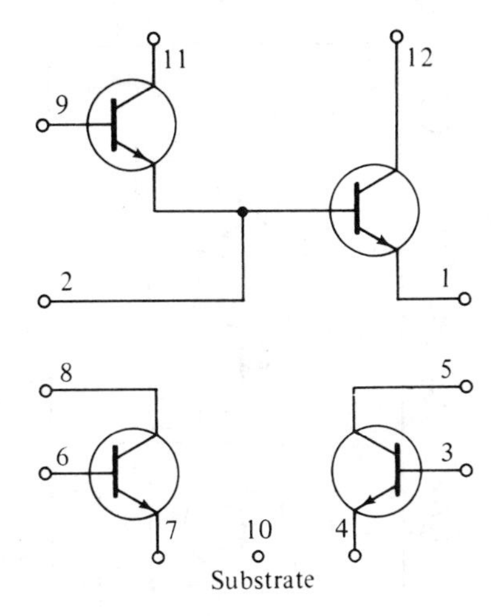

FIG. 2.8 CA3018/18A four-transistor array. *(RCA)*

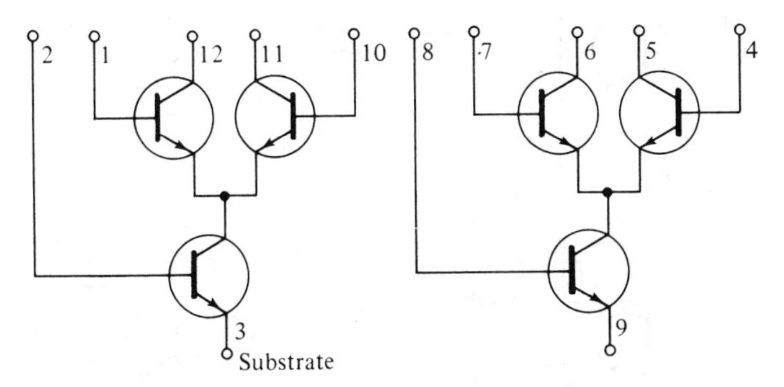

FIG. 2.9 CA3026 dual RF/IF amplifier/array. *(RCA)*

2-12. MONOLITHIC TECHNIQUES WITH TRANSISTOR ARRAYS

By observing certain precautions, the transistor array may be used to directly replace a number of discrete transistors in a conventional linear circuit. There are configurations, however, which would simply be impractical in discrete form because of required matching, which are common in modern LICs and are useful design tools in otherwise discrete component circuits. Three important examples of such configurations follow:

Accurately Matched Differential Pair

Many of the assumptions in the discussion of differential amplifiers (Chapter 3) rely upon ideal matching of the input transistor pair with respect to current gain (h_{fe}), forward base-emitter voltage (V_{be} value), and input bias

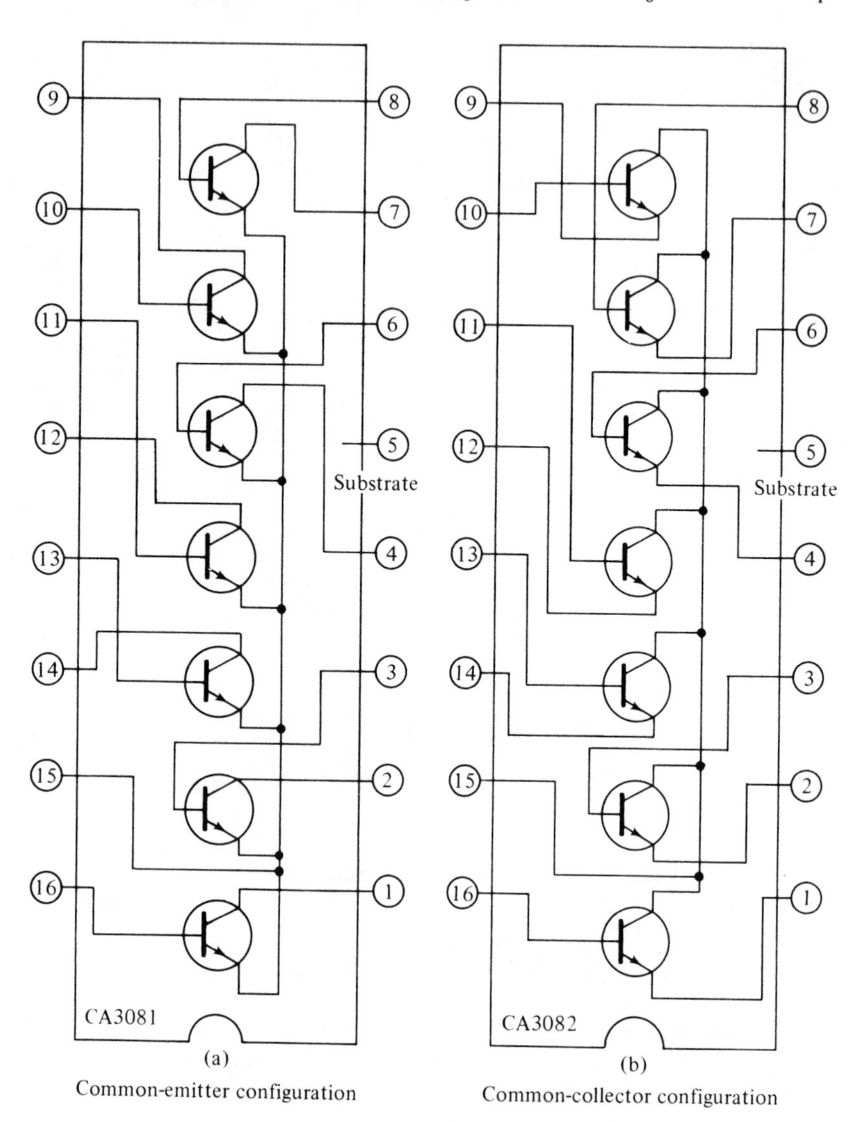

FIG. 2.10 (a) CA3081 common–emitter (7–transistor) array; (b) CA3082 common–collector array. *(RCA Solid-State Division)*

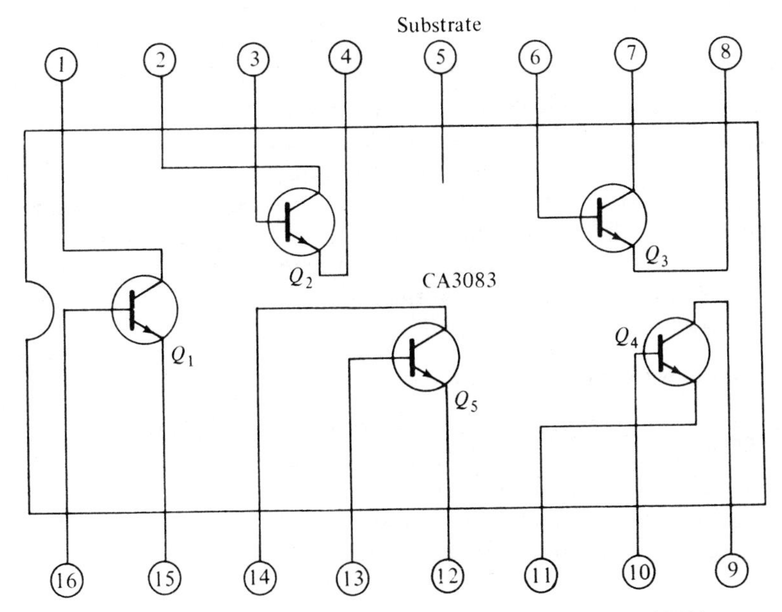

FIG. 2.11 High-current (5–transistor) CA3083 array. *(RCA Solid-State Division)*

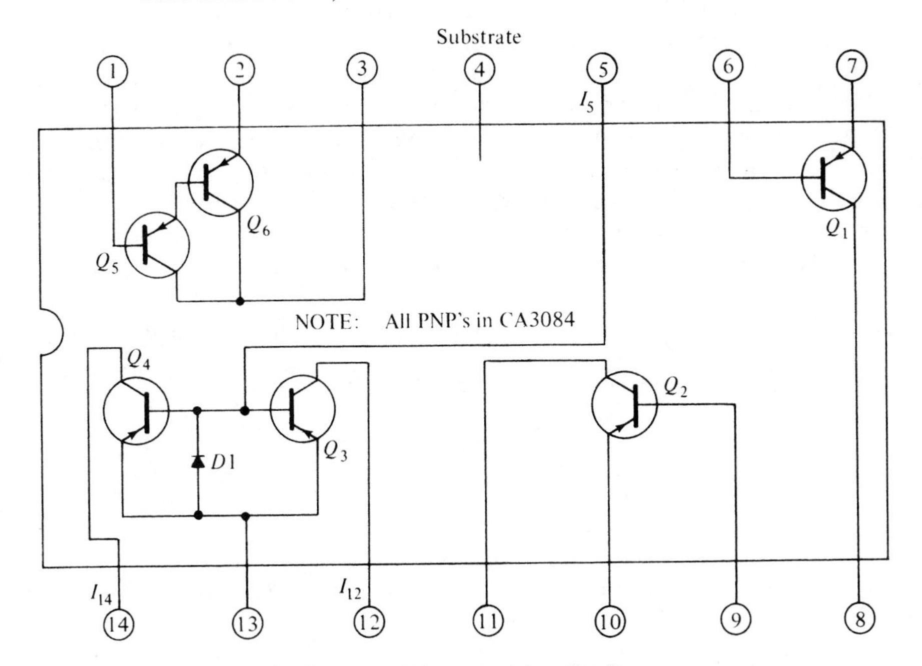

Fig. 2.12 PNP array CA3084 containing Darlington current-mirror and 2 independent transistors. *(RCA Solid-State Division)*

currents (I_B). A pair of discrete transistors randomly selected from a handful of devices, even from a single production lot from a given manufacturer, will have no better than ± 20 percent current-gain match, and V_{be} offset of ± 50 percent mV. Careful manual selection is necessary when building differential amplifiers from discrete transistors, and, even then, since the devices are in different packages, they will exhibit subsequent mismatches because of temperature differences and different rates of aging. Using a pair of transistors from a transistor array gives better than ± 5 percent current-gain match and typical V_{be} offset of less than ± 2 mV.

Constant-Current Source or "Current Mirror"

Predictable constant-current sources are needed, not only in differential amplifiers, but in many other LIC functions, such as oscillators, RC timers, and the like. Since a discrete-component current source cannot rely on matched transistors, it must use precision-matched resistors to "swamp out" current-gain and V_{be} variations seen in randomly selected discrete transistors. Such a discrete circuit is seen in Fig. 2.13. A much simpler and more effective current source is widely used in LICs, as in Fig. 2.14. For a given base–emitter structure, a fixed Base–Emitter voltage is required to force a fixed collector current. If transistor Q_1 is connected as in Fig. 2.14, with collector shorted to base, the Q_1 collector current will increase until the sum of collector and base currents equals the forced external current, reaching an equilibrium through negative feedback. Under this condition (neglecting base current, which is much smaller than collector current), the emitter–base voltage is precisely that value required to sustain a collector current equal to the forced value. If, now, a perfect replica of Q_1, called Q_2, is connected so that the equilibrium base–emitter voltage of Q_1 is applied across the Q_2 base–emitter junction, then Q_2 must also sustain a collector current equal to the original forced value.

It may be seen that operation of the LIC constant-current source

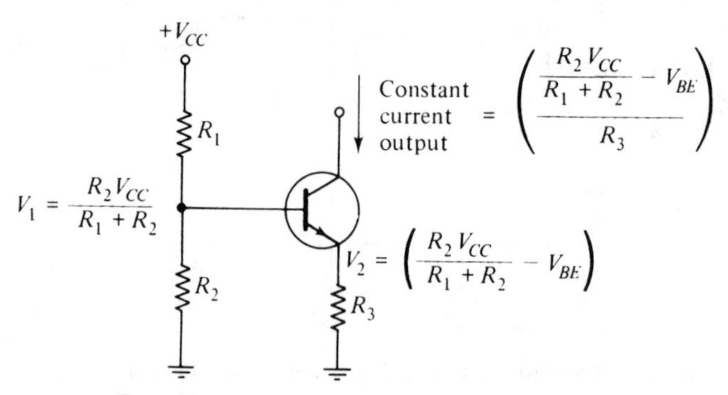

FIG. 2.13 Discrete constant-current source.

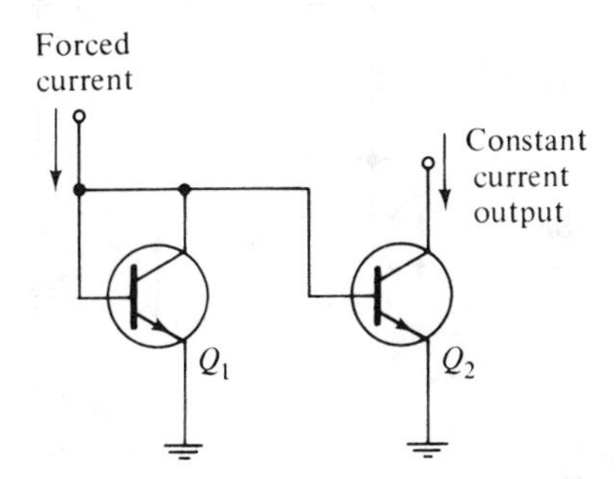

FIG. 2.14 LIC "current mirror."

depends heavily on precise matching. This convenient structure ensures equality of currents in a transistor pair (sometimes called a *current mirror*). Using minimum components, the LIC can operate within $\frac{1}{2}$ V of the negative supply, and provides a practical precision arrangement that is very difficult to obtain in discrete-circuit designs.

Biasing of Gain Stages with Predictable Diode (V_{be}) Drops

A conventional class A amplifier stage, Fig. 2.15, uses a number of resistors to establish its quiescent operating point. In particular, the emitter resistor (which requires a bypass capacitor if maximum gain is to be attained) establishes emitter current despite the unpredictable value of the forward base–emitter (V_{be}) voltage drop of a randomly selected discrete transistor. An equivalent monolithic-style amplifier, Fig. 2.16, has a grounded emitter, eliminating the emitter bypass capacitor and emitter resistor. Again, biasing such a stage by a fixed base-voltage with discrete transistors would cause unpredictable variations in collector current, whereas a transistor array used as a constant-current source can give stable bias point. Moreover,

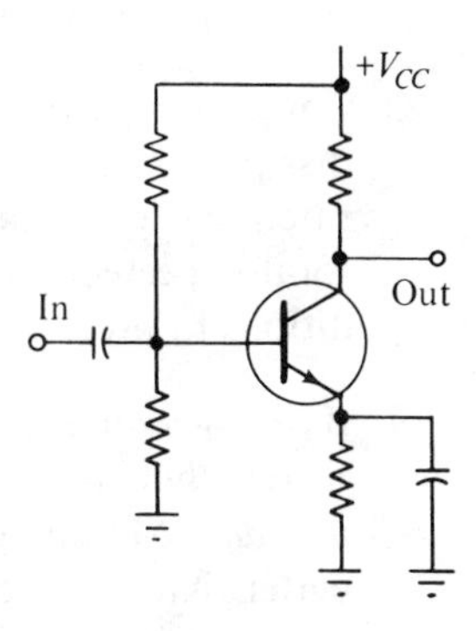

FIG. 2.15 Conventional discrete amplifier.

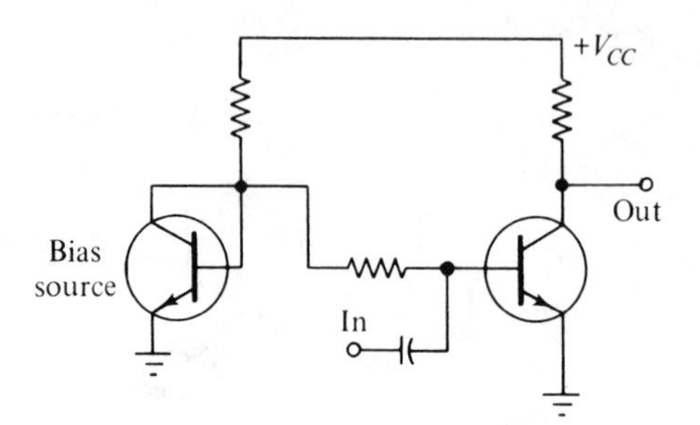

FIG. 2.16 LIC amplifier using matching transistors

if four or five such stages are to be similarly biased, a single *reference* voltage section, consisting of the transistor with collector–base short, can supply all stages simultaneously.

2-13. PRECAUTIONS IN APPLYING TRANSISTOR ARRAYS

In using monolithic transistor arrays it must be remembered that the individual transistors are not totally independent and are subject to the parasitic problems outlined previously in this chapter. The *following precautions* must be considered, in addition to those followed in normal discrete-transistor designs:

1. *The pin connected to the substrate of the array must always be the most negative dc voltage in the circuit*, or else one or more isolated regions will become forward-biased with respect to the substrate under some possible circuit conditions.

2. *Saturating designs*, in which the collector–base junction of the NPN monolithic transistor in the array becomes momentarily forward-biased, can cause the unwanted parasitic PNPN structure to turn on and latch. Destructive currents can be prevented under such conditions by using a series base resistance on the NPN transistor.

3. *Power dissipation* of a transistor array is usually limited by the total dissipation limitations of the package. While one transistor may be rated at a given maximum dissipation, it is important to calculate the total expected dissipation within the package under worst-case conditions to preclude heating problems.

4. *When several Transistor-Array packages* are used in a design, the transistors that are relied upon for matching must be within a *common package*. Matching of transistors between two separate arrays is no better than that between two discrete transistors.

2-14. MULTIPLE-FUNCTION LINEAR INTEGRATED CIRCUITS

Progressive development of newer LICs has led to the introduction of devices that combine more than one function in the same compact package. Later we shall discuss LICs that include two, three, four amplifiers housed in a single DIP package of 14 or 16 leads. In other cases there will be a number of different functions integrated on a single device, such as combinations of a voltage-controlled oscillator with an amplifier and a comparator to form an entire subsystem, all in one monolithic package.

Although in most cases the effect of companion functions may be ignored, the designer must remain aware of other cases where different LIC functions in the same package may unintentionally interact. The ways in which such interaction can take place include the following:

1. *Sharing a common power-supply connection.*—Supply currents into one function must pass through unavoidable lead resistance, causing a small, but often troublesome, voltage drop in that resistance and giving rise to undesired coupling. Such situations call for various decoupling schemes.
2. *Capacitive and inductive coupling.*—Because LIC packages have closely spaced leads, there may be appreciable capacitance between functions in the same package. Since capacitive reactance is lower at high frequencies, capacitive coupling is especially troublesome in multiple radio-frequency amplifiers; similarly, at the higher frequencies even the inductive reactance of the physical lead lengths must be considered.
3. *Thermal interaction.*—If a LIC function is temperature-sensitive, and if it shares a package with others that produce appreciable and varying amounts of internal heat, interaction will take place primarily at dc and low frequencies, since heat variation in a chip is a relatively slow process.

Despite these cautions, there is a performance advantage in complex LIC functions sharing a single package: that of matching. For example, with multiple amplifiers on a single chip, it is very likely that all operating characteristics will be closely matched and will remain matched as temperature varies, because they are in intimate thermal contact. This advantage, combined with the other integrated-circuit features of convenience and reliability, tends to favor the continual effort to produce more complex LICs that serve as subsystems and outperform their discrete counterparts.

2-15. AMPLIFIER ARRAY

Going beyond the various transistor arrays previously discussed, there is

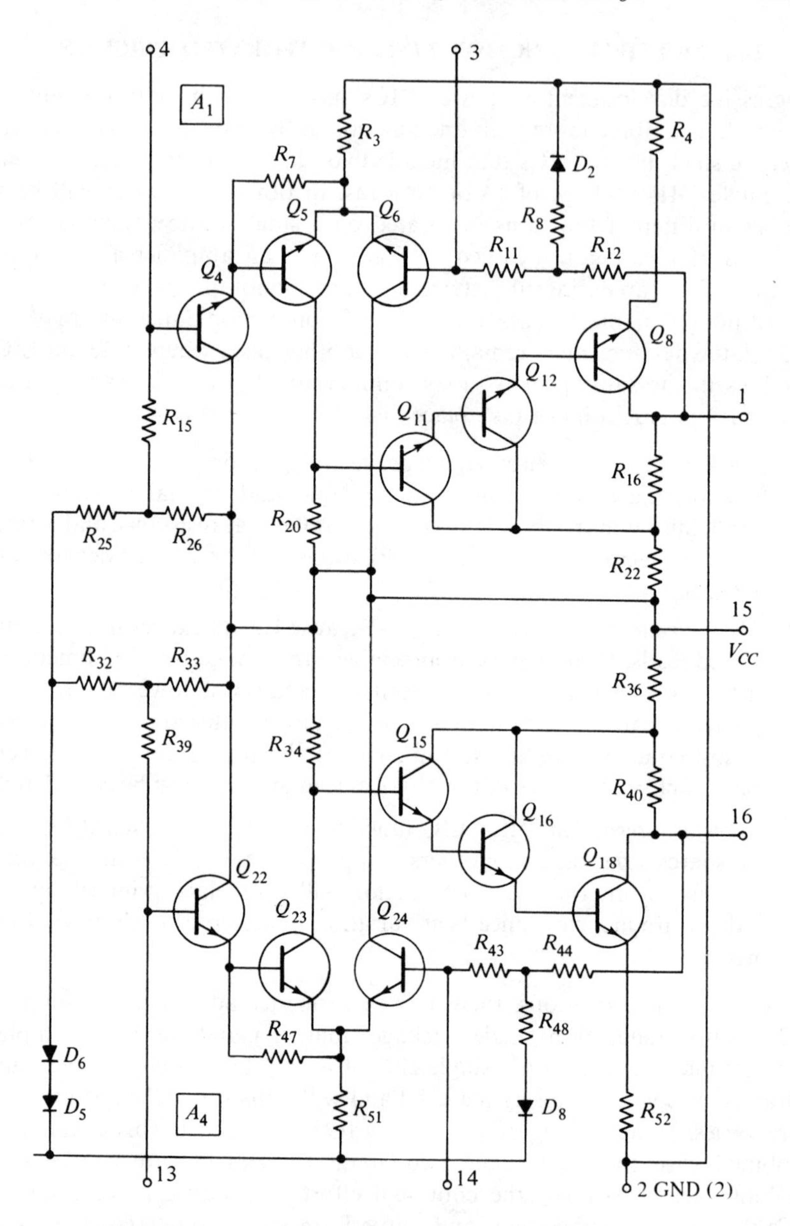

FIG. 2.17 Four-amplifier array (CA3048) showing the circuit for two of the four amplifiers (A_1 and A_4); amplifiers A_2 and A_3 are symmetrically similar, and all four are on a single-chip in a 16-lead DIP package. *(RCA Solid-State Division)*

an interesting *array of four amplifiers* in one package, as offered by the RCA *CA3048*. In this quad-amplifier array, each of the four identical amplifiers has independent inputs and outputs, all on a single monolithic chip and housed in a 16-lead DIP package.

In the schematic diagram of Fig. 2.17, two of the four amplifiers (A_1 and A_4) are shown, while the other two amplifiers (A_2 and A_3) are symmetrically similar. Referring to amplifier A_1 (at the top of the diagram), it is seen to consist of two stages of voltage gain. The input stage is basically a differential amplifier (transistors Q_5 and Q_6) preceded by a Darlington transistor (Q_4) on the left side. The output stage consists of a combination of three transistors (Q_{11}, Q_{12}, and Q_8) connected in an inverting configuration, with Q_8 as the actual output transistor. Fixed resistance between the output of Q_8 to the inverting input of Q_6 provides negative feedback, so that the closed-loop gain can be tailored by the value of input resistance connected to inverting terminal 3. Terminal 4 is the non-inverting terminal to which the input signal is normally applied. (The fact that terminal 4 is the non-inverting input can be checked by noting that there is a double inversion from the base of Q_4 to the output at the collector of Q_8.)

Each amplifier in the array has a typical open-loop gain of 58 dB and an input impedance of 90 kΩ. Together with an inherent low-noise characteristic, this combination suggests many versatile applications for various compact arrangements of the four independent amplifiers. Some suggested examples of the use of the CA3048 are as (1) an *oscillator–amplifier arrangement*, (2) a *four-channel linear mixer*, or (3) a balanced-line-driver.[1]

Other models in the *quad* OP-AMP *family* include such recently introduced types as the *National LM3900* and the *Motorola MC3301P*.

[1] Details of these and other applications are given in the RCA Application Note ICAN 4072, and the National "Quadzilla" pamphlet; see Appendix IV for list of manufacturers' addresses.

3

DIFFERENTIAL-AMPLIFIER STAGE IN INTEGRATED-CIRCUIT DESIGN

3-1. GENERAL ELEMENTS IN INTEGRATED-CIRCUIT DESIGN

In the previous introductory discussion it was emphasized that the IC can be used as if it were a single active device, and that this ease of use provides a tremendous simplification—even for nontechnical personnel—in setting up a desired circuit. However, we need to know more than just the external connections if we are to have a good understanding of the basic functioning of the little "black box." By delving a bit further into the functional design elements of the internal arrangements of the IC, we can learn how to make more effective use of its many capabilities, while—equally important—being intelligently guided as to its inherent limitations.

The general problem of translating the design of a multi-stage transistor amplifier into an integrated-circuit form can be envisioned by the three blocks shown in Fig. 3.1(a). For the input stage of the IC, it has been found that the integrated version of a pair of transistors in the form of a *differential amplifier* gives the best answer for satisfying the critical input conditions, in order to produce stable bias, minimum offsets, and the ability to have simple provision for external negative feedback. Since this type of input

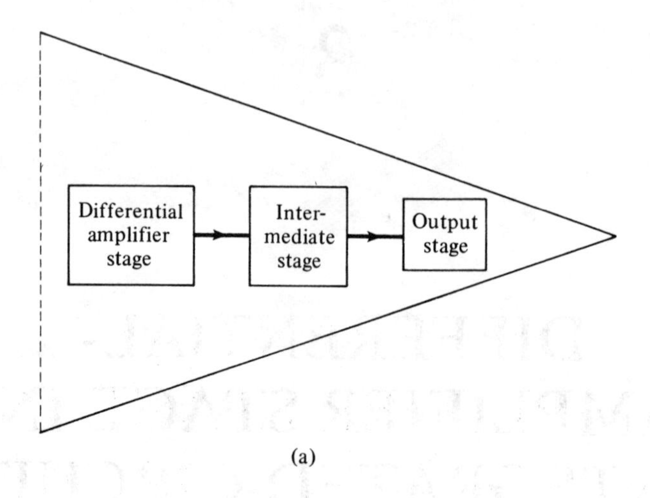

(a)

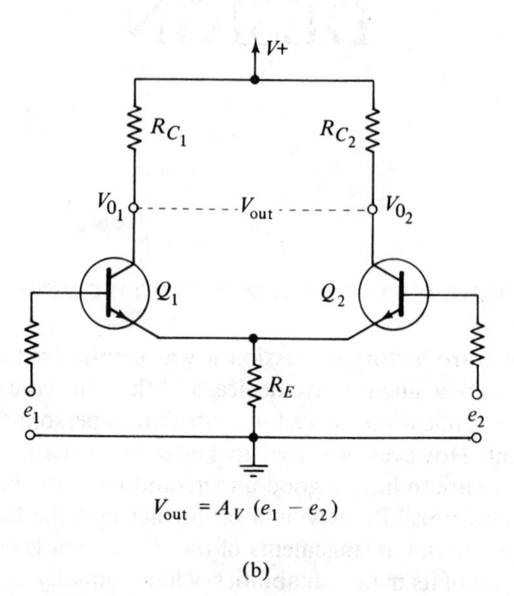

(b)

FIG. 3.1 (a) Generalized blocks of an OP AMP; (b) simplified circuit of differential amplifier input stage of OP AMP; the output $(v_{o1} - v_{o2})$ is proportional to the difference of the inputs $(e_1 - e_2)$.

stage has so great an effect in determining many characteristics of the resulting LIC, it will be discussed separately in this chapter.

3-2. CHOICE OF DIFFERENTIAL AMPLIFIER FOR INTEGRATED-CIRCUIT INPUT STAGE

In the most widely used forms the circuit arrangement of the linear IC should be able to provide a high order of amplification. While there are many versions of multi-stage ac amplifiers to accomplish the desired high gain, they necessarily include fairly large capacitors for *RC* coupling between stages—a condition that serves to rule out this circuit approach for integration. The circuit choice then narrows down to some form of multi-stage *direct-coupled amplification*, and this brings with it the attendant *problem of drift*, a characteristic that is inherent in dc amplifiers. This is so because unavoidable changes in operating conditions (such as changes in temperature, supply voltage, and so on) are accepted by a dc amplifier in the same way as it accepts the desired input signal, thus producing undesired output changes that together constitute the drift condition. Even small amounts of this undesired drift are important, since they will be magnified by the high gain of the subsequent stages. Consequently, the type of circuit that minimizes this drift, such as the differential-amplifier stage, is a preferred choice.

The *differential-amplifier stage* offers a good practical solution to the drift problem (except for the most stringent drift specifications, where a chopper-stabilized amplifier might be required). As shown in Fig. 3.1(b), it consists of a matched pair of transistors, where the drift-induced changes affecting one side of the pair tend to be canceled out by the corresponding changes on the other side of the pair. As a result of this balancing action, the differential-amplifier circuit is able to produce a substantially zero dc output under quiescent (no-signal) conditions. (This feature of "zero out for zero in" accounts for its popular use in the bridge type of electronic voltmeters.[1]) Because of this and other beneficial features, the circuit has been almost universally adopted in integrated circuits for the input stage (and often for intermediate stages as well). Variously designated as *differential-amplifier stage*, *emitter-coupled stage*, or "*long-tailed pair*," its advantageous features can be listed as follows:

1. It is *direct-coupled*, allowing both dc and ac amplification without requiring coupling capacitors.

[1] Details of its use in solid-stage multimeters (such as the FET-VM) are given in S. Prensky, *Electronic Instrumentation*, 2nd ed., Prentice-Hall, Inc., Englewood Cliffs, N.J., 1971.

2. It tends to *cancel drift conditions* that cause offset changes in the output, and simultaneously it can be made to be almost impervious to signals that are common to both inputs (such as hum pickup), because of its property of *common-mode rejection* (CMR).
3. Since its emitter resistance supplies internal negative feedback, the resulting amplifier characteristics show improvement in its *stability*, *wide-band response*, and *high input impedance.*

(It should also be noted that the disadvantage of the reduced gain, normally associated with negative feedback, is not a pressing problem in the case of integrated circuits, where it is relatively easy to fabricate additional transistors as required.)

With all these beneficial and practical characteristics, it is not surprising, therefore, to find the differential-amplifier circuit being used in the input of practically all integrated circuits as a *basic amplifier stage.*

3-3. CIRCUIT ACTION OF DIFFERENTIAL-AMPLIFIER STAGE

The performance of the differential-amplifier stage can be followed more clearly by starting with a functional circuit using a *matched pair of field-effect transistors*, as shown in Fig. 3.2. It is well to start with this circuit, because the operation of the unipolar FET requires no current in its input circuit (the conventional bipolar transistor, by contrast, requires input base-current flow as a necessary condition). Analysis of the FET circuit allows us to concentrate only on the output currents, making it much easier to grasp the relation between the output and input voltages in determining the voltage gain of the circuit. (This approach, with its concentration on the currents in the output circuit, will also be helpful in analyzing the bipolar transistor circuit, which is considered later.)

In arriving at an equivalent-circuit model for the emitter-coupled amplifier of Fig. 3.2(a), we can take advantage of the simplicity of the model[1] obtained by looking into the source resistor (R_{sc}), as shown in Fig. 3.2(b). Here the internal drain resistance (r_d) and load resistance ($R_d = R_L$), on each side of the circuit, are combined in a single equivalent resistor $(r_d + R_D)/(\mu + 1)$ and are fed by an equivalent generator $[\mu v_i/(\mu + 1)]$. We then make the simplifying assumptions that apply for good symmetry between the left and right sides; that is, ($r_{d1} = r_{d2} = r_d$, and $R_{L1} = R_{L2} = R_L$). Also, we assume R_{sc} to be much larger than the equivalent series resistors (both of these assumptions are practical ones, as will be shown later).

[1] After J. Millman and C. C. Halkias, *Electronic Devices and Circuits*, McGraw-Hill Book Company, New York, 1967, pp. 205–206.

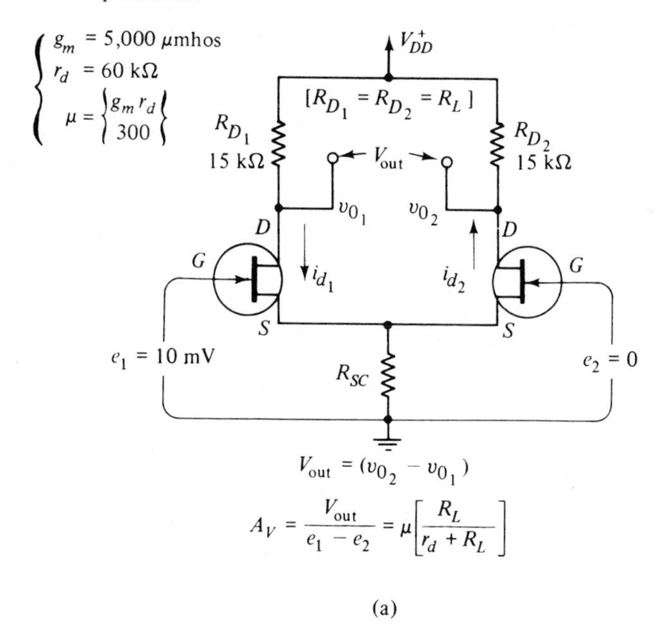

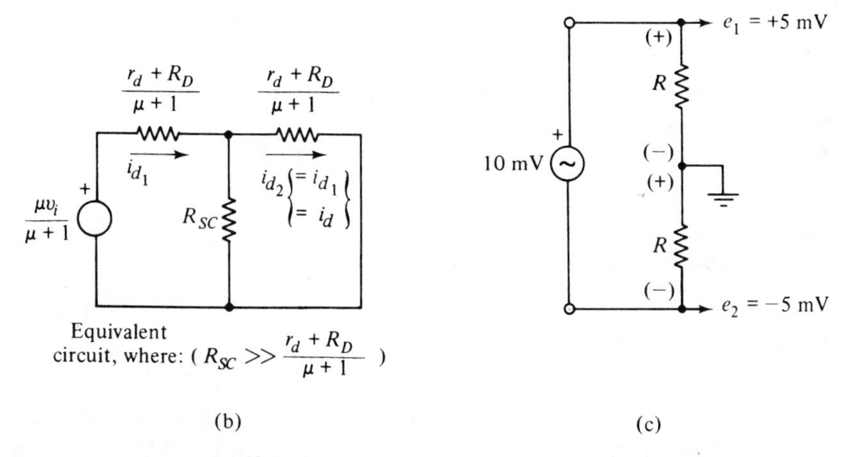

FIG. 3.2 Functional circuit of FET differential-amplifier stage: (a) actual circuit; (b) equivalent circuit looking into source S; (c) positive and negative inputs giving same results as in (a).

Accordingly, as we trace the output signal current around the equivalent circuit, we can consider negligible current in R_{sc}, and find

$$i_{d1} = i_{d2} = i_d,$$

and the corresponding output voltages are

$$v_{o1} = -i_d R_L,$$

$$v_{o2} = +i_d R_L \qquad \text{(equal and opposite to } v_{o1}\text{)}.$$

Then, for the overall output voltage, $v_{\text{out}} = v_{o2} - v_{o1}$,

$$v_{\text{out}} = i_d R_L - (-i_d R_L) = 2(i_d R_L).$$

Finding the value of i_d and substituting in the preceding equation,

$$i_d = \frac{\mu v_i/(\mu+1)}{2(r_d+R_L)/(\mu+1)}, \quad \text{or} \quad \frac{\mu v_i}{2(r_d+R_L)},$$

and

$$v_{\text{out}} = 2(i_d R_L) = \frac{2\mu v_i (R_L)}{2(r_d+R_L)} = \frac{\mu v_i R_L}{r_d+R_L}.$$

Therefore, for the gain (A_v),

$$A_v = \frac{v_{\text{out}}}{v_i} = \frac{\mu R_L v_i}{(r_d+R_L)\, v_i} = \mu \left[\frac{R_L}{r_d+R_L}\right],$$

which will be recognized as the *gain of a single FET, independent of the value of source resistance* (R_{sc}).

Output of General Differential-Amplifier Stage

Consequently, a *general expression for the output of the differential-amplifier circuit* ($v_{O2} - v_{O1}$) can be stated in terms of the gain (A_v) of a single element, and it applies equally well to any device pair, whether FETs, tubes, or conventional transistors, as follows:

$$v_{\text{out}} = v_{O2} - v_{O1} = A_v(e_1 - e_2),$$

with $e_2 = 0$ in this case; or, expressed in words, *the differential output is proportional to the difference of the two input voltages.*

Two examples will be given to illustrate this simple method for determining the approximate gain of the circuit; the first for the FET circuit and the other for a conventional (bipolar) transistor circuit.

Example 1 (Sec. 3-3)

USING THE FET MATCHED PAIR: SINGLE-ENDED INPUT TO DIFFERENTIAL OUTPUT

Using the numerical values given in Fig. 3.2(a) for the single-ended voltage input ($e_1 = 10$ mV, $e_2 = 0$), we first find the gain of a single stage (neglecting R_{sc}) as follows:

$$A_v = \mu \frac{R_L}{r_d + R_L}$$

$$= 300 \frac{15\ \text{k}\Omega}{60\ \text{k}\Omega + 15\ \text{k}\Omega} = 300 \frac{15}{75} = 60$$

Then, $v_{\text{out}} = v_{O2} - v_{O1} = 60(e_1 - e_2) = 60(10\ \text{mV} - 0)$, and $v_{\text{out}} = 600$ mV or 0.6 V as the differential output voltage, taken between the two drain resistors.

It may be noted that this is the same gain as would be obtained using the formula for the FET gain:

$$A_v = g_m R_{\text{eff}}, \qquad \text{where } R_{\text{eff}} = R_L r_d,$$

WITH DIFFERENTIAL INPUT

The input voltage of 10 mV could equally well be applied as a differential input, if desired, by arranging two equal resistors across the 10 mV generator, and connecting the center tap to ground, as in Fig. 3.2(c). Since the differential output will still be proportional to the difference of the input signals, we again have

$$v_{\text{out}} = v_{O2} - v_{O1} = A_v(e_1 - e_2)$$

$$= 60[5\ \text{mV} - (-5\ \text{mV})] \quad \text{or} \quad 60(10\ \text{mV})$$

$$= 600\ \text{mV},$$

the same differential output as before.

SINGLE-ENDED OUTPUT TAKEN FROM ONE SIDE TO GROUND

When it is desired to obtain an output voltage with one side on ground (rather than having the floating differential output), we take the output voltage from one drain resistor to ground, so that

$$v_{out} = v_{O1} \text{ (or } v_{O2}) \text{ to ground} = i_d R_L = \frac{A_v}{2}(e_1 - e_2)$$

$$= \frac{60}{2}(10 \text{ mV})$$

$$= 300 \text{ mV};$$

thus the *single-ended output* (v_{O1} or v_{O2}) *will be half the differential output voltage.*

Conventional Transistor Circuit

The differential-amplifier circuit using conventional (or bipolar) transistors for the matched pair is by far the one most commonly used in integrated circuits. Even though a high input impedance can be obtained with the FET pair, the bipolar transistor pair is able to provide a *much higher gain* while still presenting a generally satisfactory input-impedance level. In analyzing the gain of the transistor version, shown in Fig. 3.3(a), we can again concentrate on the output circuit, as we did for the FET version. Even though additional considerations arise from the flow of signal current into each base, we find that the expression for the overall gain of the circuit again turns out to be the gain of a single transistor, independent of the emitter resistor. Thus, within a reasonable approximation, the differential output, taken between the two collectors, is again $v_{out} = v_{O2} - v_{O1} = A_v(e_1 - e_2)$, where A_v is the gain of a single transistor, as if it had no resistance in its emitter circuit. The simplified expression for A_v, shown in part (b) of the figure, makes use of the additional assumptions (usually valid) that h_{re} can be neglected and that $h_{oe}R_L$ (or in this case $h_{oe}R_c$) is much smaller than unity. Accordingly, we approximate the current gain:

$$A_I = \frac{h_{fe}}{1 + h_{oe}R_L} \approx h_{fe} \qquad (\text{or } \beta),$$

and

$$A_v = h_{fe}\frac{R_L}{R_{in}} = h_{fe}\left[\frac{R_L}{R_{gen} + h_{ie}}\right].$$

Example 2 (Sec. 3-3)

MATCHED PAIR OF BIPOLAR TRANSISTORS

Using the numerical values given in Fig. 3.3(b), the approximate

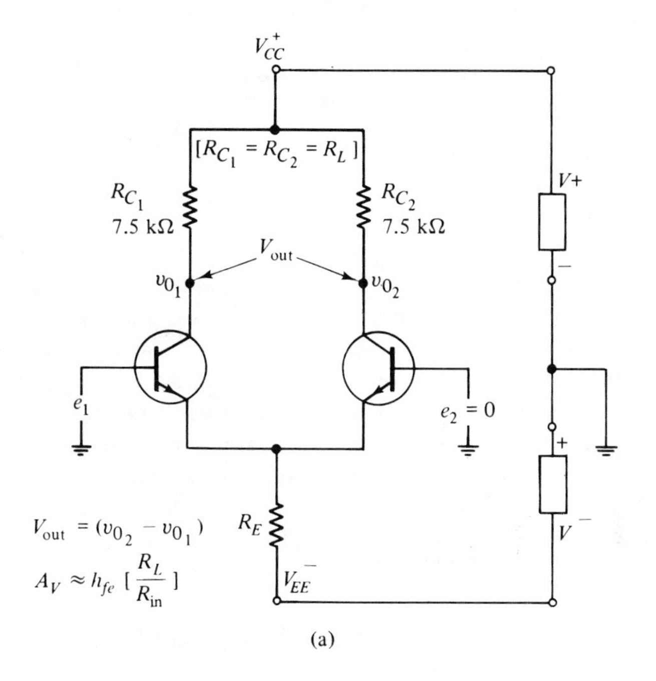

(a)

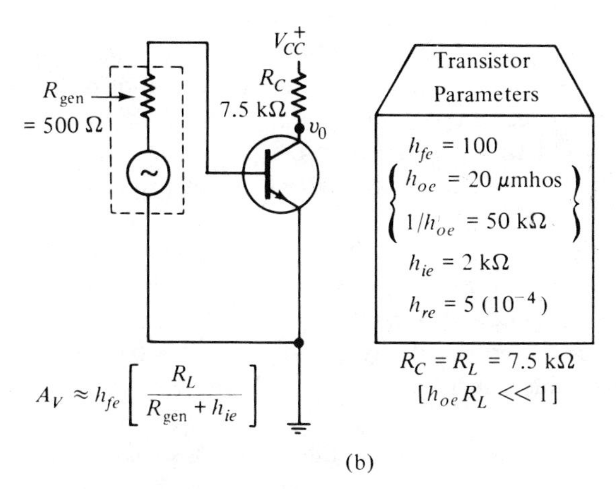

(b)

FIG. 3.3 Bipolar (conventional) transistor version of a differential-amplifier stage: (a) actual circuit; (b) equivalent diagram for stage gain, showing gain to be essentially independent of value of emitter resistance R_E.

differential gain (A_v) is as follows:

$$A_v = \frac{h_{fe}(R_L)}{R_{\text{gen}} + h_{ie}} = 100\left[\frac{7{,}500}{500+2{,}000}\right] = 100\left[\frac{75}{25}\right]$$
$$= 300,$$

which is five times as great as that for the FET pair.

In the examples cited, we have used a simplified version of the differential-amplifier circuit, along with a theoretical assumption of a very large common-emitter resistance, to establish the basic relation that the use of the matched pair of transistors provides the gain of only a single transistor, but, more importantly, that it can also produce a circuit having highly desirable characteristics for integration. The practical circuit actually used to realize these benefits involves a workable method for obtaining the assumed very high resistance in the emitter circuit. This is discussed in the next section in the form of a *constant-current device* used to produce the effect of a high resistance.

3-4. CONSTANT-CURRENT CIRCUIT

The desirability of making the common emitter resistor (R_E) as high as possible is based on a major advantage; it makes the output (v_{O1}) on one side of the pair more nearly equal and opposite to the output on the other side (v_{O2}). As a consequence of this equality, the ability to reject common-mode signals is greatly enhanced. This is expressed quantitatively as a desirably large *common-mode rejection ratio* (CMRR), which is discussed in the next section.

There is, however, a practical difficulty in choosing a very large resistor (R_E), since the large voltage drop across it would call for excessively large supply voltages to produce acceptable values of collector current. This difficulty is circumvented by adding a third transistor (Q_3) to the circuit, as shown in Fig. 3.4. This transistor acts as a constant-current source in place of R_E.

A constant-current source, by definition, corresponds to a source having an extremely large internal resistance. Such a source is obtained from transistor Q_3 by providing its base with an unvarying bias voltage, obtained in the usual case from the constant voltage across the diode, which is forward-biased from the $-V_{EE}$ supply. With the V_{BE} of Q_3 held constant in this way, the value of its collector current can be set to an appropriate value by its emitter resistor (R_3). The resulting constant collector current of Q_3 divides equally and establishes the quiescent collector currents of Q_1 and Q_2. (Looked at another way, transistor Q_3 is being operated on the almost horizontal portion

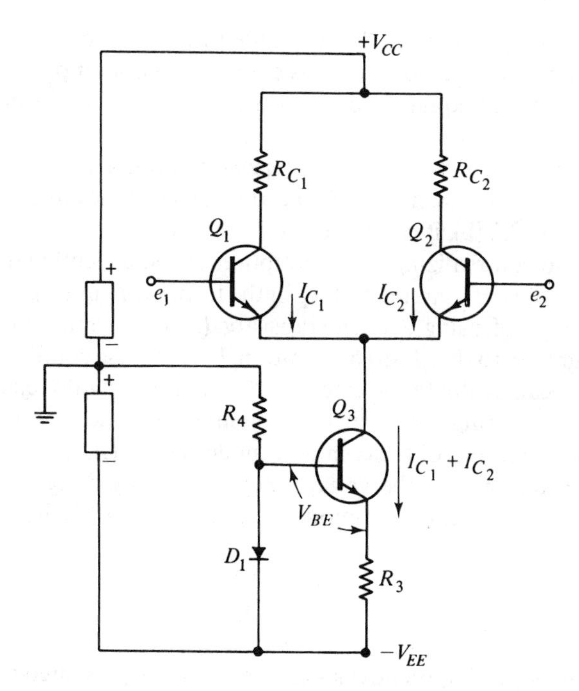

FIG. 3.4 Constant-current circuit provided by varying bias (V_{BE}) of Q_3. This produces an effective very-high resistance in place of emitter resistance (R_E) in Fig. 3.3 (a).

of its output characteristic curve, thus making its internal resistance equal to a very high resistance). In this manner, effective resistances equivalent to many megohms can be obtained without incurring excessive voltage drops in the actual circuit. Morever, when this constant-current-source arrangement is used in monolithic integrated circuits, an added advantage lies in the fact that the diode used to bias Q_3 can easily be fabricated as a diode-connector transistor on the same chip, thus supplying a good temperature-compensation match for Q_3. Since the stabilized Q_3 collector current determines the quiescent operating current, this method serves to reduce temperature drift for the whole circuit.

3-5. COMMON-MODE REJECTION RATIO

A most important characteristic required in the first stage of a high-gain amplifier was previously mentioned as its ability to suppress undesired disturbances that might be amplified along with the desired signal. The matched pair of transistors in the differential-amplifier stage has this inherent capability, based upon the fact that unwanted signals from an external source (such as hum pickup) would appear as common to both input bases, and, as such, would produce equal output voltages whose difference in the overall output would theoretically be zero.

The practical effectiveness of this rejection, however, depends upon how closely equal the currents in the left and right sides of the circuit can be held. Referring to Fig.3.2 (b), it can be seen that the equality of those currents (i_{d1} and i_{d2}) depends on realizing the assumption of a sufficiently large common resistor (R_{sc} in the FET circuit and R_E in the transistor circuit). We have seen that the strategy of using a third transistor (as a constant-current source) contributes greatly to this desired condition. Under these condition, as shown in Fig. 3.4, we can evaluate the extent of the common-mode rejection by use of the ratio comparing the desired differential amplification (A_d) with the undesired amplification of the common-mode signals (A_{cm}). (Note that A_d as used here corresponds to the voltage gain previously designated as A_v). A figure of merit for the common-mode rejection ratio (CMRR) is given as

$$\text{CMRR} = \frac{A_d}{A_{cm}}.$$

This ratio will depend upon how large a resistance is presented by the collector of Q_3 at the junction of the common emitters; its equivalent value is designated R_{cm}. The resulting CMRR can be expressed as the number of times that A_d is greater than A_{cm}, or, as is usually done, by the logarithmic ratio in decibels, as illustrated in the following example:

Example 1 (Sec. 3-5)

DETERMINING CMRR

We shall assume that measurements in the circuit of Fig. 3.4 give the following results: when tested for differential output (V_{out}) for single-ended input (V_{in}),

$$\text{input}\,(V_{\text{in}}) = e_1 - e_2 = 1\text{ mV} - 0 = 1\text{ mV};$$

$$\text{output}\,(\text{diff}) = V_{\text{out}} = v_{O2} - v_{O1} = 200\text{ mV}.$$

Thus, *differential gain* $A_d = V_{\text{out}}/V_{\text{in}} = 200\text{ mV}/1\text{ mV} = 200.$

Also, when both input terminals are tied together, assume that a *common-mode* signal of 10 mV (V_{cm}) is applied to the common input, and a common-mode output of 2 mV is obtained:

$$\text{common-mode gain } A_{cm} = V_{\text{out}\,(cm)}/V_{cm} = 2\text{ mV}/10\text{ mV} = 0.2.$$

Then CMRR $= A_d/A_{cm} = 200/0.2 = 1{,}000$ (or 60 dB).[1]

Note: The effectiveness of this fairly conservative value for the CMRR may be noted from this example; in spite of the fact that the common-mode signal is 10 times as great as the differential signal, the differential output of 200 mV contains only a negligible amount of common-mode output (2 mV, or about one percent). More typical values of 90 to 100 dB for the CMRR are obtainable, thus greatly reducing this percentage of common-mode error in well-designed differential-amplifier stages.

3-6. INPUT-STAGE VARIATIONS IN PRACTICAL INTEGRATED CIRCUITS

In examining schematic diagrams of commerical ICs, we often find that the input to each base of the matched pair of transistors is applied to a Darlington-connected arrangement for each side of the pair, as shown in Fig. 3.5. This refinement (often found in operational amplifiers) is designed to achieve a higher input impedance and so to reduce the quiescent bias currents, especially for low-level signals. Other versions might be found using FETs for the matched pair to achieve an even greater value of input inpedance. Various other stratagems using additional transistors are employed in practical

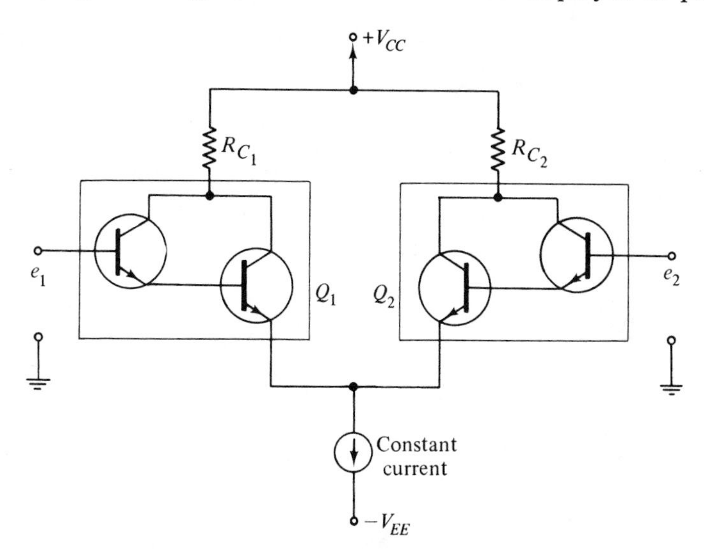

FIG. 3.5 Darlington-connected transistors, Q_1 and Q_2, serve to reduce the amoúnt of bias current required and, therefore, to increase the input resistance of the circuit.

[1] For conversion to decibels, see Appendix I.

linear ICs to achieve certain specific characteristics, and these are detailed in later chapters.

Summarizing the Differential-Amplifier Stage

The foregoing discussion may be summarized by saying that the *differential-amplifier circuit* has become practically a *standard for the input stage of an integrated circuit,* because of the greatly improved overall characteristics that it offers in a stable amplifier arrangement; its performance generally exceeds most alternative arrangements, even when using discrete components. The benefits obtained by the use of three integrated transistors for this input stage may be listed as improved values of *high input impedance* and *wide bandwidth* in a *direct-coupled amplifier* and very good *suppression of drift and undesired signals.* Another feature of integrated circuits that may be noted here is the free use of a large number of transistors that would usually not be practical in a discrete circuit. There will be many other examples of this generous use of numerous transistors fabricated on a single chip for specific objectives, as will be seen in the subsequent discussions of typical integrated circuits.

4

OPERATIONAL-AMPLIFIER CHARACTERISTICS

4-1. BASIC REQUIREMENTS FOR THE OPERATIONAL AMPLIFIER

The most prominent form of linear integrated circuit, by far, is the *operational amplifier*. In the amplification process it is generally important for the output to be a closely faithful reproduction of the input, and so *linear operation* is one of the major characteristics of this integrated circuit, accounting for the designation of "linear" in the LIC category. [The prominence of the operational amplifier types in this non-digital category explains, perhaps, the stretching of the designation of linear IC to include even some types having a non-linear output, such as comparators and regulators, in order to distinguish them from the digital (DIC) category.] Be that as it may, we can start with linearity of operation as one of the essential characteristics of the operational amplifier and proceed from there to discuss the other circuit requirements that combine to produce this very versatile form of amplifier, widely recognized by its familiar name OP AMP.

The OP AMP derives its highly attractive properties as an active device from its basic form of a very-high-gain amplifier, which also has provision for external feedback. In this form the output of the amplifier can be made to depend primarily on *externally connected passive elements*, so *its amplifying performance is virtually independent of its internal parameters*. To accomplish this highly flexible form of amplifier, there are three main requirements for the internal circuitry:

1. The *open-loop gain* A_{VOL} *of the amplifier must be very high;* (preferably well over 10,000 times, or 80 dB).
2. The *multiple stages must be direct coupled* (this allows both dc and ac amplification).
3. There must be provision for obtaining *an inverted output from which negative feedback can be obtained by a single external resistor* connected from the output to the input terminal of the amplifier (additional provision for a noninverting output is almost always present to allow positive feedback in a similar manner, when desired).

These three basic requirements are shown schematically in Fig. 4.1, in

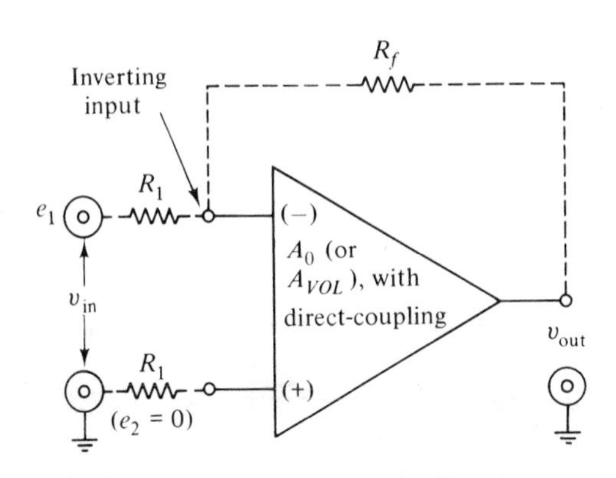

FIG. 4.1 Symbol of OP AMP, emphasizing three main requirements: (1) very-high open-loop gain, A_o or A_{VOL}; (2) direct-coupled stages; (3) external resistor (R_f) for negative feedback from output to inverting input (labeled minus).

which the triangle represents the large open-loop gain (A_{VOL}, abbreviated here as A_o); the two inputs (for dc or ac amplification) are labeled (−) for the inverting function and (+) for the noninverting function; also the feedback resistor (R_f) is shown connected externally between the output terminal and the inverting input (−) terminal to provide negative feedback. Under these basic conditions, the gain of the operational amplifier with feedback (A_f) (or, as usually stated, the closed-loop gain, A_{VCL}) can be expressed by the simple relation:

$$A_f = A_{VCL} = -\frac{R_f}{R_1} e_1 .$$

This relation also applies when the other input voltage (e_2) is nonzero (as was discussed in Chapter 3 for the differential amplifier circuit that forms the input stage of the OP AMP); in that case, the gain is expressed *in terms of the difference voltage*, as

$$A_f = A_{VCL} = -\frac{R_f}{R_1}(e_1 - e_2).$$

The analysis for tracing this relation for the gain with feedback (or closed-loop gain) is discussed in the next section.

4-2. ANALYSIS OF OPERATIONAL-AMPLIFIER ACTION

When the requirement for a very large open-loop gain is satisfied, we can analyze the circuit action of the operational amplifier in a simplified manner to readily yield the relationship for the amplifier gain with feedback. Referring to the generalized circuit of Fig. 4.2 (a), and considering the situation at the inverting input (−) of the amplifier, the voltage at that terminal (e_t) is the voltage resulting from the input source (e_1) diminished by the opposing feedback voltage (v_f), or

$$e_t = e_1 - v_f.$$

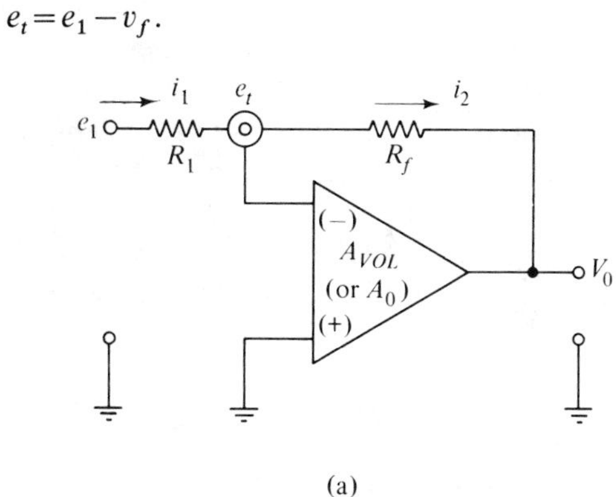

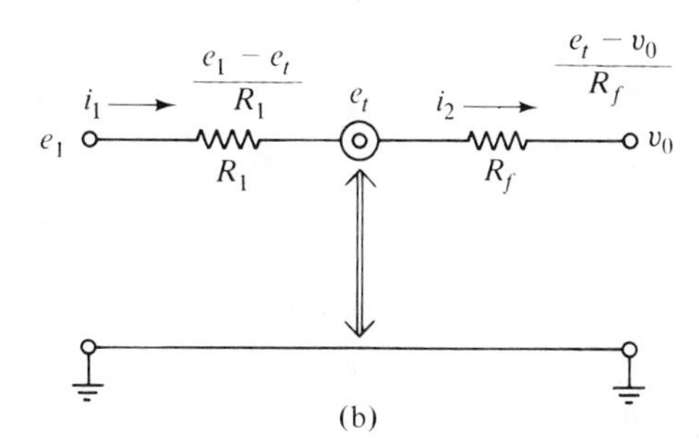

FIG. 4.2 Concept of virtual ground: (a) symbol diagram; (b) virtual ground shown as double-ended arrow.

The output voltage (v_o) can then be considered as the voltage resulting from the voltage at the terminal (e_t) multiplied by the open-loop gain, or

$$v_o = A_o(e_t)$$

and

$$e_t = \frac{v_o}{A_o}.$$

From this it can be seen that as open-loop gain (A_o) becomes larger, e_t becomes progressively smaller, so that e_t approaches zero as A_o approaches infinity. It follows, then, that in a practical amplifier with A_o sufficiently large ($> 10\text{k}\Omega$, and usually much larger) e_t can be considered as practically zero. (This remains true for all values of input voltage within the linear range of the amplifier, since larger input voltages result in correspondingly greater values of the feedback voltage.)

Virtual Ground

The situation where the voltage at the amplifier terminal (e_t) is substantially zero is known as a *virtual ground*, signifying a point that is *substantially at ground level, but not physically connected to ground.* Since we have both points (e_t and ground) at practically the same potential, we may assume that practically no current will flow between them. Any current i_1 entering terminal e_t may be considered as flowing directly to the output terminal, as current i_2.

By using this concept of a virtual ground[1] [shown in Fig. 4.2 (b) by a heavy double-ended arrow], we can readily arrive at an expression for the gain of the circuit. Here we can equate the current i_1 and i_2 as follows:

$$i_1 = \frac{e_1 - e_t}{R_1} = i_2 = \frac{e_t - v_o}{R_f}.$$

With e_t as substantially zero,

$$\frac{e_1}{R_1} = -\frac{v_o}{R_f}.$$

Thus the gain of the circuit with feedback is

$$A_f = A_{VCL} = \frac{v_o}{e_1} = -\frac{R_f}{R_1}.$$

with the negative sign indicating an inverted output.

[1] G. J. Deboo and C. N. Burrous, *Integrated Circuits and Semiconductor Devices*, McGraw-Hill, Book Company, New York, 1971.

In summary, this analysis demonstrates that the closed-loop gain of the practical inverting amplifier can be closely approximated as the *ratio of two external resistances*, based upon the action of an operational amplifier having a very large open-loop gain, and also arranged for negative feedback by means of an external resistor (R_f).

4-3. FLEXIBLE USE OF OPERATIONAL-AMPLIFIER CIRCUITS

Before entering upon the details of operational-amplifier characteristics, it is well to quickly glance at the variety of useful circuits that can be assembled quite easily by using an inexpensive general-purpose version of this versatile IC, considered as a single device, along with a few external components.

An introductory listing of 12 useful OP AMP circuits may be roughly divided into three convenient categories: general-purpose amplification (both as dc or ac amplifiers), analog computer elements, and miscellaneous uses, as follows:

A. GENERAL-PURPOSE AMPLIFICATION
 1. Inverting amplifier (for low-impedance signals).
 2. Noninverting amplifier (for high-impedance signals).
 3. Unity-gain inverter (as voltage follower).

B. ANALOG-COMPUTER ELEMENTS
 4. Summing amplifier (including a subtracting function).
 5. Integrator (for solution of differential equations).
 6. Differentiator (for pulse peaking).

C. MISCELLANEOUS USES
 7. Linear rectifier (especially useful for low-level ac signals).
 8. Active filters (low-pass, high-pass, and all-pass phase-shifting circuits).
 9. Logarithmic amplifier (for wide dynamic range).
 10. Sine-wave oscillator circuits.
 11. Multivibrator circuits (including astable, monostable, and bistable circuits).
 12. Function generator (including square-wave, triangular-wave, and ramp signals, including pulses).

The circuit applications listed are surely ample evidence of the flexibility of this op amp branch of the LIC field; moreover, it should also be borne in mind that new uses continue to be added for many particular applications. Within space limitations, many of these circuits will be described in Chapter 5; other circuit developments can be found in the references cited for further study which will be found within the chapter and at the end of Chapter 12.

4-4. GENERAL OPERATIONAL-AMPLIFIER CHARACTERISTICS

At first glance the data sheet of an operational amplifier shows a large number of characteristics that can be quite bewildering to anyone who is primarily interested in assembling a simple amplification function. But the numerous specifications—however desirable in some special instances—are by no means equally necessary for general-purpose use. For the sake of simplicity, then, it is well to first concentrate on just a few *major characteristics*, which are defined and illustrated in the next section; more refined characteristics are left for a succeeding section, in which more exacting requirements will be discussed.

4-5. MAJOR OPERATIONAL-AMPLIFIER CHARACTERISTICS

A first consideration in evaluating an OP AMP for general-purpose use is its *gain–bandwidth characteristic* for voltage amplification. Since the product of gain times bandwidth (G × BW) is generally a constant (around which gain can be traded off for bandwidth), this characteristic for each OP AMP is usually shown on the data sheet by means of a *semilog plot of open-loop gain* (A_{VOL}) *in decibels against log frequency* (f).

As shown in Fig. 4.3, one obtains the approximate bandwidth for any desired closed-loop gain by simply drawing a horizontal line from the desired value of gain to intersect the slope of the gain roll-off. The slope shown in Fig. 4.3 is the 6-dB/octave (or 20-dB/decade) slope that is typical of an internally compensated OP AMP (such as the popular **741** type and similar types). For the gain–bandwidth product of such an OP AMP, the graph shows the closed-loop gain of 100 times (40 dB) intersecting the slope at a cutoff frequency ($f_{c(-3\text{dB})}$) of 10 kHz, giving a *G × BW product of 1 MHz*. It will be noted that this agrees with the unity gain (0 dB) value of 1 MHz, giving the same product (1 × 1 MHz).

A relatively modest extension of the gain–bandwidth product is a feature of another popular OP AMP (the **748** type among others), which, for example, uses a single external compensating capacitor of 3 pF to extend the −3-dB bandwidth from 10 to 100 kHz at the same gain of 100. With this simple change of compensating capacitor, *the G × BW product is increased to 10 MHz*.

Since the graph for open-loop gain (A_{VOL} versus f) applies only for small-signal operation, a specification is usually included for *large-signal operation*. Here the value of gain is stated for the case where relatively full output swing is required ($V_{\text{out}} = \pm 10$ V). In the popular examples cited thus far, this large-signal gain is generally between 15,000 and 25,000 times.

In the customary specification data on *electrical characteristics*, the preceding major characteristics are summarized as follows:

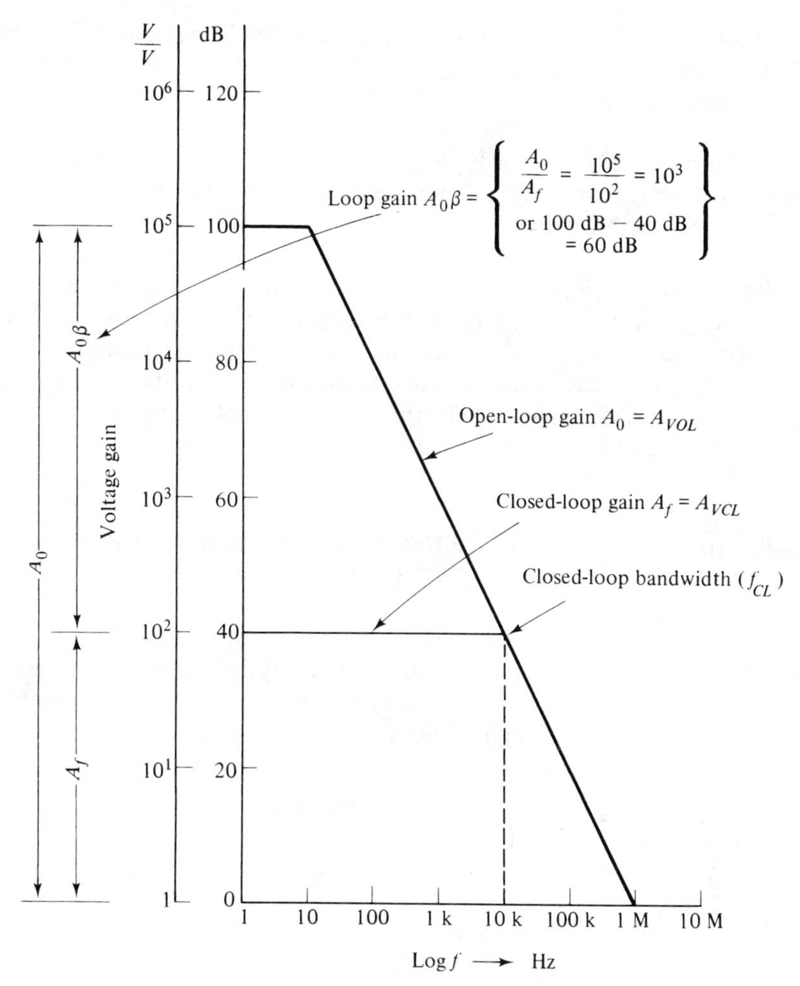

FIG. 4.3 Typical open-loop frequency-response (A_{VOL} or A_o), for internally compensated OP AMP, showing relation of closed-loop gain (A_{VCL} or A_f): $A_f = \dfrac{A_o}{1+A_o\beta} \approx \dfrac{A_o\text{(open-loop gain)}}{A_o\beta\text{(loop gain)}}$

1. *Open-loop gain (A_{VOL} or A_o).*—The voltage-amplification ratio of output to input voltage without external feedback illustrated in Fig. 4.1, with R_f disconnected:

$$A_{VOL} = V_{out}/e_{in}, \qquad \text{where } R_f = \infty.$$

2. *The G × BW product for small-signal operation.*—Obtained from the plot

of open-loop frequency response, as explained previously; further details for wide-band operation are discussed later, under the topic of frequency-compensation methods.

3. *Large-signal voltage gain* (V_{out}/e_{in}).—The ratio of the maximum output-voltage swing to the change in input voltage required to drive the output from zero to this voltage.

4. *Input resistance* (R_{in}).—Associated with the values of available gain is the input resistance that the OP AMP presents to the signal source. This depends, among other things, on the *input bias current* (at zero output). In effect, R_{in} determines the portion of the signal voltage remaining effective at the input after deducting any significant voltage drop across the internal resistance of the source.

4-6. CHART OF TYPICAL VALUES OF OPERATIONAL-AMPLIFIER CHARACTERISTICS

A comparison of some *typical values for the characteristics of some fairly standardized* OP AMPS is also very helpful in obtaining a clear view of the major features among the many refined specifications that proliferate in this highly competitive and rapidly changing field. Table 4-1 gives this chart.

TABLE 4-1 (Sec. 4-6)

Some Typical Operating-Amplifier Comparisons*

	*Device Type***									
25°C	**709**	**709A**	**741**	**748**	**101A**	**777**	**770**	**660**	**108**	**108A**
Input offset voltage (mV)	<5	<2	<5	<5	<2	<2	<4	<3	<2	<0.5
Input offset current (nA)	<200	<50	<30	<200	<10	<3	<2	<5	<0.2	<0.2
Input bias current (nA)	<500	<200	<200	<500	<75	<25	<15	<15	<2	<2
Input resistance (MΩ)	>0.040	>0.085	>1	>0.5	>1.5	>2	100 typ	>4	>30	>30
Large-signal voltage gain†	45,000	45,000	>50,000	>50,000	>50,000	>50,000	>50,000	>25,000 $R_L = 10$ kΩ	>50,000 $R_L = 10$ kΩ	>50,000 $R_L = 10$ kΩ
Full-temperature										
Input offset voltage (mV)	<6	<3	<5	<6	<3	<3	<7	<5	<3	<1
Input offset voltage (μV/°C)	NS	<25	<20	NS	<15	<15	NS	<25	<15	<5
Input offset current (nA)	<500	<250	<500	<500	<20	<10	<5	<5	<0.4	<0.4
Input offset current (pA/°C)	NS	<500	<200	NS	<200	<150	NS	<40	<2.5	<2.5
Input bias current (nA)	<1,500	<600	<800	<1,500	<100	<75	<35	<25	<3	<3
Large-signal voltage gain†	25,000 to 70,000	25,000 to 70,000	50,000	>25,000	>25,000	>25,000	>25,000	>25,000 $R_L = 10$ kΩ	>25,000 $R_L = 10$ kΩ	>40,000 $R_L = 10$ kΩ
Output voltage swing† (V_{p-p})	>20	>20	20	>20	>24 $R_L = 10$ kΩ	>24 $R_L = 10$ kΩ	>24	>26 $R_L = 10$ kΩ	>26 $R_L = 10$ kΩ	>26 $R_L = 10$ kΩ
Input voltage range (±V)	>8	>8	>10	>12	>15	>12	>12	>13.5	>13.5	>13.5
CMRR (dB)	>70	>80	>70	>70	>80	>80	>80	>80	>85	>96
Supply current (mA)	NS	<4.5	<4.5	<3.3	<2.5	<3.3	<2	<1	NS	NS
Unity-gain slew rate (V/μs)	NS	NS	0.5 typ	0.5 typ	0.5 typ	0.5 typ	2.5 typ	>0.1	0.2 typ	0.2 typ
Operating temp range (°C)	−55 to +125	−55 to +125	−55 to +125	−55 to +125	−55 to +125	−55 to +125	−55 to +125	−55 to +125	−55 to +125	−55 to +125
Operating supply voltage (±V)	9–15	9–15	9–15	15	20	15	15	20	20	20

† $R_L = 2$ kΩ unless otherwise specified.

* L. Altman, "Bridging the Analog and Digital Worlds with Linear ICs," *Electronics*, June 5, 1972 (Vol. 45, No. 12), Special Report (p. 85), copyright 1972, McGraw-Hill, Inc. All rights reserved. (*Note:* It must be understood that the examples of op amp types given in the chart are for types that were widely used as of the publication date of the chart.)

** For description of *Device Type* see Appendix III, where the extensive CROSS-REFERENCE listing identifies the type number by the italicized (or bold) part of the manufacturers' model number.

Although it represents an excellent and fairly detailed comparison of widely used general-purpose OP AMPS, it is well—certainly at the outset—to concentrate on the few major characteristics previously mentioned in this section, after first identifying the popular type numbers that are listed. (Note that the manufacturer's identifying prefixes that were listed in Section 1-7 are omitted in the chart, since *practically all these general-purpose monolithic types* are *second-sourced* by a number of manufacturers, as can be seen by the emphasized numbers in the *Cross-Reference list* of OP AMPS given in Appendix III.)

Description of Listed Op-Amp Types

Type **709** (such as μA**709**) one of the early types, is suitable for operation over a fairly wide band of frequencies, but requiring, however, different values of two compensating capacitors (C_c) for various values of gain; thus, at closed-loop gain of 100 (40 dB), a bandwidth of more than 1 MHz is obtained with $C_{c_1} = 100$ pF and $C_{c_2} = 3$ pF (as recommended for that value of gain), resulting in a *gain–bandwidth product of over 100 MHz.*

Type **741** *(such as μA***741***) is internally compensated*, and is unconditionally stable at all values of gain without any external capacitor; the resulting f_c for 40-dB gain is 10 kHz, giving a *gain–bandwidth product for this condition of 1 MHz.*

Type **748** *(such as μA***748***) offers an extended bandwidth* over the **741** type, using only a single compensating capacitor (such as 30 pF) for all values of gain when a limited bandwidth (of around 10 kHz) is satisfactory. Otherwise, using a smaller C_c of 3 pF at the same gain of 40 dB, f_c is extended from 10 to around 100 kHz, with a resulting *gain–bandwidth product of 10 MHz.*

Type **101A** (such as LM**101A**) has the same simple (30-pF) single-capacitor compensation feature as the **748** type, but with *improved input specifications.*

Type **777** (such as μA**777**) and type **660** (such as SN52**660**) are termed *"precision"* OP AMPS (as are the others that follow, indicating usefulness in circuits requiring greater accuracy); they feature *smaller offset-error values.*

Type **770** (such as SN52**770**) and type **108A** (such as LM**108A**) are *super-beta transistor* types; they both feature *high input resistance*; additionally, the **770** type achieves a *higher slew rate* (also indicating better high-frequency response), while the **108A** type exhibits still *further improvement in offset-error and drift characteristics.*

Other advanced types of OP AMPS that may be just mentioned at this point include the *FET-input types* (such as the μA**740**) and *programmable types* (such as the CA**3080**); these and other types are discussed later in Chapters 11 and 12.

4-7. INTERPRETING OPERATIONAL-AMPLIFIER DATA SHEETS

When selecting an OP AMP for a particular purpose, one is confronted with the difficulty of choosing from a large number of OP AMP types (even in this limited category of the wider field of linear ICs). For purposes of clarification, this chapter has concentrated on examples of typical specifications primarily for the widely used types (leaving highly specialized types for later); yet it is well to further subdivide these popular types according to three main kinds of applications (specific application details are covered in Chapter 5). Accordingly, in summarizing the importance of various specifications, examples of some suitable type numbers[1] are given next, along with those characteristics that are most pertinent for the *three general kinds of applications*, as follows:

A. For *voltage preamplification, within the audio-frequency band,* the main characteristics involve
   1. Large-signal voltage gain (V/V).
   2. Input resistance (R_{in}).
   3. Output-voltage swing ($\pm V_{out}$).

(EXAMPLES: TYPES **741**, **748**, **101A**, AND **770**.)

B. For *wider-band ac amplification*, additional consideration is given for the following characteristics:
   1. Open-loop frequency-response graph (for G × BW product).
   2. Frequency-compensation methods (C_c).

(EXAMPLES: TYPES **709**, **748**, **101A** AND **777**.)

C. For *"precision" dc amplification purposes* (as in instrumentation and analog computation) additional consideration is given for the following characteristics:
   1. Input-offset values of voltage and current (V_{io} and I_{io}).
   2. Input-offset drift with temperature (microvolts or picoamperes per degree Celsius).
   3. Common-mode rejection ratio (CMRR).
   4. Unity-gain slew rate (volts per microsecond).

(EXAMPLES: TYPES **660**, **770**, **777** AND **108A**.)

Keeping in mind the different requirements of these three kinds of amplification, that is, narrow-band, wide-band, and dc amplifiers, we can make a more sensible choice of a suitable OP AMP by giving proper weight to the various specifications. Accordingly, in presenting detailed application

[1] See Appendix II for "Selection Guide for OP AMPS," and also Appendix III for CROSS-REFERENCE listing of model numbers from other manufacturers.

information for the schematic diagrams that follow in Chapter 5, a particular OP AMP type is suggested with the understanding that other type numbers will also be suitable, as determined by the pertinent specifications.

4-8. APPROXIMATE VALUES OF TYPICAL OP-AMP SPECIFICATIONS

The numerical values listed for comparison in the chart of Table 4-1 can be very helpful in providing a preliminary estimate of "ball-park" figures for the various types of general-purpose OP AMPS. Referring back to the first two major characteristics (outlined previously in Section 4-5) the values of open-loop gain (A_{VOL}) and the gain–bandwidth product (G × BW) are not given on the chart, since these figures are usually obtained from the graph of the open-loop frequency response (similar to Fig. 4.4 and generally found on the in-

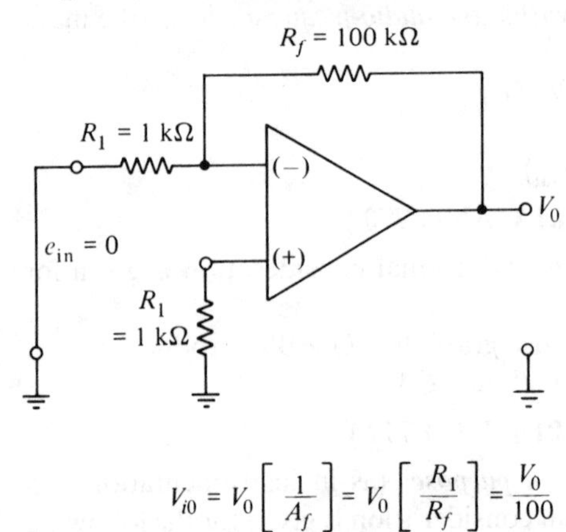

$$V_{i0} = V_0\left[\frac{1}{A_f}\right] = V_0\left[\frac{R_1}{R_f}\right] = \frac{V_0}{100}$$

FIG. 4.4 Measuring input offset voltage (V_{io}): with both inputs at ground, the input offset voltage is V_o divided by the closed-loop gain.

dividual data sheet); however, an estimate of the typical values for the G × BW product has been discussed in the previous section, where each type of OP AMP was described. (*Note:* The significance of the ratio of A_{VOL}/A_{VCL}, i.e., the *loop gain* $A_o\beta$, which is indicated on the plot of Fig. 4.4, is discussed later with regard to its effect on accuracy requirements.)

Likewise, the other two major characteristics (large-signal voltage gain (V/V) and input resistance (R_{in}) have also been estimated in the previous discussion of each type in Section 4-6). The characteristics that are discussed next have greater or smaller significance as specifications, depending on particular applications,

Offset-Error Specifications

The specifications that are given for *input offset voltage* V_{io} (or offset current

I_{io}) and the corresponding values of *offset drift* (microvolts or picoamperes per degree Celsius) are of negligible importance in the amplification of ac signals, but they do assume important significance in dc applications, *where precise dc information is required.* Such dc applications include dc instrumentation, analog computation, and, in some cases, digital interfacing.

For these dc applications, the values of offset error given in the chart present a basis of cost–performance selection. The values shown for *input offset voltage* (V_{io} *or sometimes* V_{ios}) range from around 5 mV for the **741** type down to less than $\frac{1}{2}$ mV for the **108A** type. However, a provision for the control of *offset null* is generally shown in the data sheet, so that this error can initially be adjusted to zero; consequently, greater importance is attached to the *offset-voltage drift specification.* Values for this drift characteristic are seen to range from 25 down to 5 μV/°C.

In a similar manner, the specifications for *input offset current* (I_{io}) and corresponding *drift* (in picoamperes per degree Celsius) assume greater significance when working with a high-impedance source, where it is desirable to keep the resulting voltage drop to a minimum.

Full-Temperature Operation

The values shown in the lower portion of the chart apply for all the op-amp types over the operating-temperature range of -55 to $+125$°C (the *"Military" grade*), as opposed to the room-temperature (25°C) values shown in the first portion of the chart. A lower-grade (and lower-cost) type of 0 to 70°C is generally available for these popular op-amp types, and this is indicated by various suffixes to the type number (such as C for the *commercial* grade) or by other devices. In many cases, an *industrial* grade is also offered as an in-between grade; for example, in the case of National types, we have **LM101A** (military), **LM201A** (industrial), and **301A** (commercial) as various grades of the **101A** type. It is well to emphasize here that the commercial grade (0 to 70°C) would likely be quite satisfactory for the great bulk of general-purpose OP-AMP applications, as can be determined from the manufacturer's data sheet for that type.

Output Specifications

The amount of voltage output that may be expected (over full-temperature operation) is shown as *output-voltage swing*, and is seen, in all cases, to exceed 20 V peak to peak (± 10 V), under the condition where the load resistance (R_L) is greater than 2 kΩ in most cases (but greater than 10 kΩ in the "precision" types). Since the *large-signal voltage gain* also exceeds 25,000 in all these cases, the full output voltage can be obtained for input signals as small as about 1 mV peak-to-peak. The maximum for the *input-voltage* range is around ±10 V.

The preceding values indicate the great versatility of the OP AMP in its ability to provide almost any reasonable value of undistorted voltage amplification within its *power-dissipation limits.* As a rule of thumb, the value of 500 mW for the 748 type can be taken as representative of the maximum power-dissipation value for most general-purpose OP AMPS. This indicates that this type of general-purpose OP AMP must be regarded primarily as a *voltage preamplifier*, since the amount of current in the load can be only a few milliamperes at best, in order to avoid distortion from saturation effects. In cases where greater amounts of current are required, an additional high-current (or booster) type of OP AMP (or in extreme cases, an external power transistor) is usually added in cascade with the general-purpose OP AMP, as is discussed later under the topic of power amplifiers. In the opposite situation, where the requirement for output current is not important, there are "micropower" OP AMPS, which draw very little power from the supply. These form a special type, as discussed later, for providing usable amplification from much smaller power sources, much less than the usual ±15 V and the 1 to 2 mA that are typically used in the general-purpose types.

Common-Mode Rejection Ratio

This characteristic measures the ability of the OP AMP to reject interfering signals (such as hum pickup) that are equally present at both inputs. With this ratio exceeding 70 dB in all cases, the gain for the desired signal is assured to be at least 70 dB or about 3,000[1] times greater than the gain for the unwanted common-mode signal.

Unity-Gain Slew Rate

Measuring the ability of the OP AMP in following fast-changing signals, this characteristic assumes importance in fast-rise pulse circuits and other such signals emphasizing high-speed operation (such as in many digital-interface types). This specification is discussed later at greater length for cases requiring values greater than the 100- to 500-mV/μs rates listed in Table 4-1. In the CROSS REFERENCE[2], these types are designated as "Fast" or "High Slew Rate" types.

[1] 3,160 times from the decibel conversion chart in Appendix I.

[2] See Appendix III.

5

GENERAL OPERATIONAL-AMPLIFIER APPLICATIONS

5-1. KINDS OF OPERATIONAL-AMPLIFIER APPLICATIONS

The successful introduction and acceptance of the integrated-circuit version of the operational amplifier radically changed its role from its *original functions in analog computation.* In its discrete form, whether using tubes or transistors, the operational amplifier occupied the fairly lofty position of a high-performance (and correspondingly costly) dc amplifier, capable of performing the mathematical functions of summing, multiplying by a constant, and integrating analog signals in such a manner as to accomplish a solution of differential equations. It thus served as a powerful mathematical tool in working with the highly flexible electrical analog forms that were the simulated equivalents of bulky or otherwise awkward physical processes.

With the breakthrough in techniques for the fabrication and circuit design of integrated circuits, the operational amplifier became available as a more reliable and considerably less costly device; since then it has progressed rapidly into becoming the "versatile OP AMP,"[1] closely approach-

[1] M. Kahn, *The Versatile Op Amp*, Holt, Rinehart and Winston, Inc., New York, 1970.

ing the ideal of a "universal amplifier," with its many uses as a building-block device.

In its present IC form, there are obvious physical constraints that limit the IC OP AMP with respect to such things as *high power* and *LC circuits*, conditions that generally call for external (or hybrid) additions to the circuit. Yet, even in these cases, developments in design strategy are gradually encroaching on such limitations. For example, advances have been made toward higher output-current amplifiers, and also in gyrator circuits that have been developed to simulate inductors.

The applications given in this chapter are necessarily restricted to typical examples. They illustrate some of the wide diversity of basic OP AMP uses in various *amplification and oscillating functions using general-purpose OP AMPS*. A separate chapter discusses instrumentation OP AMPS, followed by chapters discussing many other linear IC functions, such as comparators, regulators, tuned circuits, active filters, and interfacing with digital circuits.

5-2. CHOICE OF OPERATIONAL AMPLIFIER IN TYPICAL APPLICATIONS

Where a particular type of OP AMP is suggested in the application schematics, it should be noted that a number of suitable substitutions are possible. The ultimate choice will obviously also include the relative-cost factor balanced against the specifications of the chart (Table 4-1) and the relative weights attached to them, as previously discussed in the sections that follow the chart. (Possible equivalents other than those in the table may be found in the CROSS-REFERENCED list in Appendix III and the "Selection Guide" in Appendix II.)

For constructing prototypes of the applications, practical details are given in Chapter 6 for *testing and breadboarding* the application examples.

5-3. BASIC CIRCUIT CONFIGURATIONS

In the application information that follows, we start with the two configurations for *voltage amplification* (inverting and non-inverting) for which the fundamental gain relation (R_f/R_1) was discussed in Section 4-2. In addition, arrangements for analog computation include configurations for *summing*, *multiplying by a constant*, *integrating*, and *differentiating*. Also, *oscillator functions* are arranged by means of a configuration to provide positive feedback. Other *nonlinear functions* (such as *detectors*, but still done with the linear ICs) are also given. These configurations, together with their modifications, result in the typical applications that follow.

5-4. INVERTING VOLTAGE AMPLIFIER

The voltage amplifier of Fig. 5.1 is a fairly simple *general-purpose preamplifier.*

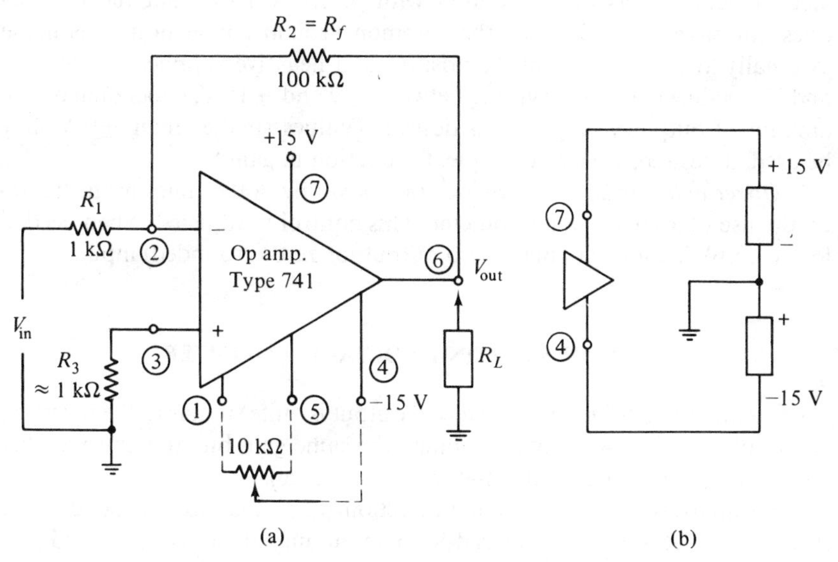

NOTE: Pin nos. apply to 8-pin package (either 8-pin metal can or 8-pin mini-DIP types).

FIG. 5.1 Inverting amplifier: (a) circuit for general-purpose preamplification (voltage gain of 100) using type 741 (internally compensated) OP AMP, with offset null adjustment shown by dotted lines; (b) arrangement for dual power supplies. *(Fairchild Semiconductor)*

The type 741 OP AMP is internally compensated, and, as such, is the easiest to use for applications *within the audio-frequency range.* We can obtain a voltage gain of any reasonable value (as low as unity gain and up to a large-signal gain of 50,000) in a very straightforward manner by the ratio of feedback resistor R_f to input resistor R_1; in the example, R_f/R_1 is 100:1 for a gain of -100. The other R_3 resistor, at the noninverting terminal 3, should theoretically be equal to the parallel combination of R_f and R_1 (R_1R_f/R_1+R_f); in this case, so close (within 1 percent) to R_1 alone, we can use the same value at both input terminals.

Input resistance for the inverting connection is fairly low, generally taken as equal to input resistor $R_1 = 1{,}000\ \Omega$ in this case; (the next section, on the noninverting connection, yields a much higher input resistance).

Frequency response, obtained from the open-loop frequency-response graph (see Fig. 4.3), will show a bandwidth or f_c (-3 dB) of approximately

10 kHz for this gain of 100 (or 40 dB). At a lower gain of 10, the bandwidth increases correspondingly, to about 100 kHz.

Power-supply connections [part (b) of the figure] show the dual-supply arrangement that is generally used with OP AMPS. Note that the IC itself does not have a terminal for the common ground connection; it is made externally at the junction of the positive and negative supplies. These (V+ and V− values) may be anything between ±9 and ±15 V, depending on the amount of output-voltage swing desired. (Values smaller than ±9 V may be used, if desired, with a consequent reduction in gain.)

Offset-null provision is provided (as shown by dotted lines in the figure) by the use of a 10-kΩ potentiometer. This control is adjusted, when needed for dc amplification, to obtain zero output for zero (grounded) input.

5-5. NONINVERTING VOLTAGE AMPLIFIER

The circuit of Fig. 5.2 for a noninverted output is quite similar to the inverting circuit of Section 5-4—providing much the same gain but at a considerably *higher input impedance* (in the high-megohm range).

By analysis (similar to that in Section 4-2),[1] the gain is found to be $A_{vcl} = 1 + (R_f/R_1)$, and the closed-loop input impedance $= Z_o(1 + \text{L.G.})$ or approximately the original open-loop input impedance multiplied by the open-loop gain (L.G.).

The use of the type 748 OP AMP in this circuit requires an external frequency-compensation capacitor (C_c), but because it has external compensation, it allows an *extended bandwidth* (beyond the audio range to around 100 kHz), which is obtained quite simply by reducing the nominal 30 pF capacitor (C_c) to 3 pF in this case. (In-between values can be obtained from the frequency-response curves given in the **748** data sheet.)

The *offset-null arrangement* (shown in dotted lines for dc applications) uses a 5-MΩ potentiometer in this case, but an alternative arrangement for a 25-kΩ potentiometer is shown in part (b) of the figure.

The *power supply* again shows a ±15 V dual supply, which can be lower (±9 V, or even less with reduced gain). Where only a single supply is available, an alternative arrangement is shown in part (c) of the figure for a 20 V supply, center-tapped at +10 V by two equal bleeder resistors (R).

Either of these widely used OP AMPS (types **741** and **748**) are generally useful for almost any reasonable value of gain, especially for ordinary *ac preamplification*, where offset-error drifts with temperature are not particularly significant.

[1] For derivation, see G. J. Deboo and C. N. Burrous, *Integrated Circuits and Semiconductor Devices*, McGraw-Hill Book Company, New York, 1971.

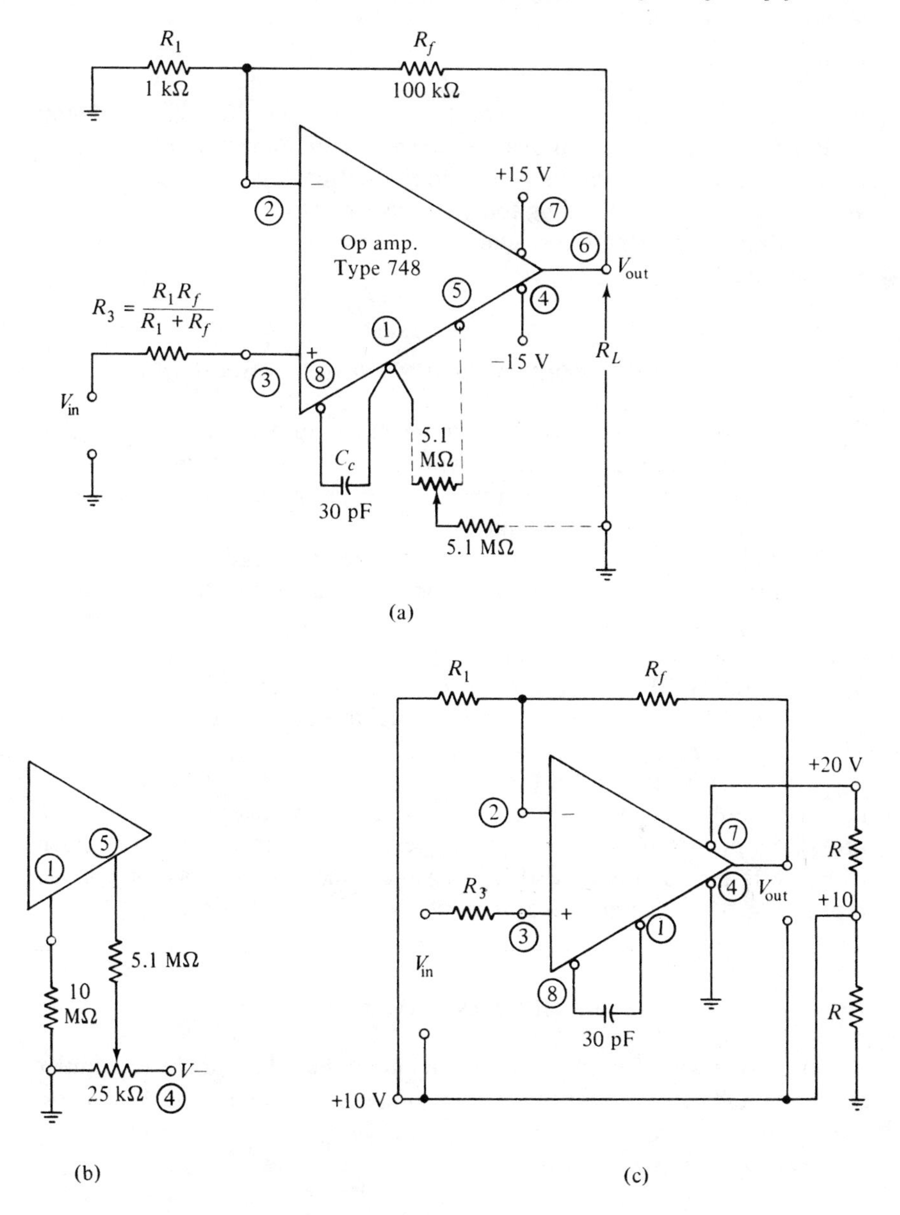

FIG. 5.2 Noninverting amplifier: (a) voltage gain of 100 circuit at high input impedance and bandwidth (≈ 10 kHz) can be extended by reducing C_c (e.g. when $C_c = 3$ pF, $BW \approx 100$ kHz); OFFSET NULL needed for dc amplification shown in dotted lines; (b) alternate OFFSET NULL arrangement; (c) circuit for single power supply. *(Fairchild Semiconductor)*

Modified Versions of Voltage Amplifier

Where there are stricter requirements for the amplifier, other refined op-amp types with tighter specifications are suggested in some typical examples. Likewise, there are instances where additional controls are desired (such as for varying gain, controlling tone, increasing current output, or the like); such additional information is summarized in Chapter 11 on Precision and Instrumentation types.

5-6. VOLTAGE FOLLOWER (UNITY-GAIN AMPLIFIER)

The circuit of Fig. 5.3(a) shows the connections for a *voltage follower* that produces unity gain (noninverted), with a high input impedance and low output impedance. Using the internally-compensated type 741 OP AMP, this very simple circuit serves quite effectively as a *buffer amplifier*, which is often employed when isolation between stages is needed.

A *fast voltage-follower* circuit is shown in part (b) of the figure, using the 101A type of OP AMP. The usual compensation capacitor ($C_{c_1}=30$ pF) of this type is used in conjunction with a second capacitor ($C_{c_2}=300$ pF, in series with a 10-kΩ resistor). This arrangement provides an *increased slew rate* of 1 V/μs to follow abruptly changing signals, and a power bandwidth of 15 kHz.

Note that both cases of the circuits in Fig. 5.3 show no provision for balancing out any dc offset errors. If this is desired, we can use the offset-null provisions previously given—for the type 741, using the 10-kΩ pot shown in Fig. 5.1, and for the type 101, using the same provision as for the 748 type in Fig. 5.2, which used a 5-MΩ pot and a 5-MΩ series resistor.

5-7. SUMMING AMPLIFIER

The OP AMP is widely used for combining a number of inputs in an additive manner, as, for example, in the analog computer. Using an inverting mode, the summing amplifier of Fig. 5.4 will produce an *output proportional to the negative of the sum of the inputs*:

$$v_o = -(k_1 e_1 + k_2 e_2 + k_3 e_3).$$

Inserting the value for each constant k, the equation becomes

$$v_o = -\left[\frac{Rf}{R_1} e_1 + \frac{Rf}{R_2} e_2 + \frac{Rf}{R_3} e_3\right],$$

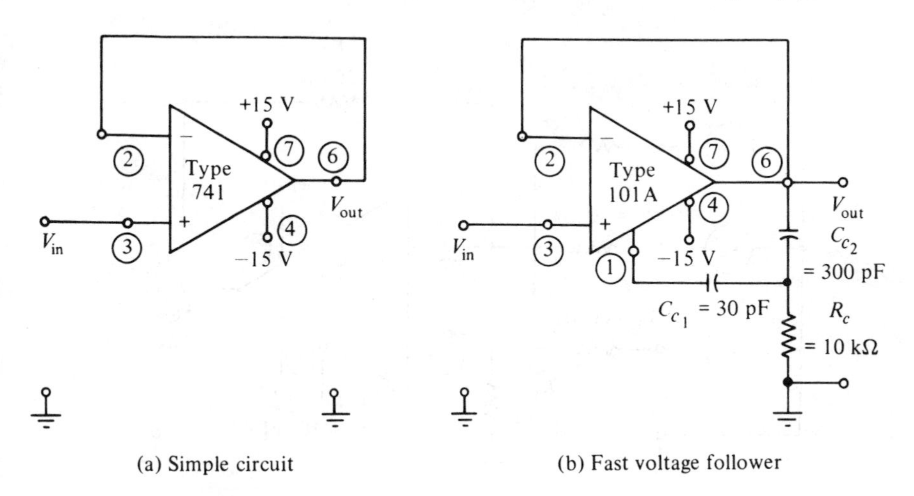

FIG. 5.3 Voltage-follower (unity-gain) circuit: (a) simple circuit suitable for most uses; (b) connections for fast voltage-follower. [(a) *Fairchild Semiconductor* and (b) *National Semiconductor*]

or, from Fig. 5.4,

$$v_o = -(e_1 + 4e_2 + 2e_3).$$

Since the junction point of the input resistors and the negative feedback resistor is forced to practically zero potential (virtual ground), the various inputs are effectively isolated from each other. In the circuit shown, a type 108A OP AMP is used by virtue of its combination of high-accuracy and high-input-impedance properties; when desired, however, a general-purpose OP AMP (such as type 741) can also be used at a small sacrifice in the accuracy of the summation.

Subtracting Function

The summing amplifier can add both positive and negative voltages, thus forming the equivalent of a *subtracting function* for general-purpose use. However, when voltages of the same relative polarity are to be subtracted, the OP AMP is generally used in the differential (or *dual-input*) mode, where input signals are applied to both inverting and noninverting terminals. This latter mode provides an output proportional to the difference of the inputs, as described later under differential OP-AMP applications in Instrumentation Amplifiers in Chapter 11.

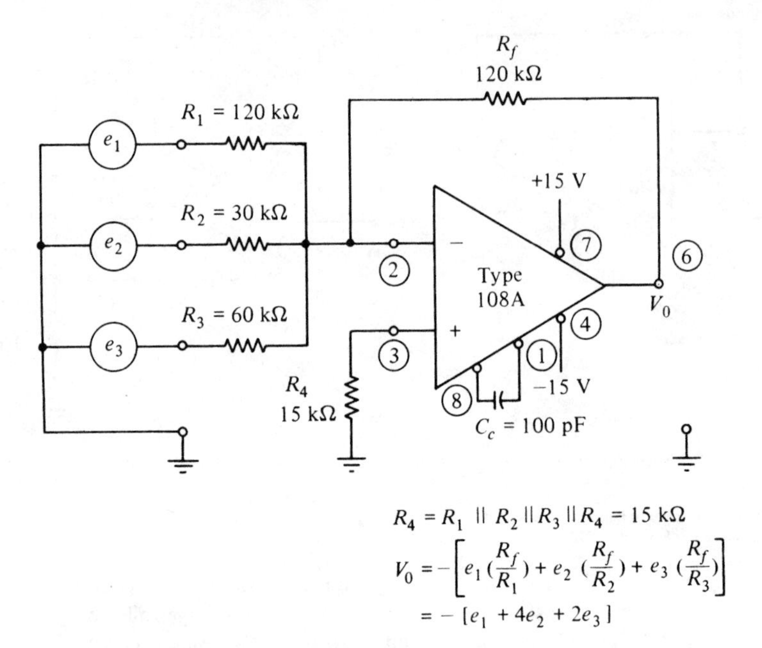

FIG. 5.4 Summing amplifier: the output (v_o) is the negative of the sum of the inputs.

5-8. INTEGRATING AMPLIFIER

The amplifier of Fig. 5.5 accomplishes the integration of an input signal by employing a capacitor (C_f) as its primary feedback impedance element (Z_f). Disregarding the shunting resistor (R_2) across it, for the time being, we may apply the basic op-amp relation in the *p-operator* form, where

$$p = \frac{d}{dt}, \qquad \frac{1}{p} = \int dt,$$

and

$$Z_1(p) = R_1, \; Z_f(p) = \frac{1}{C_p}.$$

Then, substituting the *p*-operator impedances,

$$v_o = -\frac{Z_f(p)}{Z_1(p)} e_1 = -\frac{1}{R_1 C_p} e_1$$

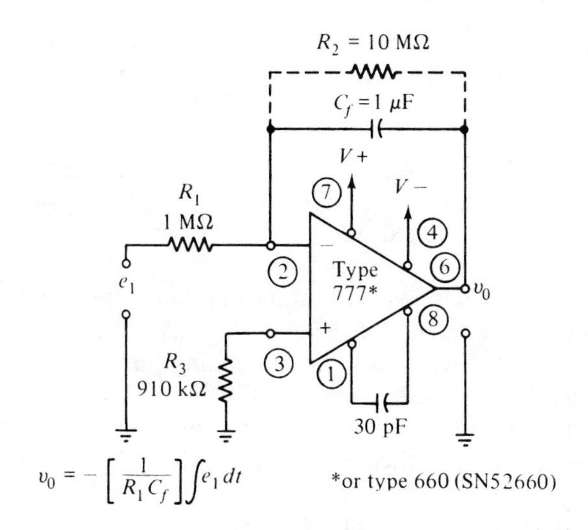

(a)

e1

v0

t

(b)

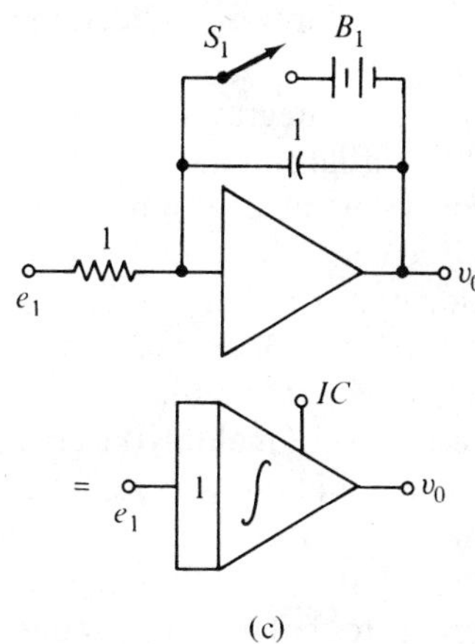

(c)

NOTE: Pin nos. apply to 8-pin package (either 8-pin metal can or 8-pin mini-DIP types).

FIG. 5.5 Integrating amplifier: (a) output v_o is the negative of the time integral of the input *e* times a constant, usually made unity ($R_1 = 1$ M, $C_f = 1$ μF); (b) triangular wave is the integral of a square-wave input; (c) initial conditions are shown, either by switch S_1 (normally closed) and voltage β_1, or in shorthand form below, by terminal IC (for Initial Condition). *(Fairchild Semiconductor)*

and, since $1/p$ signifies integration,

$$v_o = -\frac{1}{R_1 C_f} \int e_1 \, dt,$$

or the output will be the *negative of a constant times the time integral of the input*. Thus, as shown in part (b) of the figure, a square-wave input will be integrated to yield an inverted triangular-wave output with a time constant of $1/R_1 C_f$.

The resistor R_2, shunting C_f, is shown in dotted lines, to be used when needed to prevent saturation of the amplifier, especially if the OP AMP has substantial voltage offset. Since the integrating amplifier (without R_2) is operating open loop for direct current, the input offset voltage would be integrated, and the output would tend to rise in a ramp fashion to possibly saturate the amplifier (in either direction). This is avoided when the dc gain is limited by the shunt resistor R_2.

Because of the multiplying effect on the input offset voltage, a *precision type of OP AMP* is indicated for accuracy in integration. Consequently, the example for this application shows a type 777 OP AMP (or a similar precision type).

The integrator finds wide application in the analog computer for solving differential equations in simulation arrangements. For example, in equations for accelerated motion, the expression d^2y/dt^2 is easily solved for the displacement y by a double integration,[1] yielding $-dy/dt$ after the first integration and $+y$ after the second integration. In analog computation (and where saturation effects are avoided by large output-voltage swings without the use of R_2), the values of R_1 and C_f are usually taken as 1 MΩ and 1 μF, respectively, making the constant term $1/R_1 C_f$ equal to unity; this simplifies the computation. Also, when input current offsets are small enough to be neglected, it is usual to eliminate resistor R_3 at the non-inverting terminal (where it usually has the value of the parallel resistance of R_1 and R_f in order to balance the input currents, as previously discussed). With R_3 eliminated, the symbolic representation of the integrator takes either of the two simple forms shown in part (c) of the figure. The provision for *setting initial conditions* at the appropriate integrator is shown by the normally closed switch S_1, which applies the required voltage from battery B_1 and which is opened at the start of the integration.

5-9. DIFFERENTIATING AMPLIFIER

The *differentiation function* is accomplished by interchanging the input

[1] For a worked-out example, see S. Prensky, *Electronic Instrumentation*, 2nd ed., Prentice-Hall, Inc., Englewood Cliffs, N.J., 1971.

resistor and feedback capacitor of the OP AMP, as shown in Fig. 5.6. Examining the p-operator form of the basic op-amp relation, as in Section 5-8 (and neglecting the dotted-line connection), we find

$$v_o = -\frac{Z_f(p)}{Z_1(p)}\, e_1 = -\frac{R_f}{1/C_1 p}\, e_1 = -(R_f C_1 p)\, e_1$$

$$= -R_f C_1 \frac{de_1}{dt}$$

For the ideal amplifier, this relation yields the *negative of the differential of the input voltage times a constant.* It should be noted, however, that the capacitor impedance is now in the denominator of the formula, indicating that the differentiator will give an increased output at higher frequencies. Thus the *high-frequency noise problem* of the differentiator becomes *worse with increased frequency*, opposite to the integrator. For this reason, the analog computer seldom uses the differentiator arrangement of the OP AMP, preferring to arrange the simulation equations so that they require integration for their solution.

[Be careful here not to confuse the *differentiator*, discussed in this section, with the *differential amplifier* discussed at the end of this chapter; the first performs a differentiating function to produce the *mathematical derivative* of the input, whereas the differential amplifier (which is discussed later under instrument amplifiers) is an amplifier with a double-ended input which produces an output that is the *difference between the two input signals.*]

The differentiating amplifier does, however, find use in *wave-peaking* and other *wave-shaping* circuits; in such cases the use of the series resistor R_2 (shown in dotted lines in the figure) is justified to reduce high-frequency noise at the expense of attenuation. As an example of the use of the differentiating arrangement for wave shaping in pulse circuits, part (b) of the figure shows how a square-wave input produces a sharply peaked wave at the output.

5-10. SQUARE-WAVE GENERATOR (MULTIVIBRATOR)

There are many IC circuits capable of providing free-running (astable) multivibrator action, producing an output of rectangular or square-wave forms. These "relaxation oscillators" are essentially amplifiers that oscillate by means of *positive feedback, with the frequency controlled by a given RC time constant.* (Production of sine-wave output is covered in the next section.)

The effect of the positive feedback in a multivibrator circuit is to produce regenerative action that alternately affects each of two active devices, thus initiating abrupt transitions between on and off states. This results in the

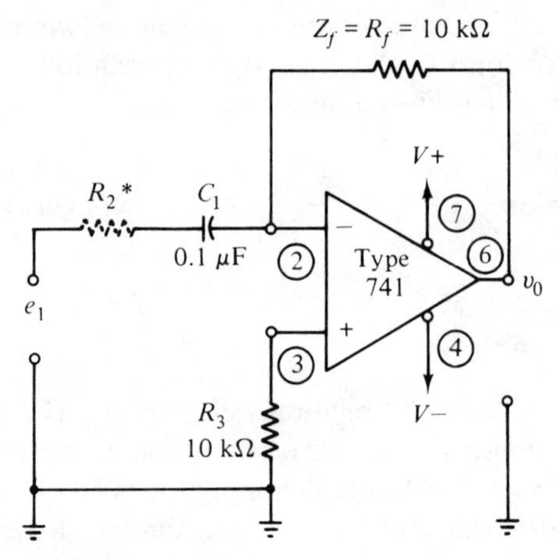

(a)

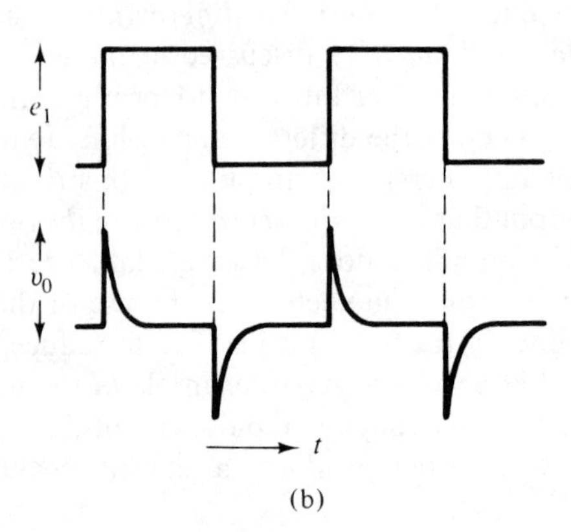

(b)

NOTE: Pin nos. apply to 8-pin package (either 8-pin metal can or 8-pin mini-DIP types).

FIG. 5.6 Differentiator amplifier: (a) the output v_o is the negative of the time derivative of the input e_1 times a constant ($R_f C_1$), which is 1 millisecond (1×10^{-3}) in this example (see text for optional R_2); (b) peaked output waveform resulting from differentiation of square wave.

fast-rising and fast-falling wave form of a rectangular wave, having a corresponding *period* of T_1 (on state) plus T_2 (off state). This type of rectangular wave form is usually called a square wave, even though asymmetric (the symmetrical square wave must have $T_1 - T_2$, and this symmetry can usually be obtained by simple adjustment of resistor values).

Although the recurrent switching (or astable) action of the multivibrator can be accomplished with digital IC gates by cross coupling, the op-amp circuits shown in Fig. 5.7 are preferred for providing a significantly greater output swing (easily up to 20 V peak to peak).

A simple circuit using a minimum of external components is shown in Fig. 5.7(a), employing a general-purpose (type 101A) OP AMP. This circuit is reliably self-starting, with the frequency adjustable by the value of C_1 within fairly wide limits.

A more complex circuit, shown in part (b) of the figure, has additional features in extending the low-frequency limits and additionally providing for a choice of either a low-impedance output or an output clamped to ± 6.2 V by the back-to-back connection of the Zener diodes. Here, again, stable operation at various frequencies can be obtained by adjustment of the C_1 capacitor.

The essential property of square waves lies in the fast rise times (and fall times) of the transitions. These abrupt step-voltage signals are the starting point for various other kinds of wave-form generation, including such wave forms as *triangular*, *ramp*, and assorted *pulse-type wave forms*. The use of linear ICs in assembling such *function generators* is covered in Chapter 12.

5-11. SINE-WAVE OSCILLATORS

The production of the basic sinusoidal waveform involves an extra consideration for amplitude stabilization of the output, to avoid distortion in the purity of the sine wave. Although this is accomplished fairly easily in a low-amplitude *LC* circuit, the examples given here are chosen to avoid the use of awkward inductors, and concentrate instead on flexible *RC* circuits that are more practical for OP-AMP applications (an exception is the use of dual OP AMPS in a *gyrator circuit* to stimulate an inductor; it is covered in Chapter 12). The *RC* frequency-determining elements employed in the examples are of two basic types, the *twin-T type* first, followed by the *Wien-Bridge* arrangement.

Twin-T Type

In the circuit of Fig. 5.8(a),[1] two T-sections in parallel (twin-T) are connected

[1] J. D. Lenk, *Handbook of Simplified Solid-State Circuit Design*, Prentice-Hall, Inc., Englewood Cliffs, N.J., 1971.

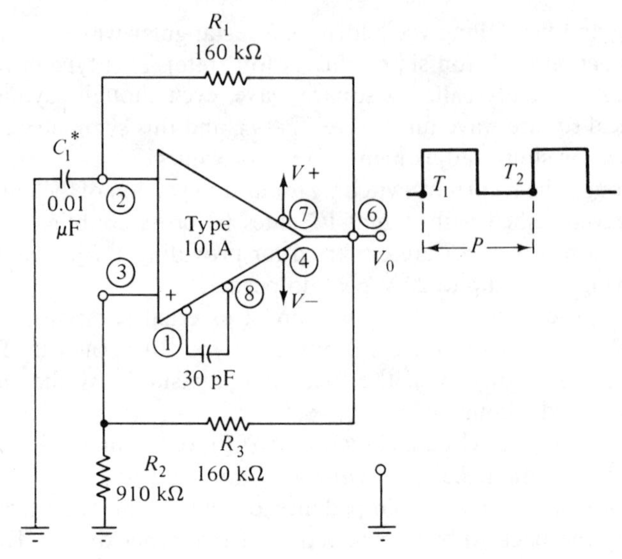

(a) Simple square-wave oscillator

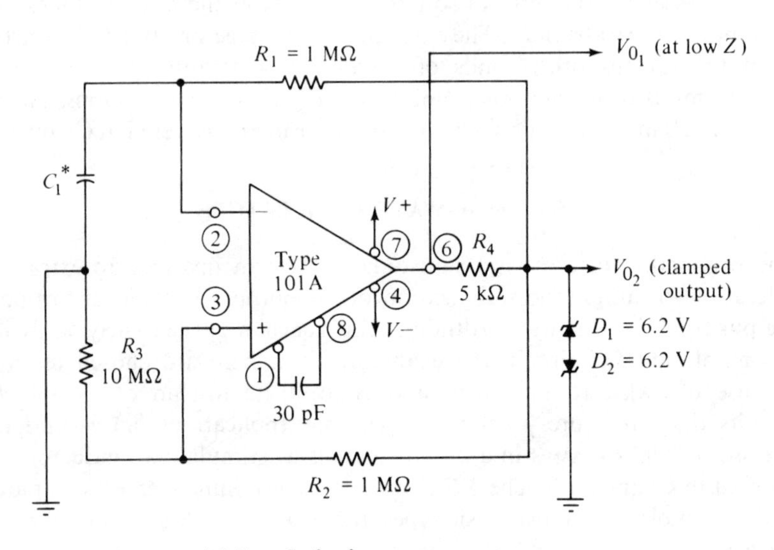

(b) Low-frequency square-wave oscillator

NOTE: Pin nos. apply to 8-pin package (either 8-pin metal can or 8-pin mini-DIP types).

FIG. 5.7 Square-wave (multivibrator) generator: (a) simple circuit using minimum components; (b) low-frequency version allows choice of low-impedance or clamped output. *(National Semiconductor)*

in the negative feedback loop. The oscillation frequency is based on the twin-T property of zero transmission at the parallel-resonance frequency, $\frac{1}{2\pi RC}$. Since positive feedback is supplied by the voltage-divider action of R_1 and R_2, oscillation takes place at this "pole" frequency, where the negative feedback through the twin-T is at its minimum.

For the values shown for R and C [RC product $=10(10^{-3})$ or 0.01], the oscillation frequency is calculated as 16 Hz. Any other combination giving this RC product can also be used for this frequency (with the provision that R should not exceed 2 MΩ). Likewise, the values for other frequencies

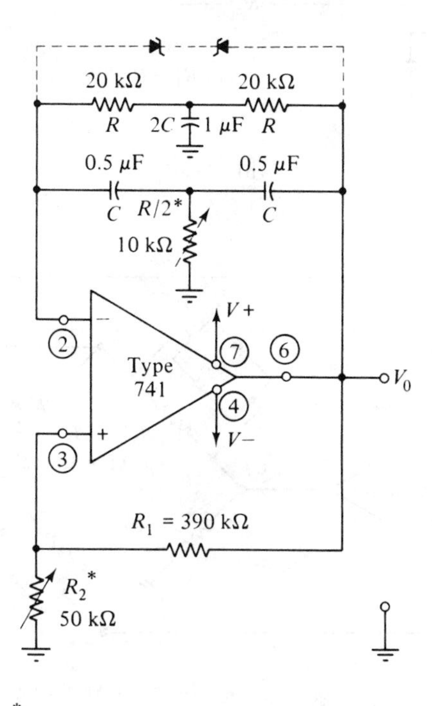

(a) Twin- T Circuit

FIG. 5.8 Sine-wave oscillators: (a) twin-T type of oscillator; (b) Wien-Bridge type of oscillator; (c) Wien-Bridge circuit of (b) redrawn to show bridge arrangement (see footnote references in text).

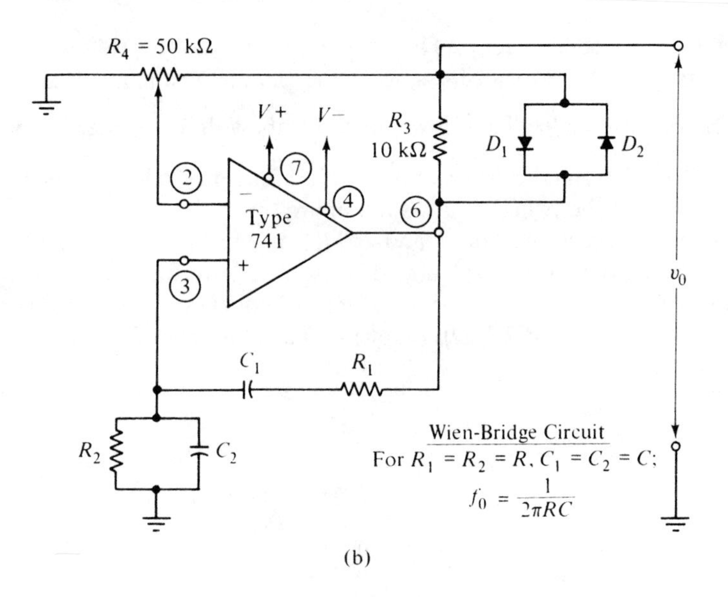

(b)

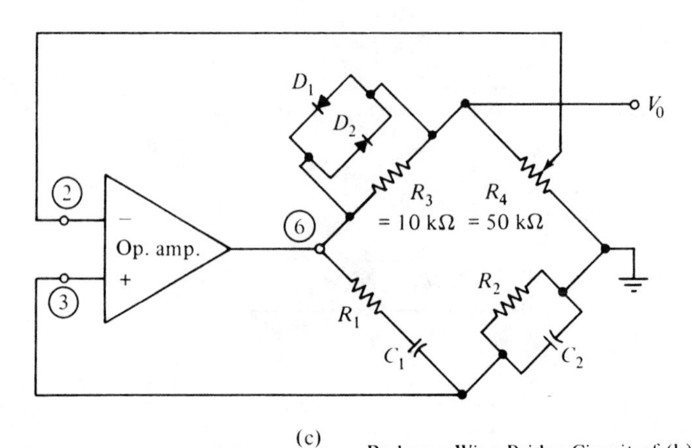

(c)

Redrawn Wien-Bridge Circuit of (b)

NOTE: Pin nos. apply to 8-pin package (either 8-pin metal can or 8-pin mini-DIP types).

FIG. 5.8 (*Contd.*)

are easily obtained from the *inverse relation of the RC product to the oscillation frequency*, as shown in Table 5-1.

TABLE 5-1

Component Values for $f = \dfrac{1}{2\pi RC}$

$f = \dfrac{1}{2\pi RC}$	R (kΩ)	C (μF)	*RC product*
16 Hz	20	0.5	$10(10^{-3})$
320 Hz $= (16 \times 20)$	10	0.05	$0.5(10^{-3}) = 10(10^{-3})/20$
1.6 kHz $= (16 \times 100)$	2	0.05	$0.1(10^{-3}) = 10(10^{-3})/100$

To assure a stable circuit that is self-starting, the value of R_2, controlling the amount of feedback, is made variable by a 50-kΩ control (nominal value is $2R$, or 40 kΩ) in conjunction with the fixed resistor R_1 having a value around $10R_2$ (here the closest standard value to 400 kΩ, or 390 kΩ).

Amplitude limiting, when necessary, is accomplished by the back-to-back connection of Zener diodes, shown in dotted lines. By making use of the nonlinear resistance of the diodes, the tendency for the amplitude to increase is opposed by the effect of decreased diode resistance, as it approaches the knee of its curve. A rule-of-thumb value for the Zener voltage is about 1.5 times the undistorted peak-to-peak value of the output sine wave.

Wien-Bridge Circuit [1]

The Wien-Bridge circuit enjoys popular use in commercial signal generators, where advantage is taken of the fact that it is conveniently tunable over a wide range of frequencies.[2] In the circuit of Fig. 5.8(b), the bridge arrangement is seen to consist of a series connection of R_1C_1 followed by a parallel combination of R_2C_2. With $R_1 = R_2 = R$ and $C_1 = C_2 = C$, the path has the property of *zero phase shift at a single frequency*, $f = \dfrac{1}{2\pi RC}$. Since this combination is used as the positive feedback element, oscillation occurs at this frequency of zero phase shift.

In its original form, amplitude stabilization in the Wien Bridge was traditionally done in the path of additional negative feedback by means

[1] G. E. Tobey, L. P. Huelsman, and G. G. Graeme, *Operational Amplifiers: Design and Applications*, McGraw-Hill, Book Company, New York, 1971.

[2] Details of operation of the basic Wien Bridge are given in Prensky, *op. cit.*

of the nonlinear resistance of a tungsten lamp. In this circuit, however, this function is obtained from back-to-back diodes, which are arranged to shunt the 10-kΩ resistor (R_3), which is in the negative feedback path, in series with the 50-kΩ distortion control (R_4). Here again, as the output amplitude tends to increase, the beginning conduction of the diodes lowers their impedance, thus opposing the increase by producing more negative feedback.

The values of R and C for obtaining the desired frequency, which were given in Table 5-1, also apply here, since the same formula applies, $f = \frac{1}{2\pi RC}$.

In part (c) of the figure the Wien-bridge circuit of part (b) is redrawn in a form to show the bridge arrangement more clearly.

5-12. LINEAR RECTIFIERS

For the rectifying function of changing ac signals to undirectional (pulsating dc) signals, the OP AMP offers a simple method, shown in Fig. 5.9, for overcoming the troublesome nonlinear properties of the rectifying diode used in this process. In the ordinary diode rectifier (whether half or full wave), diode conduction does not start until the ac voltage has risen to the breakover point, around 0.6 V (for the usual silicon diodes), and then provides current that has a nonlinear (exponential) relation to the input voltage.

Both of these difficulties are overcome by the use of the OP AMP in the linear half-wave rectifier of Fig. 5.9(a). Here, for small instantaneous negative voltages (less than 0.6 V), the amplifier is operating practically open loop, and so a relatively large positive voltage appears at the output end of the feedback elements, forcing the diode into forward conduction. In a similar manner, for various ac inputs the gain of the OP AMP automatically adjusts itself to keep the summing voltage at virtual ground. Thus the instantaneous values of $-v_o$ and e_1 are kept practically equal to each other, assuring a good linear relation. For this reason, the rectifier arrangement of the OP AMP as shown is called a "precision" or *linear half-wave rectifier.*

A similar scheme may be used to provide *full-wave rectification,* as in an ac voltmeter, as shown in Fig. 5.9(b). The meter sensitivity primarily determines the ohms-per-volt sensitivity of the resulting ac scale, with R_{cal} made variable to help in the calibration. Thus, roughly speaking, a 0- to 1-mA dc meter can provide a sensitivity around 1,000 Ω/V; progressively greater sensitivities can be obtained by using smaller full-scale dc values—coming down to a 0- to 50-μA meter for a sensitivity of 20,000 Ω/V.

A higher input impedance can be obtained from more complex circuitry, or as an alternative, by using another OP AMP to serve as a unity-gain buffer before the rectifying OP AMP. In any case, both the linearity and the sensitivity of ac rectification are improved by the amplifying ability of the OP AMP.

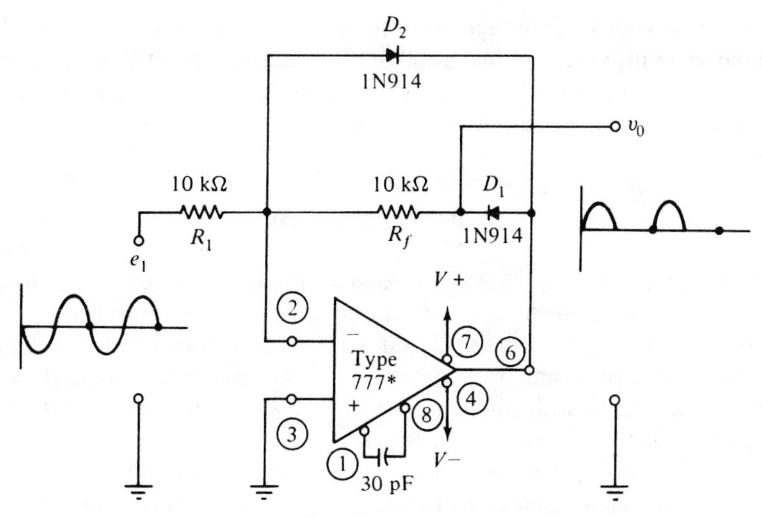

* μA 777, or type 660 (SN52660) (see Table 5-1)

(a) Linear half-wave rectifier

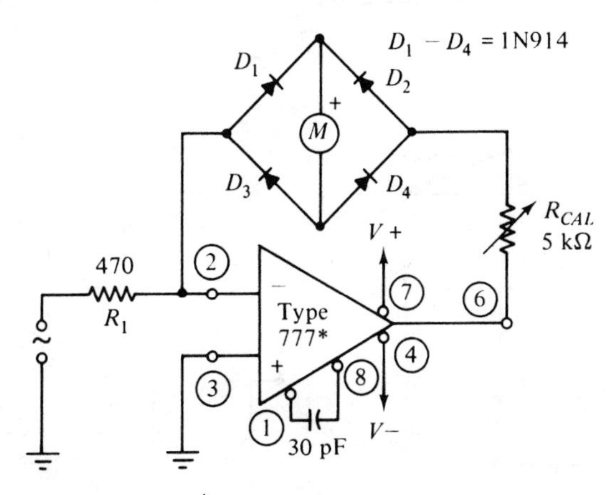

* μA 777, or type 660 (SN52660) (see Table 5-1)

(b) Full-wave voltmeter rectifier

NOTE: Pin nos. apply to 8-pin package (either 8-pin metal can or 8-pin mini-DIP types).

FIG. 5.9 Linear rectifiers: (a) half-wave rectifier for more linear rectification, using "precision" types of OP AMP; (b) full-wave rectifier for more linear scale of ac voltmeter.

This is an obvious advantage over nonelectronic voltmeters, where, for example, a multimeter having 20,000-Ω/V value for direct current comes out with a much poorer figure of around 5,000 Ω/V for ac measurements, thus necessitating a special *nonlinear* scale for the low ac ranges.

5-13. PEAK DETECTOR

It is often necessary to measure the ac voltage of signals having non-sinusoidal wave forms, and in such cases we cannot rely on the root-mean-square reading shown on the ordinary ac voltmeter, since it is generally calibrated on the basis of a pure sine-wave input. Such instances occur frequently with pulse measurements, with square and sawtooth waves, or even when dealing with greatly distorted sine waves. In such cases we are more interested in *peak measurements* (either single peak or peak-to-peak values).

An OP-AMP equivalent of a discrete peak detector is shown in Fig. 5.10. In part (a), the discrete circuit responds to the voltage developed by the fast charging of capacitor C by the positive peak, while very little voltage is lost by the relatively slow discharge of the capacitor through load resistor R_1 when the diode is reverse-biased on the negative swing, as shown by the output wave form.

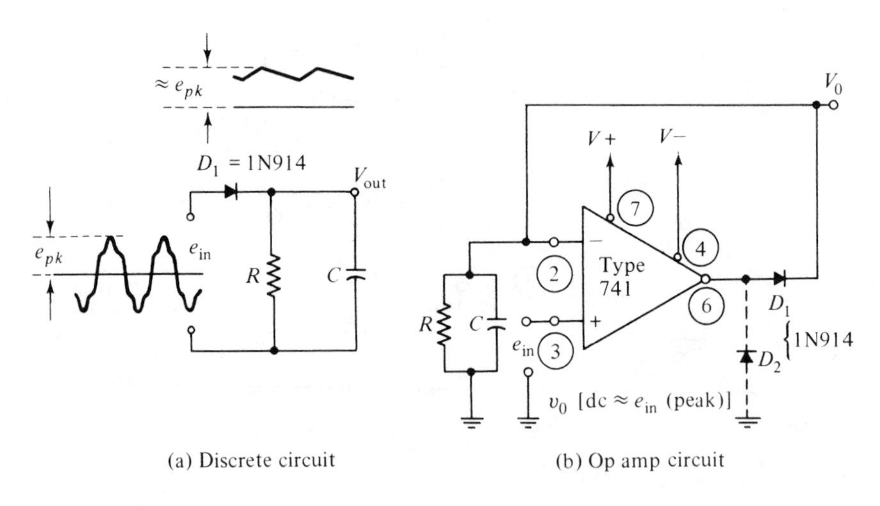

FIG. 5.10 Peak detector: (a) discrete circuit; (b) OP AMP circuit, producing a dc output voltage v_o, reasonably close to the positive peak of the signal.

The OP-AMP version of the peak detector, in part (b) of the figure, allows considerable improvement. Although a type 741 general-purpose OP AMP is shown, critical applications will profit from the use of an instrumentation type of OP AMP (discussed in Section 5-15) with advantages of an FET input and/or high slew rate. For ordinary applications, the output v_o is a dc voltage reasonably close to the amplitude of the positive peak of the input. Specific values for R and C depend upon keeping the RC time constant within a reasonable relation to the period of the input signal. (Where peak-to-peak voltage outputs are desired, the equivalent of a voltage-doubler circuit is used.)

The optional use of a second diode D_2, as shown by dotted lines in the figure, is used where it is necessary to avoid saturation of the OP AMP on the negative excursions of a large input signal. Also, when it is necessary to avoid loading the capacitor, a voltage follower may be interposed as a buffer between the peak detector and the load.

5-14. CURRENT-TO-VOLTAGE CONVERTER[1]

The current to be measured in the circuit of Fig. 5.11(a) is shown as being

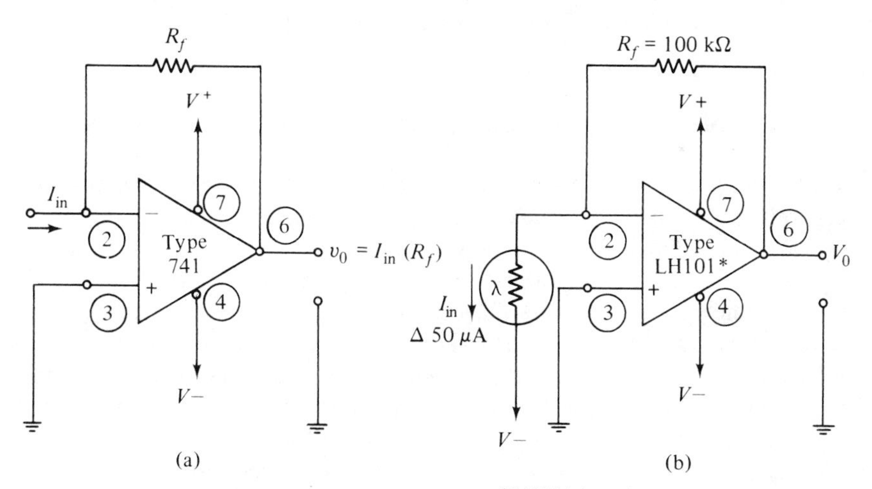

FIG. 5.11 Current-to-voltage converter: (a) general circuit; (b) converting a change in photocell current ($\Delta I_{in} = 50$ μA) to a voltage-change ($\Delta v_o = I_{in}(R_f) \doteq 5$ volts). *(National Semiconductor)*

[1] *Linear Applications*, National Semiconductor Corp.; see address in Appendix IV.

injected directly into the summing node (inverting terminal) of an OP AMP, This method of measuring a small current departs from the more conventional method of inserting a known resistor in the circuit and measuring the voltage developed across it. Another method for measuring such small currents might be that of using an OP AMP to amplify the small voltage across the inserted resistor; this, however, brings in the attendant error due to offset voltage.

The method of current-to-voltage conversion avoids these problems by making use of the basic OP-AMP property, where it forces the current through R_f continually to be equal to the input current I_{in}. The output voltage v_o thus simply becomes $I_{in}(R_f)$.

An example of the use of the current-to-voltage conversion for a photoconductive cell is shown in part (b) of the figure. Let us assume that a change from dark current to illuminated current (ΔI_{in}) is 50 μA. With a value of 100 kΩ used for the feedback resistor (R_f), the voltage output change (Δv_o) will be:

$$\begin{aligned}\Delta v_o &= 50(10^{-6})\text{ A} \times 100\text{ k}\Omega = 5{,}000(10^{-3})\text{ V}\\ &= 5\text{ V change}\end{aligned}$$

In the conversion process, the only error is the bias current (I_b), which is summed algebraically with the input current (I_{in}).

5-15. DIFFERENTIAL (INSTRUMENTATION-TYPE) AMPLIFIER

The requirements for an OP AMP used as an *instrumentation amplifier* are generally more rigid than those needed for simpler amplification purposes; in particular, the instrumentation amplifier must handle very small dc signal variations, and often these signals come through relatively long leads from remotely located sensors. This entails a combination of such properties as *high gain along with small offset errors and especially good common-mode rejection* properties that stretch the capabilities of general-purpose OP AMPS and often require the use of *premium monolithic types*, or possibly the use of the more expensive *hybrid modules* (discussed in Chapter 12). Where the requirements are not overly strict, however, the general-purpose OP AMP can be used in the *differential (or double-ended input) mode*, where the input is applied across the inverting and noninverting terminals, so that both input ends are "floating" above ground.

The connections for the differential-amplifier mode (not to be confused with the diff-amp *stage* of Chapter 2 are shown in Fig. 5.12(a).[1] The signals e_1 and e_2 usually come from a bridge arrangement of transducers [such as

[1] Tobey et al., *op. cit.*

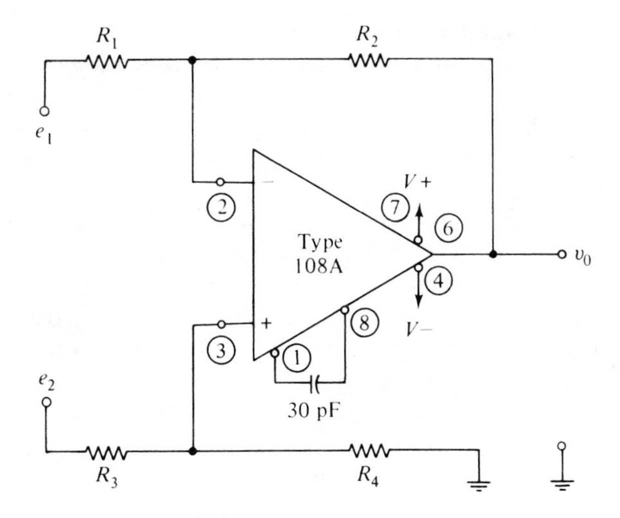

(a) Differential input to difference amplifier

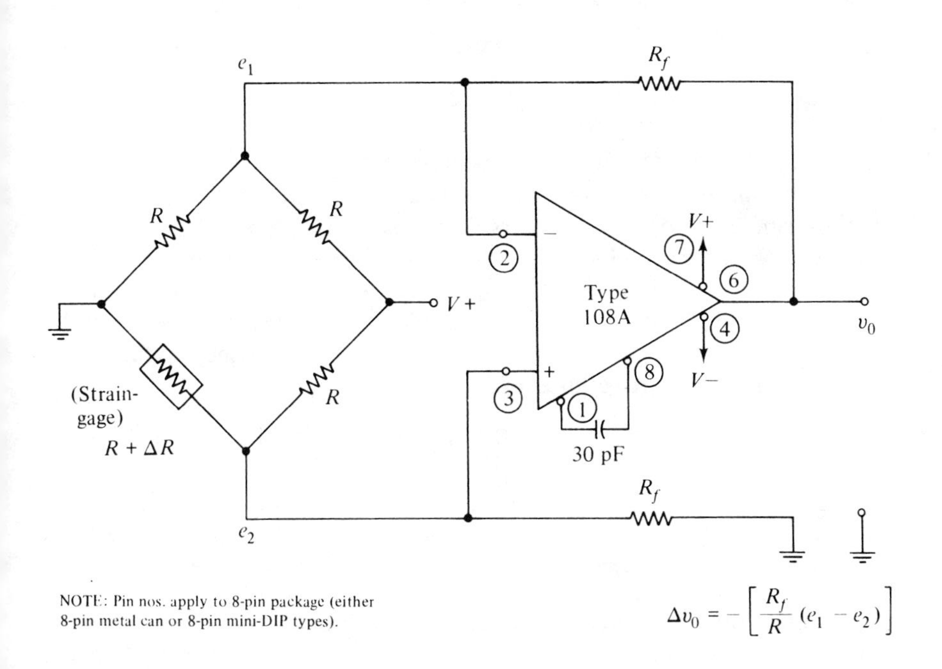

(b) Strain-gage bridge amplifier

FIG. 5.12 Differential (double-ended input) amplifier: (a) double-ended input for amplification of difference $(e_1 - e_2)$; (b) example with strain-gage transducer bridge. *(National Semiconductor)*

the strain-gage amplifier in part (b) of the figure], and the signal to be amplified represents the *difference* between the voltages at opposite corners of the bridge $(e_1 - e_2)$. For this reason, *the dc differential amplifier* is also known by such names as a *difference amplifier* or *error amplifier*. As previously noted, in the differential connection, neither e_1 nor e_2 is at ground potential.

In this differential-amplifier connection, with the OP AMP, one retains the valuable property of a high common-mode rejection ratio (CMRR) and also keeps the offset errors very small by the selection of precision resistors to provide a proper balanced arrangement of the four resistors. The arrangement for providing a good initial balance is obtained by using four well-matched resistors to equalize the resistor ratios, as follows:

$$\frac{R_2}{R_1} = \frac{R_4}{R_3}.$$

In this way, the output of the differential amplifier responds only to the difference of the input signals $(e_1 - e_2)$ and is thus able to cancel out common-mode signals and also equal-and-opposite offset tendencies. As a result, the OP AMP can amplify the signals from the unbalanced bridge with the least interference from undesired signals.

Bridge Amplifier

An example of the OP AMP in an instrumentation application is given for the *strain-gage amplifier* of Fig. 5.12(b).[1] (This bridge type of instrumentation applies similarly to other sensors, such as thermistors, thermocouples, photocells, and the like.) It will be noted in the circuit diagram that the bridge supply is grounded, and that each signal voltage (e_1 and e_2) is above ground by approximately the value of R. By using equal values for the feedback resistors (R_f) and equal arms for the bridge (R), we satisfy the condition for good common-mode rejection, since

$$\frac{R_2}{R_1} = \frac{R_4}{R_3}$$

becomes

$$\frac{R_f}{R} = \frac{R_f}{R}$$

Additionally, we retain initial equality of bias currents, since the resistance seen at each input terminal is the parallel combination $R_f \parallel R$.

The signal to be amplified comes from the change in resistance of the

[1] *Ibid.*

strain gage from its original value of R to a new value $(R+\Delta R)$. As an idea of the numerical values involved, we may assume an equal-arm-bridge value of $R=100\ \Omega$, and a change (ΔR) of $2\ \Omega$, for a 2 percent change ($\delta=\Delta R/R$ $=2$ percent). With a supply voltage of $+12$ V, a Thèvenin-equivalent analysis[1] for the approximate bridge output gives

$$e_1-e_2=V\frac{\Delta R}{4R}\text{ V},$$

or

$$e_1-e_2=12\cdot\frac{2}{400}=0.06\text{ V}\quad\text{or}\quad 60\text{ mV}$$

To obtain an amplified output for this change in the range of volts, let us assume that a gain of 50 is needed. Since the approximate gain of this simplified circuit is $-R_f/R$, we could use the following values:

$$R_f=5\text{ k}\Omega,\qquad R=100\ \Omega.$$

Then amplifier output v_o is, then,

$$v_o=-\frac{R_f}{R}(e_1-e_2),$$

and

$$\Delta v_o=-50\,(60\text{ mV})$$

$$=3\text{V output change for a 2 percent change in } R.$$

In instances where unequal bias currents dictate the use of different values for the four resistors, the modified formula becomes more complicated, as follows:

$$v_o=-\left[\frac{R_2}{R_1}(e_1)\right]+\left[\left(\frac{R_4}{R_3+R_4}\right)\left(\frac{R_1+R_2}{R_1}\right)e_2\right]$$

When it is desired to change the amount of gain, it is necessary to alter both ratios of the combination of resistors (rather than a single resistor) in order to preserve the balanced arrangement.

5-16. ADDITIONAL OPERATIONAL-AMPLIFIER CIRCUITS

The dozen or so circuit diagrams given as examples in this chapter by no

[1] Prensky, *op. cit.*

means exhaust the list of applications using OP AMPS. The examples given have been chosen to illustrate a fair sampling of typical OP AMP uses; additionally, however, we find OP AMPS used in many other circuits, where linear ICs are employed under specific names, and these other applications are covered later in separate chapters; they include such functions as *comparators*, *regulators*, *active filters*, *communication-type amplifiers*, *digital-interface devices*, and *hybrid IC modules* in specialized system applications.

While the field of op-amp applications continues to widen, it is well at this point to emphasize the unifying theme that the operational amplifier in the integrated-circuit form has evolved into a basic building block of great versatility. As a consequence, the OP AMP has presented the circuit designer with a revolutionary device for constructing applications that represent significant advances in both traditional and modern circuitry.

6

TESTING AND BREADBOARDING INTEGRATED-CIRCUIT OPERATIONAL AMPLIFIERS

6-1. LEVELS OF TESTING LINEAR INTEGRATED CIRCUITS

The measurements made in testing linear ICs will vary, according to the intended purpose, from relatively simple *manual tests on an individual IC*, as in the laboratory, to the much more complicated and extensive *automated test systems* used for production testing in industry. The test methods discussed here will concentrate on the *OP AMP as a typical linear IC*, with emphasis on the relative importance of the parameters to be measured. In this way those parameters of particular importance to other linear ICs can be identified and, with suitable modifications, can also serve to indicate tests for other groups of linear ICs. In addition, details of *practical means for setting up breadboard arrangements* in the laboratory for performing these tests will be discussed.

6-2. BASIC LABORATORY TESTS

The chart of OP-AMP characteristics, given previously in Table 4-1, includes

TABLE 6-1

Basic Op-Amp Properties for General Use

Open-Loop Properties	*Closed-Loop Measurements*
Voltage gain (A_{VOL} or A_o)	Large-signal gain (A_{VCL} or A_f)
Input resistance (R_{in})	R_{in}, inverting mode
	R_{in}, non-inverting mode
Frequency-response curve: (f_c at -3 dB)	Bandwidth (BW, at stated gain)

specifications that cover a wide range of properties that might enter into industrial uses. For testing in the laboratory, however, we list in Table 6-1 the relatively more important *basic properties for general-purpose use* (with symbols as shown in Fig. 6.1).

Other properties, as applicable depending on circuit use, include input voltage offset (V_{io}), input bias current (I_B), voltage and current drift errors with temperature, output resistance (R_o), and effects of load resistance (R_L).

Test conditions, unless otherwise stated, are standardized as follows:

$$\text{Supply voltage (dual)} = \pm 15 \text{ V},$$

$$\text{Room temperature } (T_A) = 25°\text{C},$$

$$\text{Load resistance (when used)} \geqslant 2\text{k}\Omega.$$

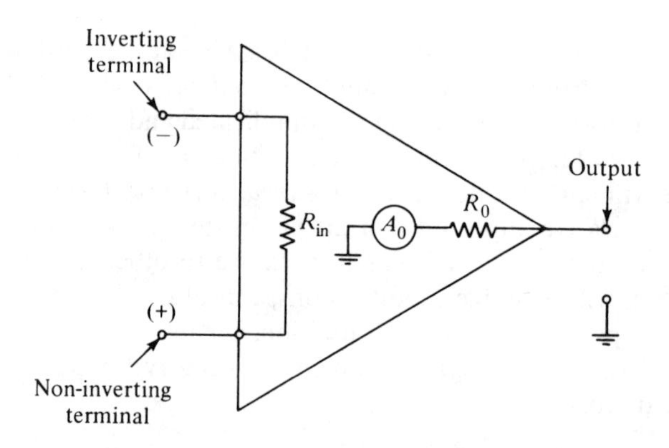

FIG. 6.1 Simplified functional equivalent model of OP AMP: the open-loop gain (A_o or A_{VOL}), the input resistance (R_{in}) and output resistance (R_o) are intrinsic properties of the OP AMP, before any external components are connected.

6-3. VOLTAGE GAIN (TESTS FOR OPEN-LOOP, CLOSED-LOOP, AND LOOP GAIN)

Open-Loop Gain

A high value of the open-loop gain (A_{VOL} or A_o) is assumed in practically all the simplified formulas for the OP AMP. This assumption is validated in all the typical OP AMPS (as listed in Table 4-1), and is usually given in the data sheets, where a plot of the *open-loop frequency response* shows values of A_{VOL} centering around 100 dB (or 100,000 times) at the low and medium frequencies.

Because of the extremely high gain possible in open-loop operation, the test for A_{VOL} is quite difficult to perform, and it is usually sufficient to take its value from the frequency-response plot of the data sheet. The value may be checked when desired, in the circuit of Fig. 6.2,[1] by the magnitude relation

$$A_{VOL} = \frac{v_o}{e_{\text{in}}}$$

The use of the voltage divider for attenuation in Fig. 6.2[1] is made necessary by the high gain and is used to ensure a sufficiently low input voltage (e_{in}) to avoid saturation at the output. For similar reasons, shielded input should be used, and precautions must be taken to *null the initial dc input-offset voltage*, which could cause amplifier saturation, even with zero ac voltage input.

Importance of Loop Gain (LG)

The specified value of open-loop gain for practically all OP AMPS is consistently high enough to justify the simplified relation for the resulting gain in an actual closed-loop circuit, as being simply the ratio of two external resistors (R_f/R_1). A parameter, more important than A_{VOL} for determining the *accuracy of circuit gain*, however, is the *loop gain (LG)*, which is the ratio of A_{VOL} to the actual closed-loop gain A_{VCL}, or

$$\text{LG} = \frac{A_{VOL}}{A_{VCL}} \quad \text{or} \quad \frac{A_o}{A_f}$$

This value depends on the amount of negative feedback used in the actual circuit, as discussed next for closed-loop operation.

[1] J. Millman and C. C. Halkias, *Integrated Electronics*, McGraw-Hill Book Company, New York, 1972.

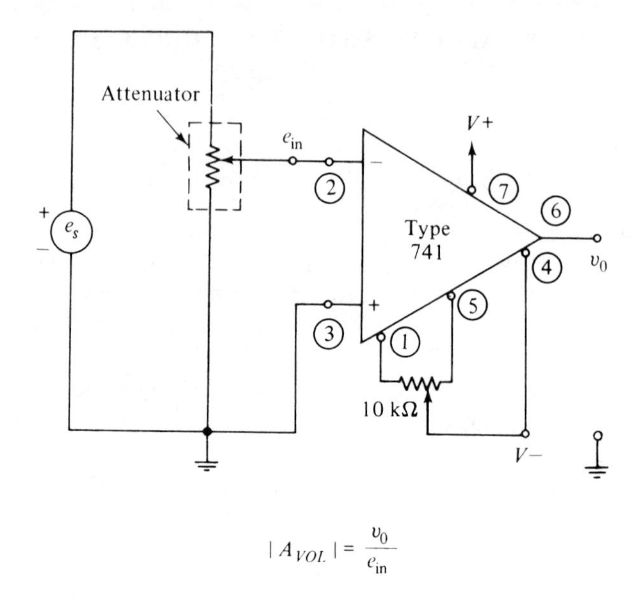

FIG. 6.2 Test for open-loop gain (A_{VOL} or A_o): control for OFFSET NULL (10 kΩ) must be set for zero dc output at no-signal condition; e_s is an ac signal, attenuated to avoid saturation of the OP AMP.

Closed-Loop Gain (A_{VCL} or A_f)

The measurement of closed-loop gain in the circuit of Fig. 6.3 is the more practical measurement, since the OP AMP is very seldom used as open-loop. For simplicity, the figure shows the circuit for the internally-compensated type 741 connected for a nominal gain of 100. Under the conditions shown, the measurement should closely approximate the simple relation:

$$A_{VCL} \approx \frac{-R_f}{R_1}$$

The actual value obtained in this measurement is affected by two main approximations implied in this simplified circuit. First, the value used for R_1 should include the source resistance R_s. (In this case, the 50-Ω resistance of R_s may be neglected within 5 percent; otherwise, we either include the known resistance of the generator, or, if unknown, we can use a voltage divider across the source to reduce the equivalent internal resistance to a 50-Ω value). Second, the value of R_3 at the non-inverting terminal should be the parallel

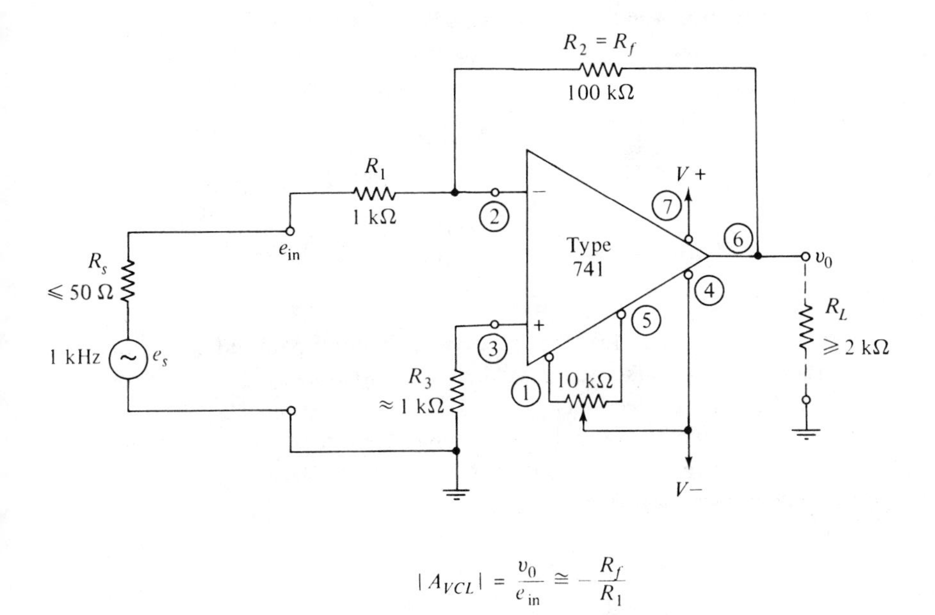

FIG. 6.3 Test for closed-loop gain (A_{VCL} or A_f), for dc and frequencies within the audio band; for the actual gain to closely approximate $R_f/R_1 = 100$, R_1 must include R_s, and R_L must be sufficiently high to have negligible loading effect, as shown.

combination $R_f \parallel R_1$ to equalize the bias currents at each terminal; here the round value of 1k for R_3 is within 1 percent and voltage offsets are assumed to be negligible for large-signal operation. In a similar way, the effect of using a load resistor (R_L) equal to or greater than 2 kΩ will not be appreciable, compared to the no-load condition.

Frequency-response curves at a desired gain within the audio range can also be obtained from the circuit of Fig. 6.3.

Example (Sec. 6-3)

LOOP-GAIN CALCULATION

Since the value of loop gain ($\text{LG} = A_o/A_f$) has important significance in determining accuracy and stability considerations, it is well to evaluate it for a given circuit, once the closed-loop gain (A_{VCL} or A_f) is found. Using a typical value for the open-loop gain (A_{VOL} or A_o) of 100 dB, or 100,000 times, we can easily find the loop-gain value for the circuit of

Fig. 6.3, where A_{VCL} or $A_f = 100$ or 40 dB, as follows:

$$\begin{aligned} \text{LG} &= \frac{A_o}{A_f} \\ &= 100 - 40 = 60 \text{ dB, or} \\ &= \frac{100{,}000}{100} = 1{,}000 \text{ times.} \end{aligned}$$

This value of 1,000 (or 60 dB) for the loop-gain indicates the gain-reduction factor (sometimes called the "throw away" factor) introduced by the negative feedback.

It should be noted that the loop gain obtained in this fashion is the same gain-reduction factor that comes from the familiar generalized feedback relation

$$A_f = \frac{A_o}{1 + A_o\beta},$$

where β is the fraction of the output that is returned as negative feedback. Thus the loop-gain is the expression in the denominator ($1 + A_o\beta$, simplified by neglecting to add the 1), or

$$\begin{aligned} \text{LG} &\cong A_o\beta \\ &= 100{,}000\left(\frac{R_1}{R_1 + R_f}\right) \\ &= 100{,}000\,\frac{1\ k\Omega}{101\ k\Omega} \\ &\cong 1{,}000, \qquad \text{as before.} \end{aligned}$$

A large value of loop-gain is generally advisable, since it can be shown that it reduces, by the same factor, the effect of the possible spread of parameter variations from one device to another. Thus the *value of loop-gain is significant in determining the expected accuracy of the closed-loop gain.* It is also a significant multiplying factor in obtaining a high input impedance in the noninverting mode, as will be seen in the next section, where both the gain and input impedance for the noninverting mode are discussed.

6-4. INPUT-RESISTANCE TESTS

Open-Loop (R_{in}, Intrinsic)

The property of input resistance (R_{in}) of an OP AMP under open-loop conditions is intrinsic to the device and accordingly is shown on the data sheet. It can be checked by the circuit of Fig. 6.4.

Values of R_{in} encountered in typical OP AMPS are much greater than the few thousand ohms that might be expected from the conventional input resistance ($2h_{ie}$) of an ordinary differential-amplifier input stage.

Again taking the internally compensated type 741 as our example, for the sake of simplicity, we find that the value of the intrinsic R_{in} ranges from 300 kΩ to the megohm range, this relatively high value being obtained by the use of a modified Darlington circuit at the input. Another scheme for increasing R_{in} is the use of super beta transistors, as in the type **LM108A**; such specially fabricated transistors can provide a beta of around 5,000 at only 1 μA of collector current, producing a resulting R_{in} of around 30 MΩ. And where still higher values for R_{in} are important, the example of the type 770 is given in the OP-AMP chart (Table 4-1) as providing up to 100 MΩ.

Still larger values are available with the use of special-purpose OP AMPS with FET inputs (such as type 740), providing a listed value of 1 million MΩ (10^{12} Ω).

Reverting to the type 741 of Fig. 6.4 for simple laboratory testing, R_{in} is measured as the differential resistance of the OP AMP by inserting a high resistance (R) in series with the input terminal to act as a voltage divider. (Note that for open-loop conditions only, either input terminal may be used). As a result, the signal voltage e_s divides, with the fraction e_{in} appearing across R_{in} as

$$e_{in} = e_s \frac{R_{in}}{R + R_{in}}.$$

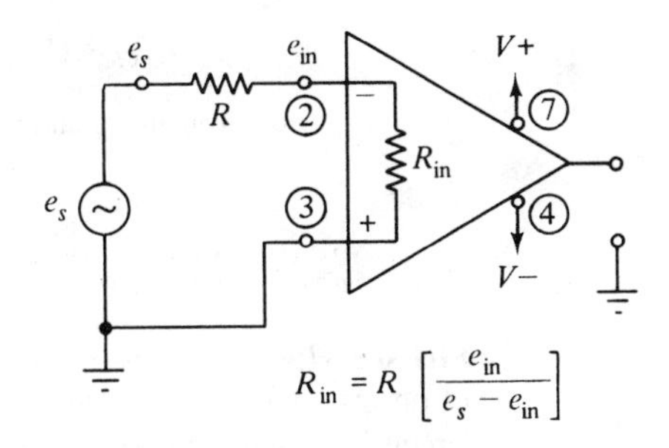

FIG. 6.4 Measuring intrinsic input resistance (R_{in}) of open-loop OP AMP: external resistor R is added to form a voltage divider with R_{in}, across signal voltage (e_s).

Solving for R_{in}, gives

$$R_{in} = R\,\frac{e_{in}}{(e_s - e_{in})}.$$

(Note that if R is a variable resistor and is adjusted to make $e_{in} = \frac{1}{2}e_s$, then $R_{in} = R$.)

Closed-Loop R_{in}

1. *Inverting Mode.* When the OP AMP is used closed-loop in the inverting mode (as was shown in Fig. 6.3), the *input resistance of the circuit is radically reduced by the loop gain* (as in shunt voltage feedback), so that the approximate expression becomes, for closed-loop, approximately:

$$R_{in} \text{ (for inverting mode)} \approx R_1$$

2. *High-Impedance Non-inverting Mode.* When the input to the OP AMP is connected to the non-inverting (+) terminal under the closed-loop conditions of Fig. 6.5, the *input impedance of the circuit is radically*

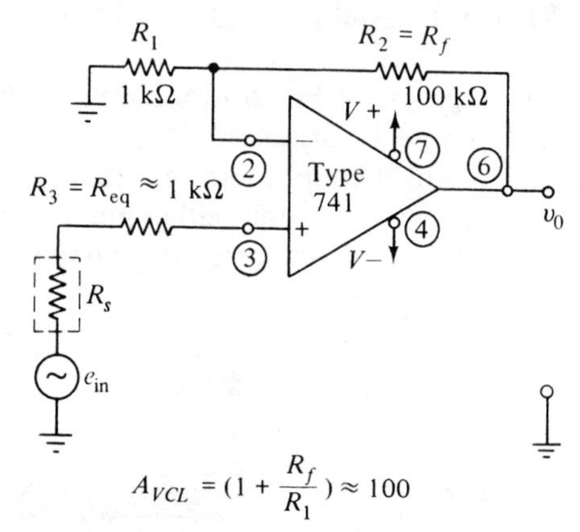

FIG. 6.5 High input resistance (R_{in}) for the noninverting mode of OP AMP operation: while the closed-loop gain (A_{VCL}) of this circuit is roughly the same as in the inverting mode of Fig. 6.3, the input resistance (R_{in}) has increased from 1 k to about 300 megohms. *(Fairchild Semiconductor)*

increased by the loop-gain factor. [The gain, however, is substantially the same, changing only slightly to $(1 + R_f/R_1)$.] Resistor R_{eq} (including R_s) should equal the parallel resistance $R_f \parallel R_1$ to equalize bias currents for minimum dc offsets. The effect on the intrinsic resistance is now similar to changing from shunt to series feedback, so that the expression for the new input resistance is

$$R_{in}\text{ (for non-inverting mode)} = R_{in\,(intrinsic)}\,(\text{LG})$$

Since loop gain (LG) is defined as

$$\text{LG} = \frac{A_o}{A_f} \qquad (\text{or } A_o\beta),$$

then

$$R_{in}\text{ (for non-inverting mode)} = R_{in\,(intrinsic)}\,\frac{A_o}{1+\dfrac{R_f}{R_1}}.$$

Using the example of *loop gain of 1,000* (given in Section 6-3 for a closed-loop gain of 100), we can expect a multiplying factor of 1,000 times the intrinsic R_{in} (given as a minimum of 300 kΩ on the data sheet for the type 741).[1] This calculation gives a theoretical value of 300 MΩ in the noninverting mode, which agrees well with the stated value of 280 MΩ for R_{in} (as given on this data sheet for a closed-loop gain of 100).

6-5. BANDWIDTH (ESTIMATING AND TESTING)

The open-loop frequency response of an OP AMP is given in its data sheet and is shown for a typical case in Fig. 6.6(b). The type 101A illustrated (similar to the 748 type) is known as an "extended bandwidth" OP AMP. Here, the bandwidth (BW) is made more flexible by allowing the use of different values for the external compensating capacitor (C_1), rather than the situation where the bandwidth is restricted in the internally compensated type (such as type 741).

Although the OP AMP is rarely used open-loop, we can *easily estimate the closed-loop performance from the open-loop response curve* by employing the concept of constant GAIN × BW product. Thus in the plot of Fig. 6.6(b), we estimate the bandwidth for a closed-loop gain of 100 (or 40 dB) by simply

[1] Fairchild μA741 data sheet.

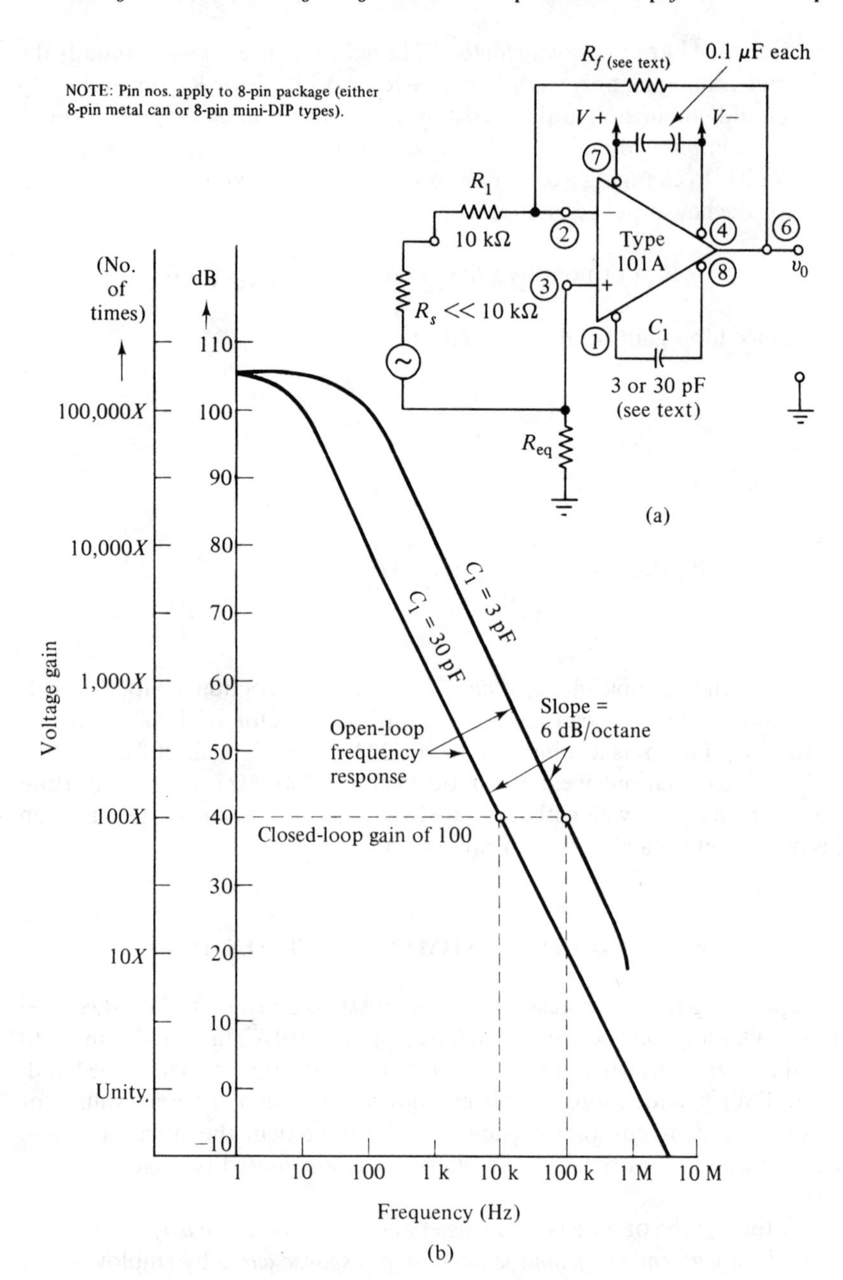

FIG. 6.6 Obtaining bandwidth of closed-loop gain, for different values of frequency-compensating capacitor C_1: (a) for circuit gain of 100, $R_f = 1$ MΩ, $R_{eq} = R_f \| R_1$; (b) two open-loop frequency response curves. *(National Semiconductor)*

moving horizontally to the first curve ($C_1 = 30$ pF), and finding an approximate *breakpoint (or f_c) at 10 kHz*, after which the gain rolls off at 6 dB octave (or 20 dB decade) for increasing frequency. This form of roll-off is known as "single-pole compensation." (The circuit for two-pole compensation is also generally given in the data sheets.)

The value of 30 pF provides adequate compensation (or protection against oscillation) for all gains down to unity gain, where the quoted amount of compensating capacity is needed. For larger gains, smaller amounts of compensating capacity are satisfactory and the resultant bandwidth is "extended." Thus, at the same gain of 40 dB (100 times), the horizontal extension to the curve for $C_1 = 3$ pF intersects at 100 kHz, indicating a *tenfold increase in bandwidth.* A point of interest in this connection is the fact that in choosing an OP AMP for fairly high gain in the complete audio-frequency range, 748, the 101A or similar types are preferred over the internally compensated 741 types, whose bandwidth is similar to the 30-pF curve (BW = 10 kHz at a gain of 100).

Test Circuit

The diagram [Fig. 6.6 (a)] shown for testing the bandwidth (or frequency response) in closed loop is tailored for the desired gain by the ratio R_f/R_1 (thus $R_f = 1$ MΩ for a gain of 100). The input resistor R_1 is given arbitrarily as 10kΩ to allow for the signal-source resistance R_s, since the published curves apply only if R_s is less than 10 kΩ. The value for R_{eq} is the equivalent resistance of the parallel combination $R_f \parallel R_1$ (a 10-kΩ resistor would be satisfactory here with most signal generators). The 0.1-μF capacitors are not critical and are used to bypass the positive and negative power supplies.

The bandwidth of the circuit is tested by increasing the input frequency to *the point where the output drops to 70.7 percent (−3 dB) of the low-frequency value.* The bandwidth would then extend from direct current to this point. However, if the dc gain is to be measured, the offset-null circuit must be included to prevent the amplified input-voltage offset from appearing in the output.

It should be kept in mind that wide-band amplifiers are available as special linear ICs to provide much wider bandwidths than the general-purpose OP AMPS discussed here, and they are covered in a separate chapter.

6-6. OUTPUT-RESISTANCE (R_o) TEST

The output-resistance specification for an OP AMP is generally of little importance in voltage amplification, especially when reasonably high load resistances (>2 to 10 kΩ) are used. It assumes greater importance when relatively high-current outputs are required, as in booster amplifiers, which are discussed separately later.

The following discussion indicates the nominal values of R_o to be expected under the three operating conditions of open-loop and of inverting and non-inverting modes of closed-loop operation.

Open-Loop R_o

The measurment of intrinsic output resistance (open-loop) follows the same plan of using a voltage-divider arrangement as was used previously for the intrinsic input-resistance test. In this case, as seen in the equivalent model of Fig. 6.7, load resistance (R_L) acts to reduce the original open-loop output voltage (V_{OL}) before loading to a smaller value (V_o) after R_L is connected. Again here, as in previous open-loop measurements, the difficulty in working with the very high open-loop gain suggests the advisability of using the value of R_o given in the data sheet as an initial figure for specific calculations. This value centers around 100 Ω.

When it is desired to check the value of R_0 in open loop, the voltage-divider relation shown in Fig. 6.7 can be used as follows:

$$V_o = V_{OL} \frac{R_L}{R_o + R_L}.$$

from which

$$R_o = R_L \frac{V_{OL} - V_o}{V_o}.$$

(Note that when V_o is made equal to $\frac{1}{2}V_{OL}$, then $R_o = R_L$.)

Closed-Loop R_o.

The output resistance is reduced by negative feedback in both the inverting

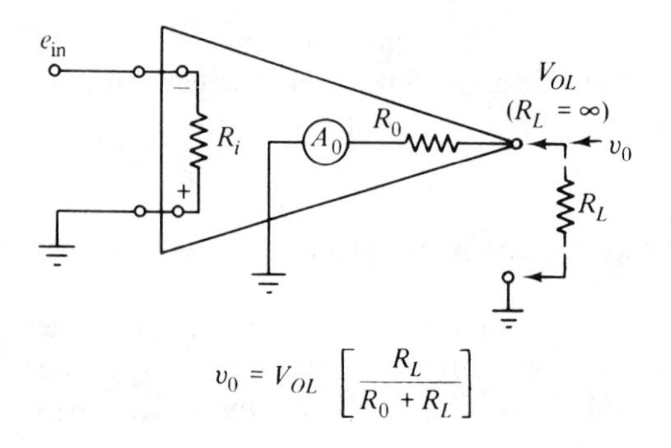

FIG. 6.7 Testing intrinsic output resistance (R_o): the unloaded open-loop output voltage (V_{OL}) is reduced to V_o, when the load resistor (R_L) is connected.

and non-inverting modes. Consequently, the expression for R_0 is identical in both configurations:

$$R_{o\,(\text{closed loop})} = \frac{R_{o\,(\text{intrinsic})}}{\text{LG}}$$

$$= R_{o\,(\text{intrinsic})} \frac{A_f}{A_o}.$$

For the example used before of loop gain = 1,000, we find the satisfyingly small value of R_o to be around 100/1,000, or 0.1 Ω.

6-7. COMMON-MODE REJECTION TESTS

An outstanding advantage of the theoretical operational amplifier is that there is *zero output for common-mode signals* (i.e., for equal signals on both inputs to the differential-amplifier stage). This theoretical infinite rejection is modified by the practical consideration that both sides of the input stage cannot be expected to be exactly identical, and so the practical common-mode rejection ratio (CMRR), as a measure of the actual unbalances, is defined as a ratio, comparing differential gain (A_d), that is, the gain to the difference signal ($e_1 - e_2$), to the gain for common-mode signals (A_c), or

$$\text{CMMR} = \frac{A_d}{A_c}.$$

Thus, particularly in noisy locations, a large value of this ratio is desired within the limits of the magnitude of allowable common-mode voltages that are specified.

In the CMRR rest circuit[1] of Fig. 6.8, the signal voltage (V_s) is essentially the common-mode voltage (V_c) applied simultaneously to both of the OP AMP input terminals. Under the conditions of matched resistors, $R_{f1} = R_{f2} = R_f$ and $R_1 = R_2 = R$, the relation for the common-mode rejection becomes

$$\text{CMRR} = \frac{A_d}{A_c} = \frac{R + R_f}{R} \frac{V_s}{V_o}.$$

For a numerical example, let $R = 100\ \Omega$ and $R_f = 100\ \text{k}\Omega$, for a nominal gain of 1,000. If a signal voltage V_s of 1 V is applied and produces an output

[1] Millman and Halkias, *op. cit.*

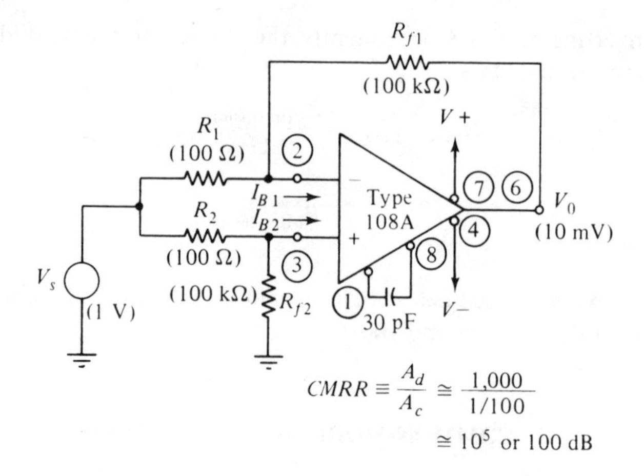

FIG. 6.8 Measuring common-mode rejection ratio (CMRR): a 1 volt ac signal applied equally to both input terminals yields an output V_o of 10 mV ($A_c = 1/100$), while the circuit has a differential gain (A_d) of 1,000, resulting in a ratio of 10^5 or 100 dB for CMRR.

V_o of 10 mV, then

$$\text{CMRR} = \frac{100 + 100{,}000}{100} \, \frac{1{,}000 \text{ mV}}{10 \text{ mV}}$$

$$\approx 1{,}000(100)$$

$$= 100{,}000 \text{ times} \quad \text{or} \quad 100 \text{ dB}.$$

This value of 100 dB is fairly typical of "precision" OP AMPS (such as type μA**725** or type LM**108A**), providing the measurement is made within the specified common-voltage swing.

6-8. INPUT-OFFSET ERROR (I_B, I_{io}, AND V_{io}) TESTS

The factors contributing to the unbalances that give rise to offset errors may be summarized as five in number, each one perhaps negligibly small, but potentially significant, depending on the application. These factors are listed in Table 6-2 with typical magnitudes for two types: the internally compensated 741 type and the super-beta 108A type, as given in Table 4-1 (Sec. 4-6):

TABLE 6-2 OFFSET FACTORS

	Type 741	*Type 108A*
Input bias current (I_B)	<200 nA	<2 nA
Input offset current (I_{io})	< 30 nA	<0.2 nA
Drift of I_{io}	<200 pA/°C	<2.5 pA/°C
Input offset voltage (V_{io})	< 5 mV	<1 mV
Drift of V_{io}	20 μV/°C	5 μV/°C

In this list of offset factors, we can concentrate on the ones most commonly measured, *input bias current* I_B, and *input voltage offset* V_{io}, since these are dominant under ordinary laboratory temperature conditions.

Accordingly, the test circuit of Fig. 6.9 shows the measurement for the input offset voltage (V_{io}); also, placing a meter in series with each input terminal will measure input bias current (I_B) as the average of I_{B1} and I_{B2}, and determine I_{io} as their difference ($I_{B1}-I_{B2}$). But the effect of I_B or I_{io} is generally negligible, assuming significance only when large input resistors are used such that the voltage drop $\Delta I_B R$ (or $I_{io}R$) becomes comparable to the input offset voltage.

In the circuit of Fig. 6.9 the unbalance producing the offset voltage at the input is amplified by the closed-circuit gain of the OP AMP, so that the

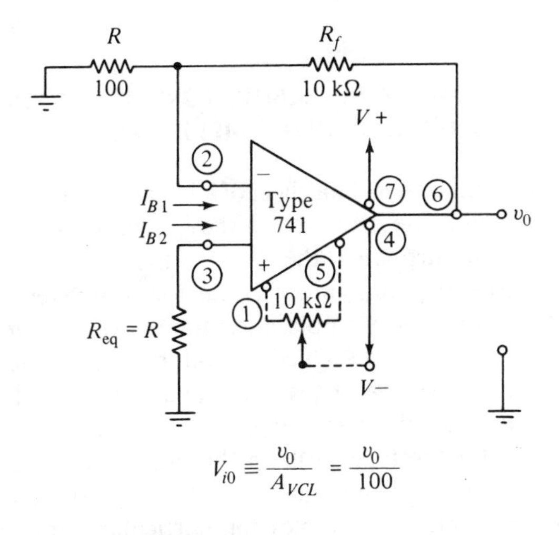

FIG. 6.9 Measuring input offset voltage (V_{io}); also, bms current $I_B=(I_{B1}+I_{B2})/2$, and $I_{io}=I_{B1}-I_{B2}$ may be measured, as described in the text. The dotted connections are for OFFSET NULL control, when used to cancel V_{io} in dc applications.

value of V_{io} is obtained as follows:

$$V_{io} = \frac{v_o}{A_{VCL}}.$$

For a numerical example using the closed-loop gain of 100 in the figure and the typical offset values mentioned previously, we have the data shown in the example below:

Example (Sec. 6-8)

	Type 741	*Type 108A*
$V_{io} = \frac{v_o}{A_{vcl}}$	$= \frac{500\text{ mV}}{100} = 5\text{ mV}$	$= \frac{100\text{ mV}}{100} = 1\text{ mV}$
Contribution of $I_{io}R$	$= 30\text{ nA}(100) = 3\ \mu\text{V}$	$= 0.2\text{ nA}(100) = 0.02\ \mu\text{V}$

For general-purpose dc applications (such as with the type 741 OP AMP), the *offset-null provision* allows cancellation of the initial offset error by manual adjustment. However, in precision instrumentation dc applications, it is preferable to use *premium* OP AMPS (such as the type **LM101A/108A**, μA**725**[1]/777, or SN52**660/770**), because of the much smaller offsets, as shown in the specifications of Table 4-1 (Sec. 4-6).

6-9. SUMMARIZING LABORATORY TESTS FOR OPERATIONAL AMPLIFIERS

The specifications given on the data sheet of a linear IC will generally include other parameters than those discussed in the previous sections. The task of testing for production purposes rather than in the general laboratory must take all these additional points into account for completeness in covering diverse applications. For one example, a test figure for *slew-rate* is important in an OP AMP that is specifically designed for voltage-follower action involving large and abrupt voltage swings. Thus, where the unity-gain slew rate for the general-purpose OP AMPS given in Table 4-1 is consistently below 1 μV/s (except for 2.5 μV/s for the type 770), on the other hand, the type **LM110**, a type particularly useful for unity-gain application, is specified as having a slew rate of 30 μV/s. Other examples for particular parameters, such as power-supply rejection ratio (PSRR), are discussed, where pertinent, in Chapter 12.

[1] The offset specifications for the μA725 (not given in Table 4-1) are generally similar to the type 101A; designed particularly for low-level instrumentation, it features a large-signal voltage gain of 1,000,000 (120 dB), and CMRR of 110 dB; "precision" performance is similarly offered in the CA3100S (RCA).

6-10. CURVE-TRACER METHOD OF TESTING AMPLIFIERS

A fast and convenient method for determining satisfactory operating condition of an OP AMP makes use of a curve-tracer method, where the transfer function of the device under test (DUT) is displayed on an oscilloscope. This method involves using a sawtooth signal to sweep the input voltage through a stated range of values, and applying the output to the vertical terminal of an oscilloscope, while the same sawtooth is connected to the horizontal terminal. The resulting trace is an input–output graph of the transfer function on the scope screen. The trace can be used as a visual GO/NO-GO indication, but, more importantly, it also provides speedy information on three main OP-AMP characteristics; *gain (either* A_{vol} *or* A_{vcl}*), input voltage offset* V_{io} *and maximum output* V_o *limits.*

The details of the test method are very similar to those for a simple diode curve tracer, with the active OP AMP taking the place of the diode being tested. The circuit requirements for the tester (shown in Fig. 6.10) are even simpler than the circuit needed for a transistor curve-tracer, since, as a two-port device, the OP AMP does not require a stepping voltage. The OP-AMP tester circuit of Fig. 6.10 was developed[1] to give the user "a degree of confidence that the units are functioning satisfactorily."

The drive circuit, in parts (a) and (b) of the figure, uses a portion of the 60-Hz voltage (from the 12-V ac secondary winding of the power transformer) and applies it (at point *A*) to the normally off transistor. This allows the *RC* combination to freely charge toward the +15-V dc power supply. With a sufficiently long time constant (relative to the 60-Hz supply), the charging curve reaches a sufficient amplitude, while deleting the more curved portion of the exponential curve. When the input sine wave reaches a peak of about 7 V ($2\,V_{BE} + V_Z$), the transistor is turned on hard, discharging the timing capacitor. The resulting wave form, as shown, has the charging time T_1 much longer than the discharge time T_2, and thus the fast retrace is barely visible.

The connections to the device under test are shown in part (c) of the figure. The input to the noninverting terminal of the OP AMP (from point *B*) has been attenuated by the factor of 1,000 (51/50,000), so that the highly amplified (open-loop) output applied to the scope vertical will have a reasonable value relative to the horizontal value of the same wave form, as set by the 5-kΩ AMP ADJ control.

[1] "An Operational Amplifier Tester," Application Note AN 400, Motorola Semiconductor Products, Inc.

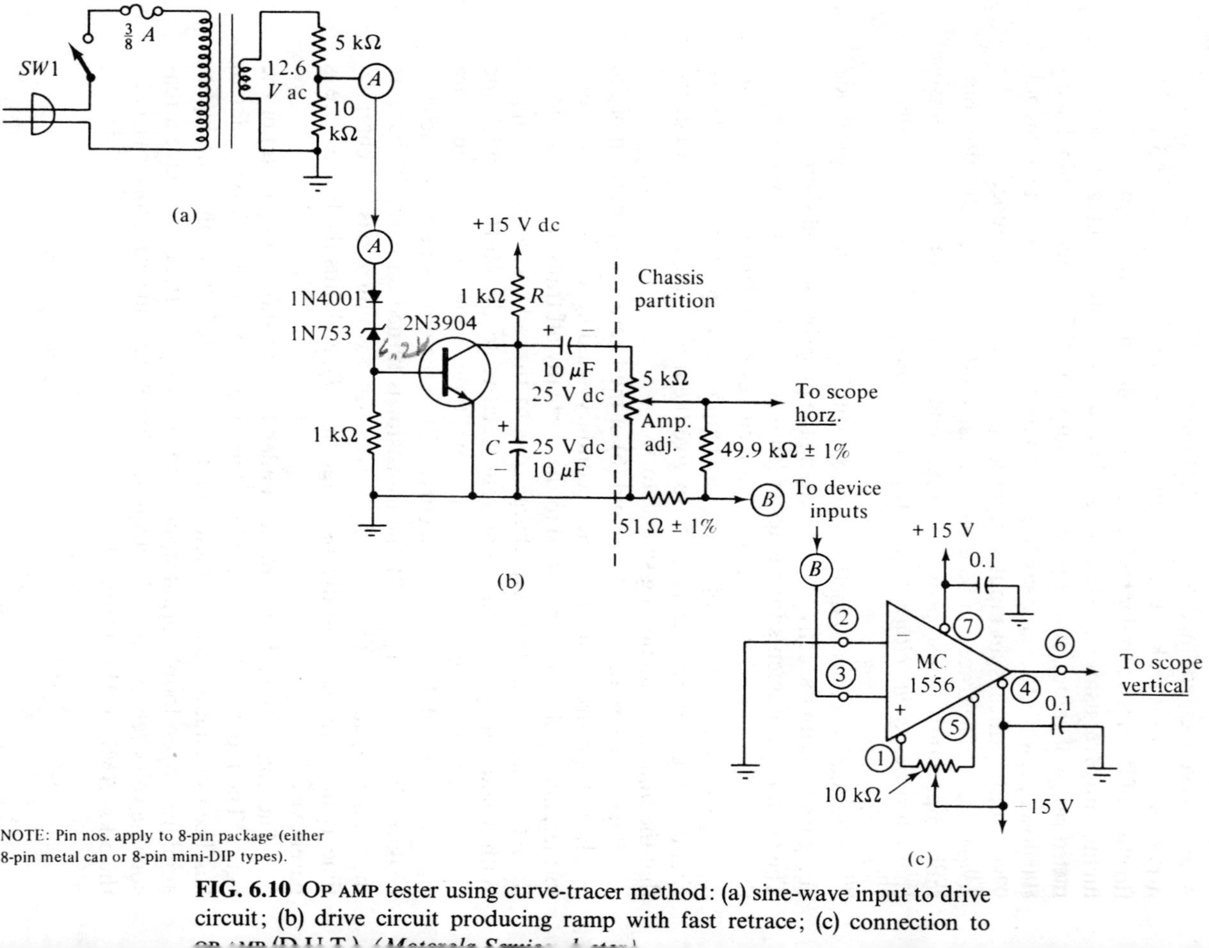

FIG. 6.10 OP AMP tester using curve-tracer method: (a) sine-wave input to drive circuit; (b) drive circuit producing ramp with fast retrace; (c) connection to OP AMP (D.U.T.)

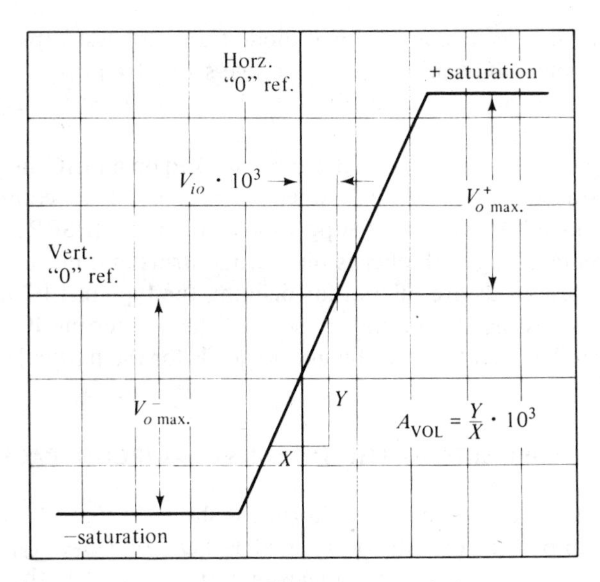

FIG. 6.11 Interpreting oscilloscope trace of IC curve-tracer.

Interpreting the Oscilloscope Waveform

A typical transfer function is illustrated in Fig. 6.11, with the main features as noted.

Initially, the oscilloscope inputs are grounded. The oscilloscope dot is then centered; this establishes the vertical and horizontal references. After the device to be tested has been connected, turn up the AMP ADJ until the unit is in deep saturation. Read directly positive and negative $V_{o\,(max)}$. The open-loop gain (A_{VOL}) is the calculated slope of the transfer curve times 1,000. The input offset voltage (V_{io}) is the horizontal displacement from the horizontal reference to the point where the function crosses the vertical reference divided by 1,000. The straight line sloping between the saturation points is also indicative of the device's linearity. In reading V_{io} and A_{VOL} from the scope presentation, it is advisable to increase the horizontal sensitivity, so that the resolution is increased for better accuracy.

This kind of tester, when constructed with appropriate input jigs, forms a very handy do-it-yourself instrument in the laboratory for quickly resolving any doubts concerning a questionable OP AMP device.

Commercial Form of Curve-Tracer Testing

A commercial type of an LIC testing scheme, producing a number of *oscilloscope-trace displays*, is offered as *test fixture 178*, which attaches to the Tektronix 577 Curve Tracer. With this connection, the curve-tracer measurement system adds to its conventional diode-transistor test capabilities, and provides for linear IC testing with a storage-scope (D_1) display, which accommodates the very slow sweep rate that must be used to display many LIC characteristics.

The typical characteristics that can be checked on an LIC amplifier with this curve-tracer system include open-loop gain (A_{vol}), common-mode rejection ratio (CMRR), power-supply rejection ratio (PSRR), input and supply currents, along with checks on thermal drifts and noise.

Plug-in test cards are offered for defining the type of LIC that can be tested; one for testing IC amplifiers and the other for testing IC regulators. Each card can be quickly set up by jumper leads for the particular linear IC under test.[1]

6-11. BREADBOARDING THE INTEGRATED-CIRCUIT PACKAGE

The IC package comes in three main forms, as shown in Fig. 6.12: one form is the *can type* in part (a), (usually designated as TO-5 or TO-99 types), having 8, 10, or 12 leads; another is the *flat pack* in part (b); and then the third main form is the *dual-in-line plastic (DIP) type*, having 14 or 16 leads in its full form in part (c); a variation of the DIP is the increasingly *popular Mini-DIP* package with 8 leads, shown in Fig. 6.12 (d).

In commercial assembly practice, the accepted method for mounting the IC package is on a prepared printed-circuit (PC) board, and making the connections presents no particular problem using commercial printed-circuit techniques. But for laboratory purposes there may be considerable difficulty in the use of commercial sockets, calling for connecting the many closely spaced leads (from 8 up to 16 leads) with the attendant likelihood of possible short circuits between adjacent leads.

Of the many possible ways for connecting the IC on an experimental breadboard, *laboratory practice strongly prefers a plug-in socket arrangement with the multiple leads brought out to numerous posts.* It is preferable also to allow for easy change over of circuit configurations, wherever feasible. For this purpose, a perforated board is generally used, and Fig. 6.13 (a) shows one such arrangement, using a *Vector board*[2] with associated spring-clip terminal posts. When making connections in the laboratory to the multiple pins of the IC package, a number of special methods are available for avoiding the difficulty of soldering the closely spaced leads. *Sockets with long wire-wrap leads*, which may then be bent to spread out the spaces between the leads, is one suggestion; another is the use of *adhesive-backed conductors*, which have

[1] *Tekscope*, Nov. 6 issue from Textronix, P.O. Box 500, Beaverton, Ore. 97005.

[2] Vector Electronics, Gladstone Ave., Sylmar, Calif. 91342.

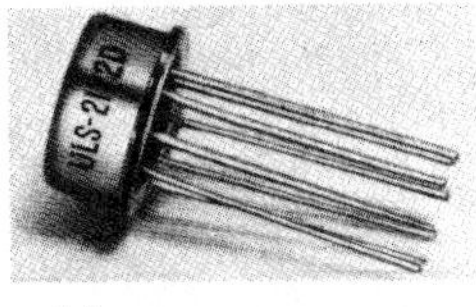

(a)

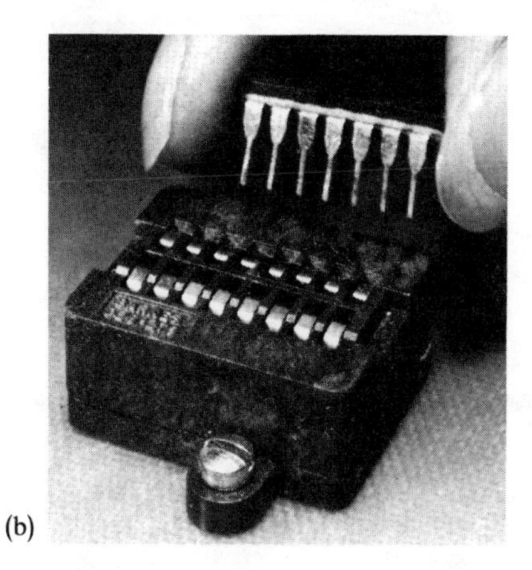

(b)

(c)

FIG. 6.12 IC packages: (a) metal can (TO-5 or TO-99 cans); (b) 14-lead dual-in-line plastic (DIP) in Barnes test socket; (c) 8-lead Mini-DIP package. [(a) *Sprague Electric* and (b) *Barnes Engineering*]

a printed conductive pattern extending the socket-pin contacts to more widely spaced terminals.

An effective alternative to the regular socket is the use of *multiple-terminal connection strips*, which, if desired, require no soldering at all. Such a multiple-terminal connection board, as shown in Fig. 6.13 (b), acts as a socket to accept both DIP and TO-5 type packages, and simultaneously provides multiple terminals for each pin of the package. This is done by simply inserting the bared end of an insulated wire into any one of four holes, whose contacts are wired in parallel with each pin connection, thus providing for up to four additional connections (without soldering) to each of the package leads (EL Socket, type SK-10).[1]

[1] EL Instruments, Inc., 61 First St., Derby, Conn. 06418, who also supply a kit for a "Universal OP-AMP Breadbox."

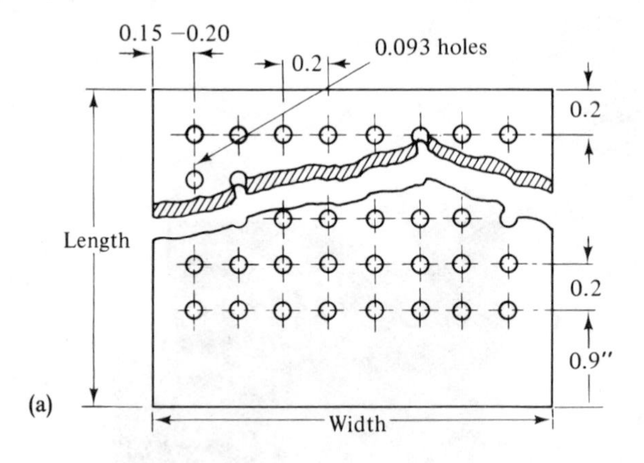

(a) Vector board

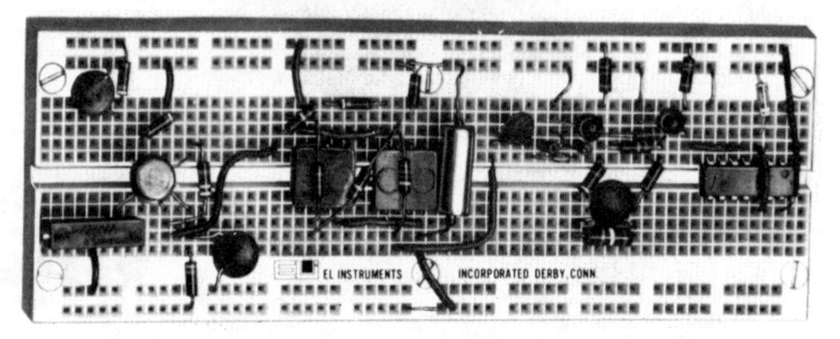

(b)

FIG. 6.13 Convenient (no-soldering) socket and breadboard arrangement: (a) Vector perfboard (.093 holes) takes Vector spring clips; (b) EL Instruments, socket SK-10 accepts DIP packages (and metal-cans with formed leads) and provides 4 parallel-connected receptacles for each lead.

In the breadboard example shown in Fig. 6.13 (a) a Mini-DIP package has been inserted into a shortened portion of the multiple-terminal block. The original block, as furnished (a bit over 6 in. long), has been cut into thirds. Wires (#20 or #22) have been inserted into the contact receptacles, and connect to the Vector spring terminals. Since the spring terminals also require *no soldering*, this arrangement of the breadboard conveniently provides posts for connecting all necessary input, output, and power-supply connections, while also allowing any desired resistors (or other external components) to be rapidly substituted for the purpose of altering the circuit as desired.

This arrangement on a perf-board avoids any possible difficulty in soldering closely spaced leads, while allowing components to be easily interchanged. The use of the multiple-terminal block, together with the

Vector spring clips that fit into the perf-board holes, combines to make a highly convenient and flexible type of breadboard. (A similar type of multiple-terminal block is manufactured by AP, Inc.[1])

In practical use, as in the illustrated breadboard arrangement, it is a very simple matter (without soldering) to make such circuit alterations as *changing the gain* of the circuit (by substituting a new resistor for R_f), *changing from inverting to non-inverting mode* (by shifting the signal-source and ground connection), and even *changing from an amplifier to an oscillator circuit* (by simply substituting a capacitor for a single resistor). It thus offers a particularly useful and convenient breadboard arrangement for experimental use in the laboratory.

6-12. AUTOMATED LINEAR INTEGRATED CIRCUIT TESTING

The testing of linear ICs in quantity, as in production testing, calls for a tester capable of speedy and automatic operation. The industrial testing instruments range from the *semi-automatic types*, where both manual and automatic modes are provided, to the highly developed *computer-controlled automatic types*.

An example of a highly versatile and speedy tester, which though automatic does not require connection to a computer, is illustrated in Fig. 6.14, the *General Radio GR 1730 Linear Tester*. It can be set up to provide as much information as desired, from a simple GO/NO-GO indication of a series of up to 18 tests, to a detailed account of each test, displaying any or all test results by a $3\frac{1}{2}$-digit display, plus decimal point and unit of measurement.

It tests a wide variety of linear ICs, including not only OP AMPS, but also comparators, regulators, and other such low-voltage linear IC devices. Some of the parameters tested are listed in Table 6-4.

TABLE 6-3

Test Parameters for Automated Testing

Gain–bandwidth product (G × BW)	Input offset voltage (V_{io})
Gain (with and without load) (A_v)	Input offset current (I_{io})
Common-mode limits ($\pm E_{cm}$)	Power-supply rejection ratio (PSRR)
Output impedance (Z_o)	Maximum output [V_o(max)]
Bias current (I_B)	Quiescent operating current ($\pm I_{cc}$)

[1] AP, Inc., 72 Corwin Drive, Painesville, Ohio 44077.

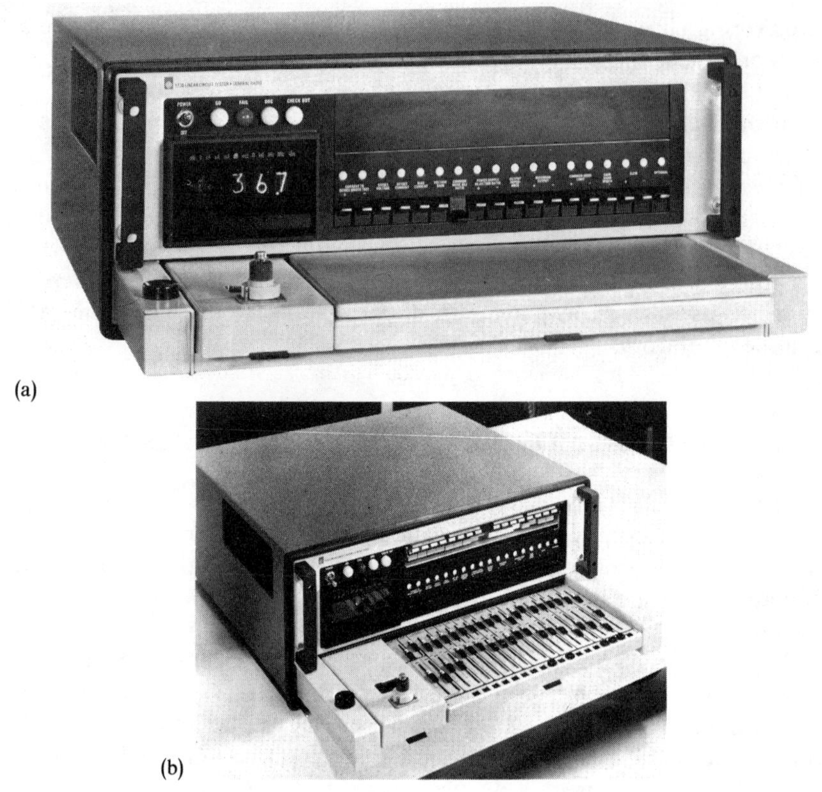

(a)

(b)

FIG. 6.14 Automatic linear IC tester: any number of tests and digital readouts (up to 18 tests) can be set up by the forty switches shown in the pull-out drawer in (b) after which one button initiates all tests. *(General Radio, model GR 1730)*

Circuits can be tested as fast as they are connected to the device-adaptor boards, and one button initiates all tests once the desired conditions have been established. The proper conditions are set on an interchangeable "memory panel" consisting of 40 slide switches for the selection of all tests; also skip tests and skip limits are included to make the tests as comprehensive or simple as desired. Together with various options, as, for example, printing the results, the tester provides a wide range of tests conveniently arranged for both simple or comprehensive automatic testing.

6-13. ENGINEERING EVALUATION (SEMIAUTOMATIC) TESTING

The two forms of testing examined in the previous sections constitute two extreme testing techniques: the *manual form*, where the test for each parameter can be set up separately, suitable for student laboratory work on individual ICs, and the automatic form, where quantities of ICs are to be tested for production or quality-control purposes in an automatic sequence type of operation, and calling for facilities for the requisite complex pre-programming of the varied tests provided by "memory panels" or, in some cases, by a computer. For the in-between cases where a modest amount of ICs is to be evaluated by the user, whether in school or industrial laboratories, a more suitable form is a third kind of *semiautomatic testing.*

The operational amplifier tester illustrated in Fig. 6.15 *(Teledyne/Phil-*

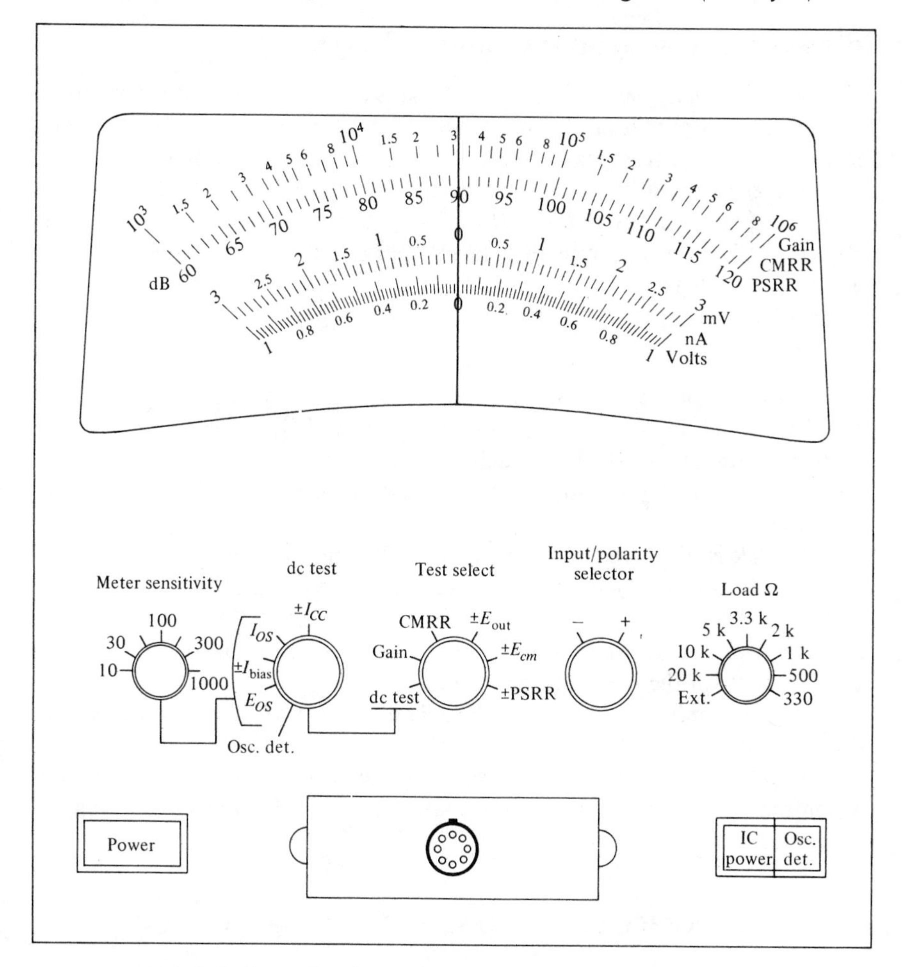

FIG. 6.15 Operational amplifier tester (semi-automatic): tests all OP AMPS (discrete and hybrid) for engineering evaluation, with meter readout for direct magnitude. *(Teledyne/Philbrick, model 5102)*

brick, model 5102) is of the semiautomatic type, designed for conveniently obtaining engineering evaluation of the dynamic and the static characteristics of OP-AMP ICs in both discrete and hybrid types. There are just five selector switches here to be set, after the proper amplifier socket has been plugged in. For testing any given parameter, the setting of three of the controls determines the automatic functioning of the proper circuit conditions for that parameter, while the operator simply sets the other two controls (METER SENSITIVITY and LOAD switches) for the evaluation at the desired operating conditions, as indicated in the tests given next.

Dc Static Tests (As Set by DC TEST Switch)

The tests for *offset errors* (V_{io}, I_{io}, and I_B) are performed with the required zeroing for output done automatically. Tests for *current drain* ($\pm I_{cc}$) are measured under no-signal condition, giving a direct reading of current from either the positive or negative supply as selected by *INPUT POLARITY switch.* A check for *oscillation* (*OSC DET*) triggers a red light to indicate any IC oscillation that exceeds a 10 mV peak, 100 Hz to 10 MHz, with the amplifier connected for unity gain (100 percent feedback).

Dynamic Tests

The following parameters are determined dynamically under ac conditions by the use of low-frequency square waves and synchronous detection, with indications as noted for each setting of the *TEST SELECT switch*:

Gain (A_{vcl}) at full output, under load, with readout in decibels and volts per volt.

CMRR is displayed both logarithmically and in decibels.

Common-mode limits ($\pm E_{cm}$) are determined automatically for the voltage limits.

Peak output voltage swing ($\pm E_{out}$) is determined automatically and read directly on the meter for either polarity.

Power-supply rejection ratio (PSRR) is determined automatically with direct readout in decibels.

Together with the proper test socket *accessories*, the tester thus provides a combination of manual and automatic operation for convenient evaluation of the linear IC device.

6-14. "QUICK-CHECK" TEST METHOD FOR OP AMPS[1]

For the purpose of checking widely-used OP AMPS (of the general-purpose

[1] Article '"Quick-Check" Test for OP AMPS with Your Multimeter' by the author (scheduled for Apr. 1974 issue of RADIO-ELECTRONICS).

variety), the circuit given in Fig. 16.16 offers a simple, yet practical method that *rapidly verifies its functioning ability.* In view of the complicated circuitry and the detailed operation involved in performing complete tests on LICs, this checking method gives a quick answer for determining confidence in the operating condition of a doubtful OP AMP, and, as such, can well serve as a rough but practical *GO-NO-GO check.*

The checking circuit employs the OP AMP as a multivibrator, generating a square-wave output. The simplicity of the system stems from the fact that the circuit uses only five active leads, out of the industry-standard 8 pins in the popular forms of OP AMPS. Thus, the connections shown will apply to dozens of OP AMPS, where the 8-pin package (whether metal can or Mini-DIP), and the terminal numbering have become standardized by industry for these five active leads, as follows:

Inputs (inverting and non-inverting) = ② and ③;
Output = ⑥, and ±V supply = ⑦ and ④ respectively.

Aside from these connections and the RC components, the circuit requires only a multimeter (preferably of the FET variety) and a small (2″) loudspeaker as accessories. With the push-button (PB) switch open, the meter reads the p-p value of the square-wave output in a nominally-loaded condition. As a rule of thumb, this (p-p) value should read at least 2/3 of the supply voltage (or at least 12V (p-p) from a ±9 V supply). Pressing the push-button

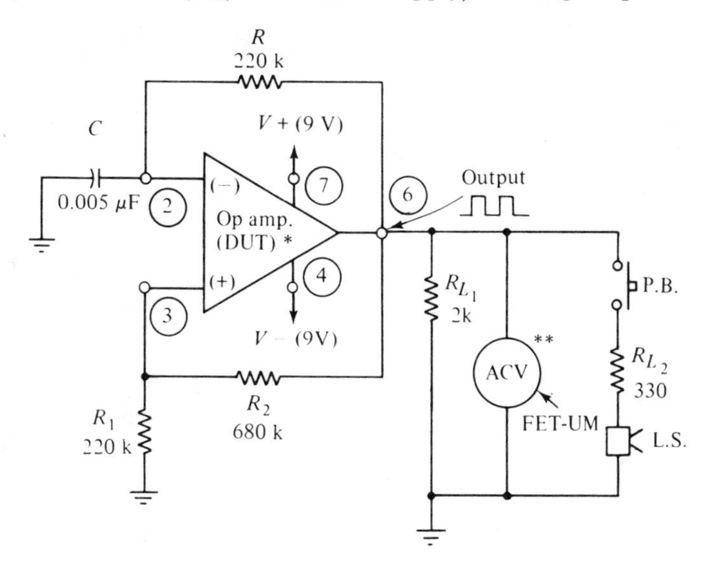

* Circled pin nos. apply equally to 8-pin metal can or min-DIP;
** FET-VM preferably for p-p output readings

FIG. 6.16 Circuit diagram of "quick-check" for functional operation of many 'standardized' types of OP AMP device-under-test (DUT)

switch calls for the OP AMP to furnish a reasonable flow of output current to produce an audible tone from the loudspeaker (around 1 kHz in this case); the output under this speaker-loaded condition should not drop to less than one-half of the previous nominally-loaded value [6 V (p-p) for this 18 V supply].

These two (p-p) meter readings, (plus the audible tone), serve to establish the basic ability of the OP AMP to produce reasonable values of voltage and current output, and are thus able to verify the operating condition of the OP AMP in a simple manner.

As an added convenience, the *connections for the five active leads remain the same*, when checking practically all general-purpose types of OP AMPS, in spite of the various type numbers encountered from different manufacturers; this fortunate condition is due to the beneficial industry-standardization of these popular devices. (Further details of construction and interpretation of test results are given in the article reference at the beginning of this section.)

7

POWER AMPLIFIERS: DIRECT-CURRENT AND AUDIO

7-1. POWER LIMITATIONS IN INTEGRATED-CIRCUIT OPERATIONAL AMPLIFIERS

The previous discussions of general-purpose IC OP AMPS emphasized their most general use as *voltage amplifiers*; as such, the OP AMP ordinarily is not called upon to deliver any substantial amount of current, thus limiting its output power capability to a great extent. As an example of this limitation, we may estimate the power output obtained in the usual voltage-amplifier operation of a general-purpose OP AMP (such as the 741 type). With the customary dual power supply (± 15 V), and with a recommended load resistor of 2 kΩ, we can expect a nominal voltage output of ± 10 V (or 20 V peak to peak); this indicates an output current of only ± 5 mA (or 10 mA peak to peak). Converted to rms values, this represents approximately 7 V at 3.5 mA, or *about 25-mW power output for the typical* OP AMP.

Of greater concern than this small value of output power is the *limited capability for substantial output current.* This is important in some dc applications, especially so in audio applications, where it is desired to avoid the use of an output transformer by feeding the current-operated voice coil of a speaker directly.

In cases where only a relatively small increase in output current is desired, there are two fairly simple options available in the OP-AMP type, without resorting to the power-amplifier type whose output is generally considered as exceeding 1 W. One method is the use of a *high-output-current monolithic type* of OP AMP, or, alternatively, one can use the *power boosters* described next; however, for appreciable amounts of power (over 1 W) we must turn to linear ICs that are designated as *power amplifiers*, which are described later in this chapter.

7-2. HIGH OUTPUT-CURRENT OPERATIONAL AMPLIFIERS

Some OP AMPS offer a bit more in output-current capability than the typical OP AMP. For example, the *Analog Devices AD 512* offers a ±10-V output feeding a load resistor of 1,000 Ω, thus supplying ±10 mA, which is double the amount of current specified for this voltage output in the typical 741 type (the latter specifies a ±10 V output feeding loads equal to, or greater than, 2 kΩ). When feeding a lower-resistance load of 400 Ω, a maximum power output of up to 125 mW can be obtained. As a monolithic OP AMP, packaged in an eight-lead TO-99 (can)-type case, the *AD 512* would be quite suitable for certain limited-power dc or ac applications.

For a further increase of *output current up to as much as 1 A*, there is the *Fairchild μA791* OP AMP, which is available as a 10-lead device in a case similar to the familiar TO-3 case of power transistors to handle the large current output. Since it combines the input characteristics of the type 741 op amp as a preamplifier with a direct-coupled power amplifier in one package, it is discussed later as a power operational amplifier, useful for both dc and ac amplification. Similarly, the *CA3094/A* programmable power switch/amplifier provides high output current (to 300 mA peak to peak), and power output to 600 mW as a class A amplifier.

7-3. POWER BOOSTERS

Another method for increasing the current output, as a supplement to a circuit already using a general-purpose OP AMP, is that of adding a "power booster." Here the use of Darlington emitter followers in a separate package serves to greatly increase the current gain of the overall amplifier; it provides unity voltage gains, thus retaining the voltage-gain characteristics of the general-purpose OP AMP.

One example of a convenient power booster is the *Burr–Brown 3329/03* type. (Although not a regular DIP package, it still fits the regular DIP socket.) When used to follow the OP AMP, it increases the output current capability of the combination to ±100 mA at ±10 V without the need for a heat sink. Since its characteristics allow it to be included within the feedback loop of an OP AMP (as shown in Fig. 7.1), it forms a very stable circuit for driving low-impedance loads. Its class B output stage ensures a minimum of quiescent power-supply drain, while its low open-loop output impedance (10 Ω) enables it to drive a 50-Ω cable with its associated

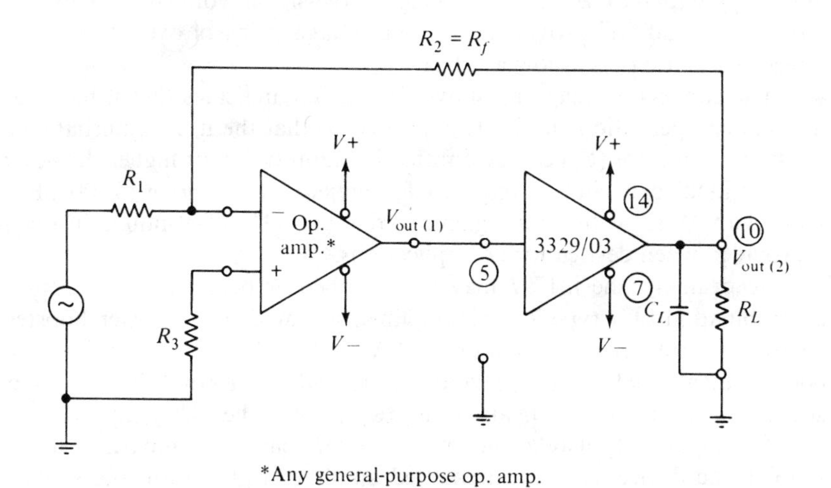

FIG. 7.1 Power booster type (BB 3329/03): used with OP AMP for increasing output-current capability as in feeding a 50-ohm cable load (R_L). *(Burr–Brown)*

capacity without degrading the frequency response of the OP AMP being used. Such an output, incidentally, is quite suitable for driving a small (45-Ω) loudspeaker to a satisfactory low-volume listening level.

A similar power-booster type is produced by *Motorola (MC1438R)* for output currents up to ±300 mA with a 1.5-MHz bandwidth (also at unity gain). This comes in a nine-pin package (looking like a small TO-3 power-transistor can), and requires a special nine-pin socket.

7-4. POWER OPERATIONAL AMPLIFIER

A device with an output-current capability of a full ampere is the *Fairchild μA791* OP AMP, packaged as a 10-lead device in a case similar to the familiar TO-3 case of power transistors. It combines a type 741 preamplifier and a power amplifier in the one package. As shown by the equivalent circuit of Fig. 7.2, it is direct coupled throughout and can therefore be used as a servo amplifier at very low frequencies, for dc uses, and in audio-amplifier applications with proper heat-sink arrangements, as discussed later. It is thermal and short-circuit-protected, having a current-sensing terminal allowing selection of the sensing resistor (R_{SC}) for limiting the amount of short-circuit current to 500 mA (when $R_{SC} = 1.5\ \Omega$) or to 1 A (with $R_{SC} = 0.7\ \Omega$). With R_{SC} equal to zero, the specified large-signal voltage gain exceeds 20,000 V/V, and will provide an output voltage swing of over ± 11 V into a load resistance (R_L) as low as 11 Ω.

The connection diagram, shown in Fig. 7.3, indicates that it has some internal compensation similar to type 741, so that the usual external compensating capacitor (C_C) can be omitted for gains of 100 or higher; however, an additional capacitor is required for *output compensation* (3,300 pF in series with 3.9 Ω). Moreover, again like the type 741, offset-null adjustment is provided when desired for dc applications.

Accordingly, the *μA791* may be described as being the equivalent of a combination of a type 741 preamplifier, followed by a "power booster" having an output-current capability of 1 A. (It should be noted that the power booster mentioned in the previous section, *Burr–Brown 3329/03*, can be added to any type of OP AMP and is not restricted to the 741 type.)

The input–output relations of the *μA791* may be summarized by describing the device as a direct-coupled amplifier with an open-loop input impedance of 2 MΩ and with an output capability of ± 11 V into 11 Ω, or a power output of 5.5 W under maximum operation with heat sinking. This high-gain and high-power-output capability is obtained in this power OP AMP at a relatively low quiescent (zero-signal) supply current of 30 mA from a dual ± 15-V supply.

7-5. AUDIO POWER AMPLIFIERS (1 TO 2 WATTS)

In pushing the power of amplifiers up to outputs of 1 W or higher, the general difficulty lies in problems of heat buildup in the semiconductor junctions. This is particularly important in integrated circuits, because of their compactness, where the *generated heat is necessarily confined on a very small silicon chip.* Satisfactory methods for meeting this problem have been developed, resulting in a variety of package and heat-sink forms; presently, both monolithic as well as hybrid versions have been made available, going beyond the previous levels of 1 or 2 W and providing devices of up to 4- and 5-W outputs. These devices are in a form compatible with the compact layouts of the standardized OP AMPS. Finally, linear power-amplifier hybrid systems of 15 and 100 W are discussed.

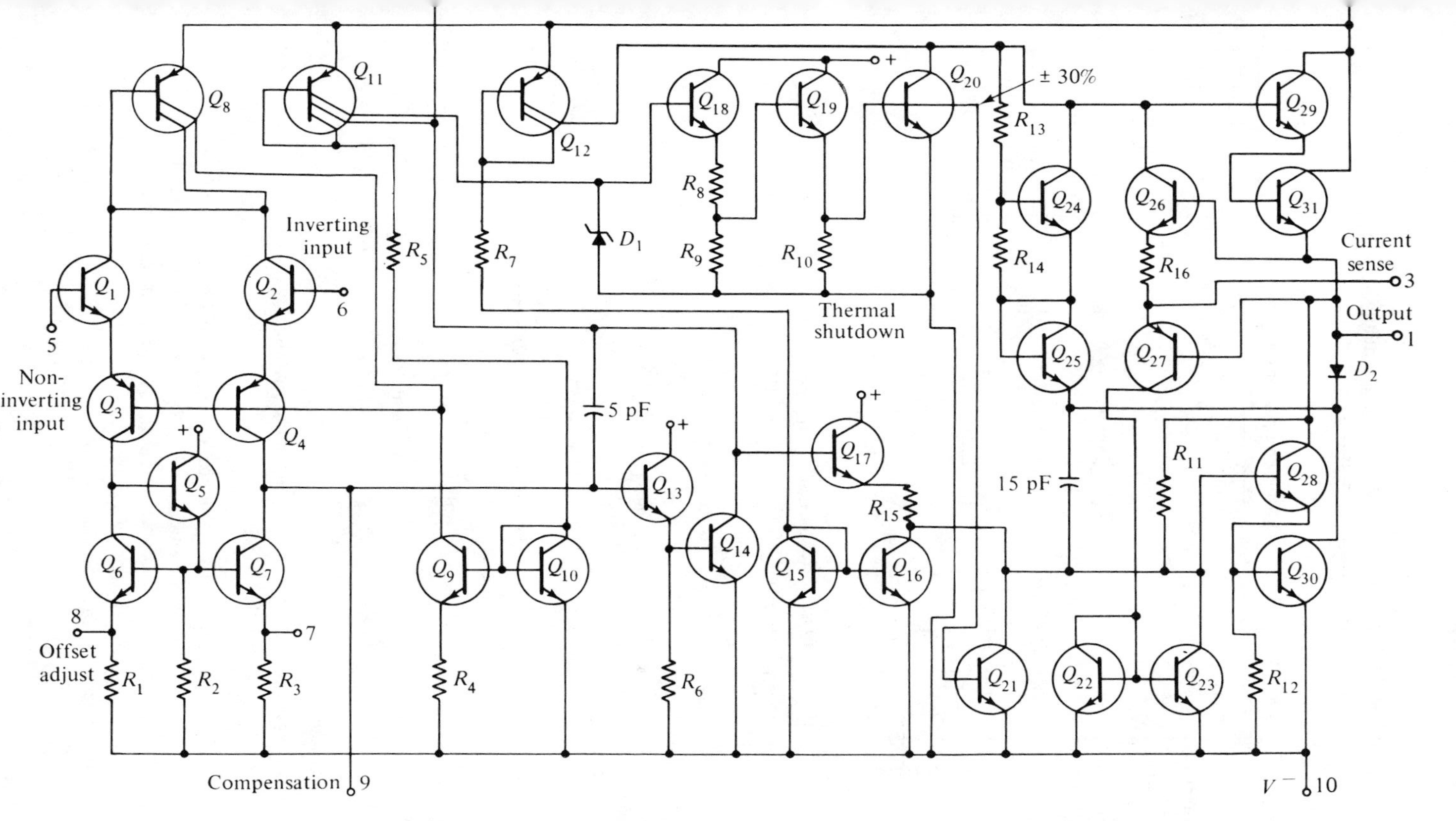

FIG. 7.2 Equivalent circuit of power operational amplifier, µA791: It combines the input characteristics of a 741 type OP AMP as a preamplifier with an output current capability of ±1 A at ±11 V, while retaining its dc and ac characteristics as an OP AMP. (*Fairchild Semiconductor*)

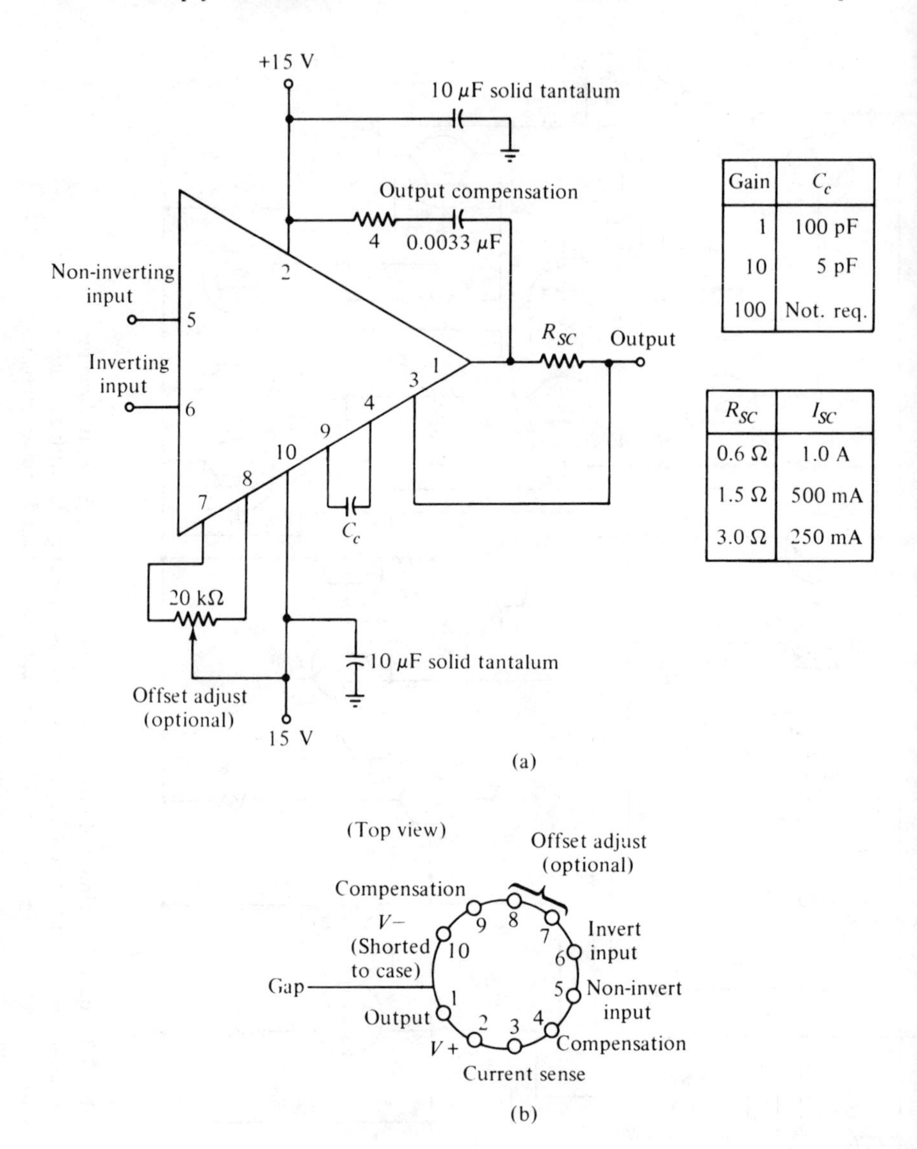

Gain	C_c
1	100 pF
10	5 pF
100	Not. req.

R_{SC}	I_{SC}
0.6 Ω	1.0 A
1.5 Ω	500 mA
3.0 Ω	250 mA

Note: Power supply decoupling capacitors and compensation networks must have short leads and must be located at the amplifier pins.

FIG. 7.3 Connection diagram for μA791 power operational amplifier: (a) current-sensing resistor (R_{SC}) limits the short-circuit current and output compensation of 3300 pF is in series with a 3.9 Ω resistor; (b) socket connections for 10-lead, TO-3 type package. *(Fairchild Semiconductor)*

1- and 2-Watt Amplifiers

As a start in tracing the trend toward higher powers (and passing by the *MC1306*, $\frac{1}{2}$-W level), we can start with the early audio amplifiers at the 1-W power level. A typical example at this level is the *Motorola MC1554G/1454G*, which comes in a 10-lead, metal-can package. It is designed to amplify signals to 300 kHz, delivering 1 W to a typical load of 16 Ω, either direct or capacitively coupled. Gain options of 10, 18, or 36 V/V are provided at single-supply operation (V+ = 16 V), or, alternatively, at a split supply of ±8 V. For consumer audio systems in this 1-W class, there are also available plastic DIP versions; examples are the *Motorola MFC8010* (1 W) and the *Texas Instruments SN76011* (with heat-sink tab) for 1 W into an 8-Ω load, operating from a 4- to 13V supply.

The 2-W power level is often used for relatively inexpensive audio systems, such as for a phono amplifier. A typical IC audio amplifier for this level is the *Motorola MFC9020*. (Previously popular units of this type from *General Electric*, such as the *PA237*, are no longer being manufactured.)

The appearance of the DIP package of the *MFC 9020* is shown, with the heat-sink tab, in Fig. 7.4(a). This tab is designed for a minimal heat-sinking requirement for 2-W operation at ambient temperature (T_A)=55°C. For a larger power requirement (or for a greater ambient temperature requirement), the tab must be soldered to an effective copper-surface area of at least 1 in.^{2}, either on one side of the printed-circuit (PC) board or as external wings. (Linear derating requirements are given on the data sheet for up to 150°C operation.)

The typical circuit connection diagram [Fig. 7.4(b)] shows a 16-Ω speaker load connected to the power supply (V+), along with volume and tone controls. The circuit, as used with a ceramic phono cartridge, for example, exhibits an input impedance of approximately 800 kΩ, with an input sensitivity of about 250 mV for the rated 2-W output, with around 1 percent total harmonic distortion (THD) in the audio-frequency range.

Another version of a 2-W amplifier is the *National LM380*, which aims at a low parts count for consumer applications when feeding an 8-Ω speaker. In the diagram for a phono amplifier application (Fig. 7.5), the connections for the 14-lead DIP package of the *LM380* are shown. In this case, the heat-sink requirement is met without the use of a separate tab; instead, three pins in the middle of each side of the standard DIP package are soldered together (pins 3, 4, 5, and 10, 11, 12). When mounted on epoxy-glass board, the maximum junction temperature of 125°C is controlled by soldering the six pins together to the copper foil of the mounting board (minimum surface of 6 sq. in.). The quiescent supply current for the *LM380* has been brought down to 7 mA.

For stereo amplifiers there are dual versions, having both 2-W channels

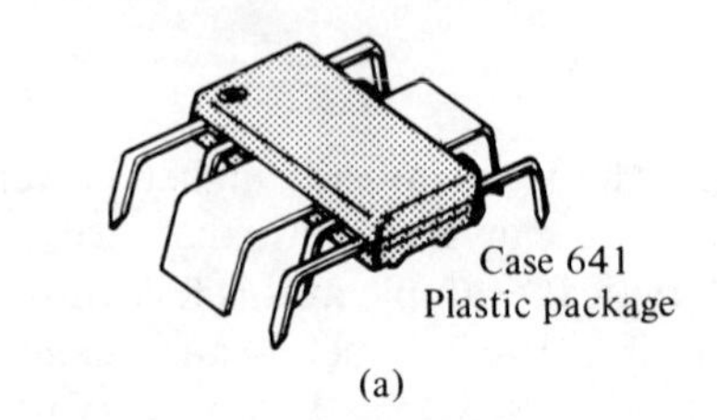

(a)

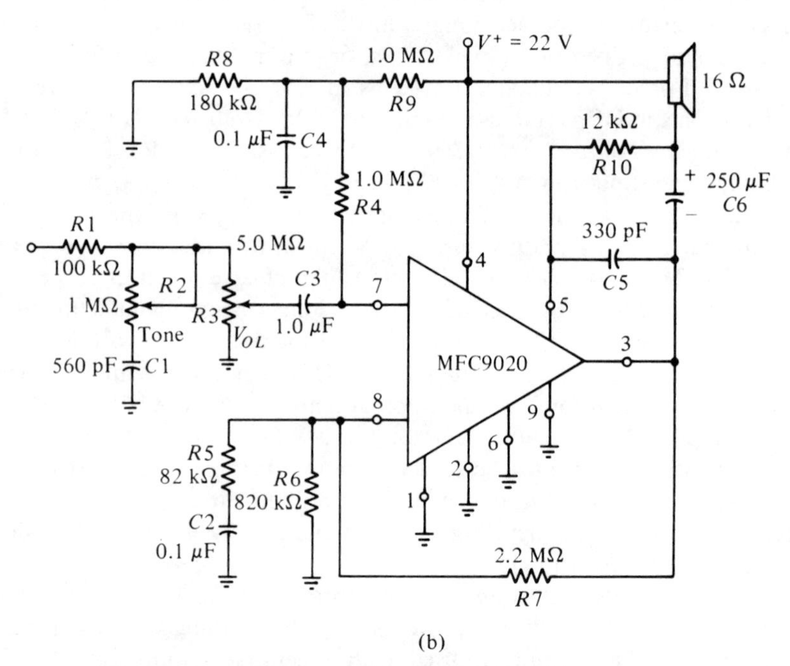

(b)

FIG. 7.4 Two-W audio amplifier, MFC9020: (a) DIP package with heat-sink tab; (b) typical circuit connections, as for a phono amplifier with 16-ohm speaker load, and showing volume and tone controls. *(Motorola Semiconductor)*

in a single 14-lead DIP package. The *National LM377* is an example of such a type, where each channel feeds an 8-Ω speaker, returned to ground from a supply voltage of from 9 to 30 V.

To accompany dual audio amplifiers for stereo use, there are companion *dual preamplifiers*, containing both preamp channels in the one DIP package. Examples of such dual preamplifiers are the *Motorola MC1303* and the *National LM381* types, both designed for stereo use. A circuit diagram using half of the *MC1303* to preamplify a *magnetic phono cartridge* with RIAA equalization is shown in Fig. 7.6.[1]

[1] B. R. Rogen, "Op Amps at Work: 10 Audio Circuits," *Radio Electronics*, Dec. 1972.

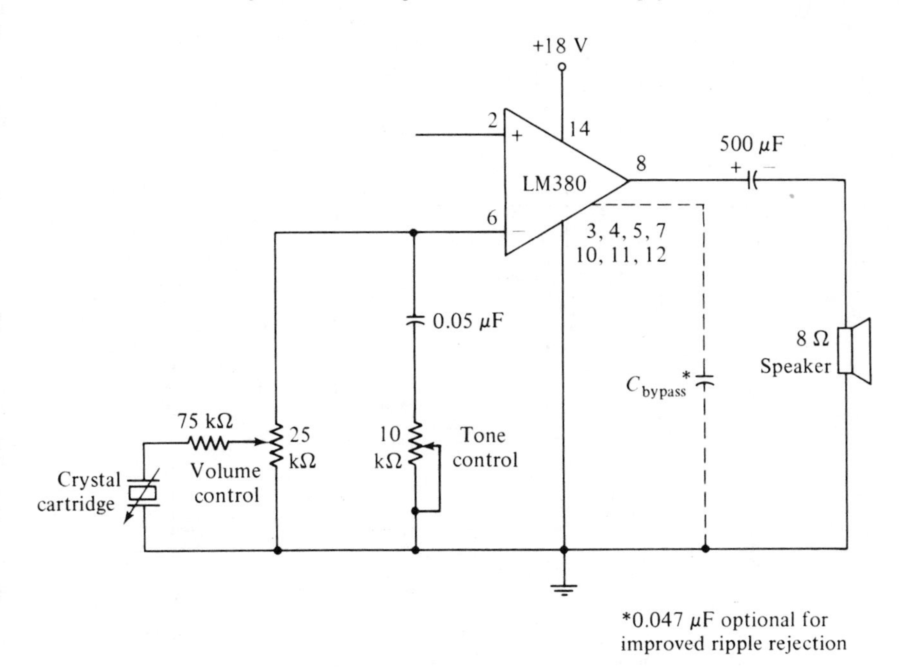

FIG. 7.5 Phono amplifier circuit of low component count: 2-watt amplifier (LM380), with loud speaker connection returned to ground; the middle pins on each side of the DIP package (pins 3, 4, 5, and 10, 11, 12) are soldered together for heat-sinking. *(National Semiconductor)*

7-6. HIGHER-POWER INTEGRATED-CIRCUIT AUDIO AMPLIFIERS (TO 5 WATTS)

Monolithic versions of audio power amplifiers up to the 5-W level find use in such consumer applications as auto radios and TV receivers, and also in some industrial applications. While still in the favored DIP form, the IC package is necessarily modified to accommodate heat-sink considerations at this power level. However, the developments in IC fabrication have succeeded in producing compact packages that are vastly more convenient than any of the discrete forms of 5-W amplifiers.

A typical 5-W IC audio amplifier, *Fairchild μA706*, is shown in schematic form in Fig. 7.7. It consists basically of a *preamplifier and power amplifier*. The preamp has high-gain transistor Q_2 acting as a common emitter with a high-impedance collector load provided by current source Q_3. Transistor Q_1 acts as an input-buffer transistor, thus providing a high input resistance (3 MΩ) referenced to ground.

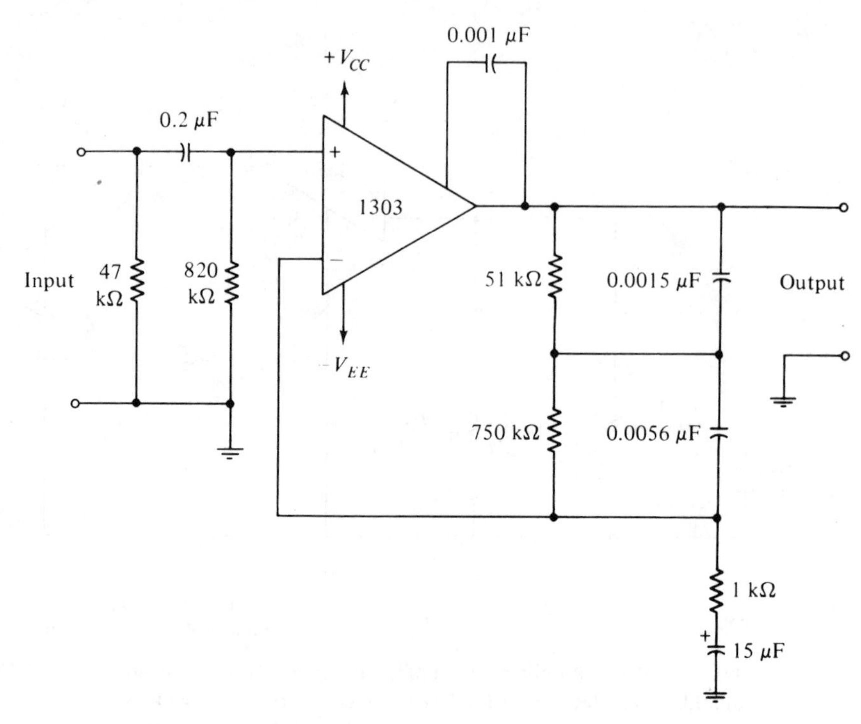

FIG. 7.6 Preamplifier circuit for magnetic phono cartridge; uses ½ of dual stereo preamp MC1303, with RIAA equalization *(Motorala)*.

The *power-amplifier* section feeds power transistors Q_{13} and Q_{14} as a *quasi-complementary push–pull amplifier stage.* (The term "quasi" is used to indicate that the circuit operates in complementary form, even though both power transistors are NPN, instead of the usual complementary combination of NPN and PNP types.)

The overall closed-loop voltage gain of the amplifier is determined initially by the internal negative feedback applied through resistor R_6 from the output to the emitter of Q_2. This internal feedback reduces the high open-loop gain by a factor of about 50 to a value of $A_v = 200$ or 46 dB (obtained from the ratio of R_6/R_3). A gain-control option is provided at terminal 8 for the external connection of a gain-reducing resistor (R_B in series with a 100-μF capacitor). Thus, as shown in the data box of Fig. 7.8 with $R_B =$ 100 Ω, the gain is reduced from 46 to 34 dB (or from 200 to 50 V/V).

Operating from a single power supply (V+ = 14 V), the 5-W output can feed a 4-Ω speaker connected between output and supply, as in Fig. 7.8(a); alternatively, with the addition of a 47-Ω resistor and 220-μF capacitor, the speaker can be connected between output and ground as shown in part (b). In the latter grounded-speaker connection, an optional tone control

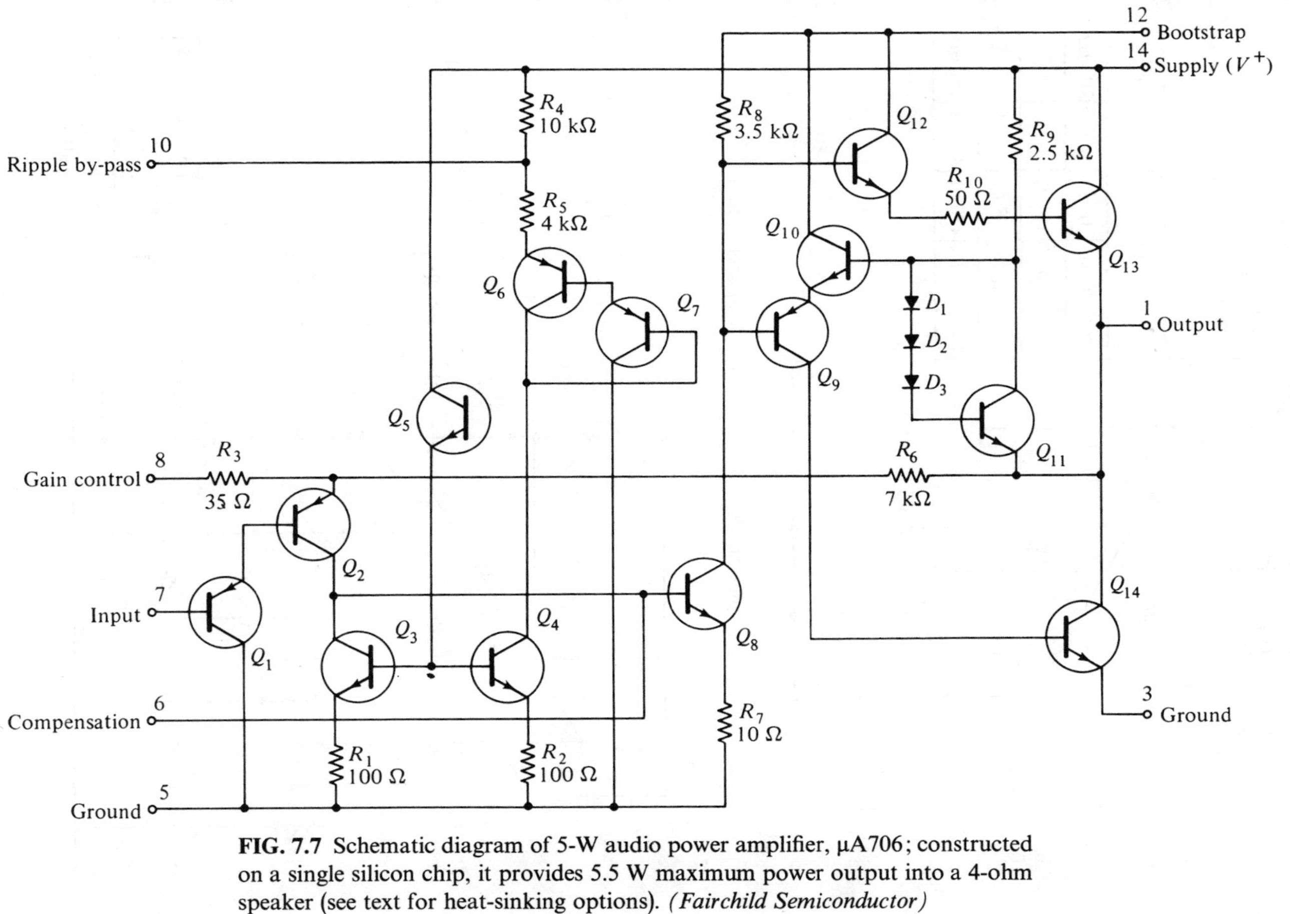

FIG. 7.7 Schematic diagram of 5-W audio power amplifier, μA706; constructed on a single silicon chip, it provides 5.5 W maximum power output into a 4-ohm speaker (see text for heat-sinking options). *(Fairchild Semiconductor)*

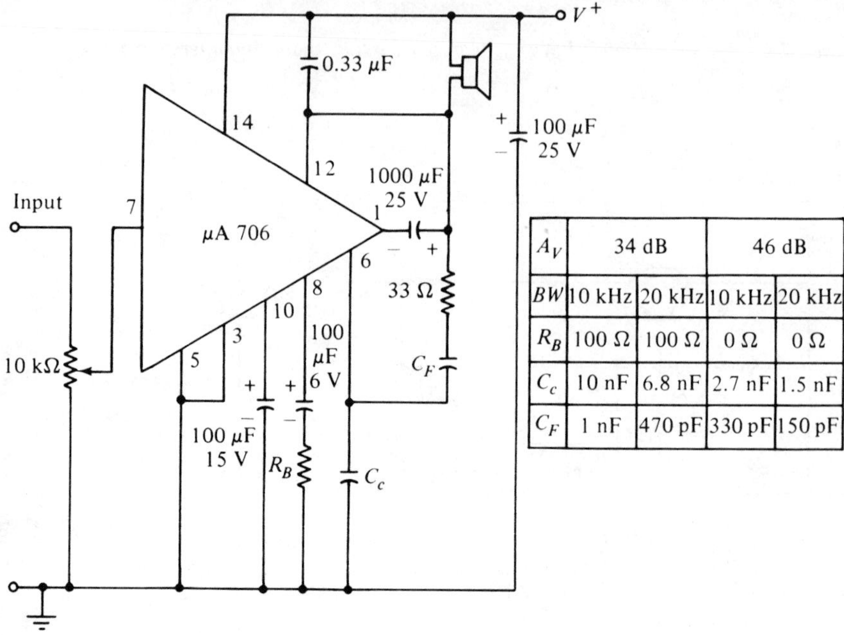

A_V	34 dB		46 dB	
BW	10 kHz	20 kHz	10 kHz	20 kHz
R_B	100 Ω	100 Ω	0 Ω	0 Ω
C_c	10 nF	6.8 nF	2.7 nF	1.5 nF
C_F	1 nF	470 pF	330 pF	150 pF

(a) 5-watt audio amplifier with minimum component count

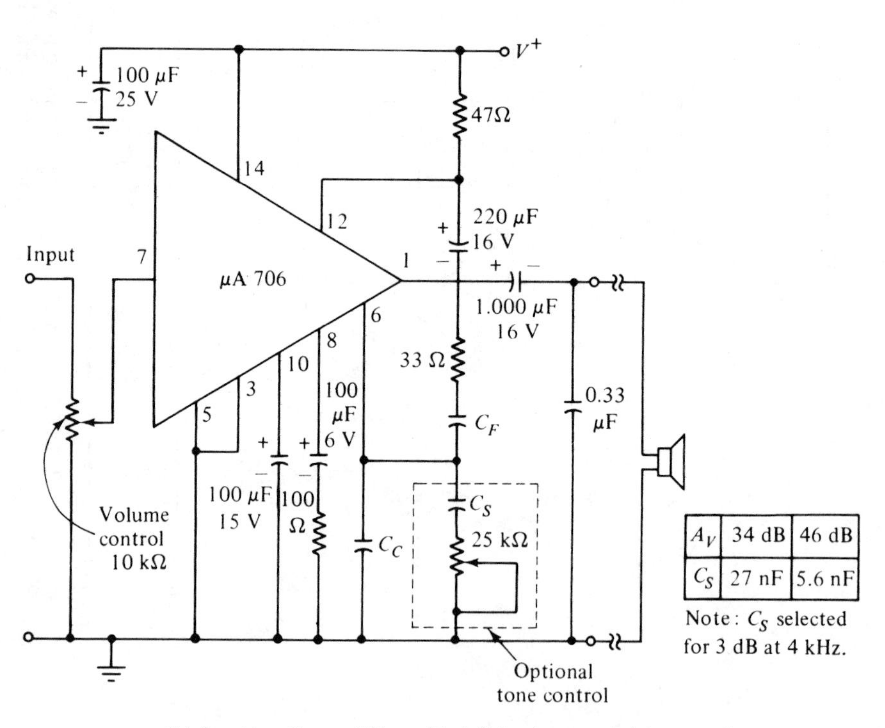

A_V	34 dB	46 dB
C_S	27 nF	5.6 nF

Note: C_S selected for 3 dB at 4 kHz.

(b) 5-watt audio amplifier, with 4 Ω load connected to ground

FIG. 7.8 Connections for μA706 audio power amplifier (5.5 W max): (a) with loudspeaker connected to V+ for minimum component count; (b) with loudspeaker connected to ground (see text for explanation of capacitor symbols). *(Fairchild Semiconductor)*

is included. Under test-circuit conditions, a value of C_S equal to 5.6 nF (5,600 pF) provides for a 3-dB falloff in gain at 4 kHz.

Examining the purpose of the other external components in Fig. 7.8(a), there are the two filter (or ripple) capacitors of 100 μF each, the output coupling capacitor of 1,000 μF, and the boot-strapping capacitor of 0.33 μF. In addition, we have compensation capacitor C_c at terminal 6 in conjunction with capacitor C_F at the same terminal. The test-circuit values for these (at the maximum gain of 46 dB) are given as $C_c = 1.5$ nF (or 1,500 pF) and $C_F = 150$ pF; corresponding values of C_c and C_F are given for reduced gain (A_v) or reduced bandwidth (BW).

Using the test-circuit values (46-dB gain and 20-kHz bandwidth at a supply voltage of 14 V), the device can deliver 5.5 W into the 4-Ω load, with less than 10 percent total harmonic distortion (THD). This high value of distortion is caused by clipping; however, it is radically reduced to a THD value of less than 2 percent at an output-power level of 4 W, and still further reduced to below 1 percent at power levels of 3 W or less.

As a device designed for consumer applications, the package has been kept compact in the familiar 14-lead dual-in-line plastic form, but this convenient DIP form has been modified to accommodate various heat-sink requirements. In the basic (package A) form, a copper slug is exposed at the top of the package, providing a thermal conduction path from the junction to the case. To improve the thermal junction-to-ambient thermal resistance (Θ_{J-A}) of package A (73 °C/W), a copper bracket soldered to the topside is optionally offered as package B, reducing this thermal resistance to 55°C/W. Various heat-sink devices are commercially available for attachment to the bracket of package B to further reduce the thermal resistance, as required by specific applications, to satisfy the relation

$$\Theta_{J-A} \leqslant \frac{T_d - T_A}{P_d},$$

where T_d, maximum junction temperature = 150°C,

T_A = ambient temperature,

P_d, maximum device dissipation = 3 W.

For ambient temperatures reasonably below the maximum of 85°C, a satisfactory thermal resistance of 22°C/W can be obtained by soldering the bracket of package B to the copper side of a two-sided printed-circuit board. Moreover, at power-output levels smaller than the maximum of 5.5 W, satisfactory operation of either package A or B can be obtained in particular situations without employing additional heat sinking.[1]

[1] Fairchild Application Note 317 illustrates commercially feasible layouts for both a single 5-W μA706 amplifier and also for two devices in an FM stereo receiver.

Similar audio power amplifiers up to 4 or 5 W, but which still retain the convenient package form of monolithic LICs, are exemplified by such devices as *Texas Instruments SN76024* (4 W) and *Sprague ULX2285* (5 W). As another example, single amplifier *National LM383* (5 W) is also available as a dual audio amplifier *LM378* (4 W/channel) for use in stereo amplifiers.

7-7. HYBRID POWER AMPLIFIERS

Power amplifiers at levels beyond 5 W are generally integrated in hybrid packages. The packages vary according to the wattage levels of the power transistors that are integrated in the hybrid package (and their corresponding heat-sink arrangements), and they *extend up to 100 W*. Two examples will be cited here, the first for a 15-W audio amplifier and then for a linear 7-A amplifier capable of 100-W output.

15-Watt Amplifier

This thick-film amplifier *(type EAA0015* from *Electronic Associates, Inc.)* provides continuous operation at output powers up to 15 W into a 3.2 Ω speaker load. It is offered in a 10-lead rectangular package (approximately 2 in. long by 1 in. wide), as shown in Fig. 7.9(a). A ceramic substrate forms the bottom of the package, and must be heat-sinked for output powers greater than 2 W (at full-power output, this is typically equivalent to an aluminum sheet 6 by 6 by $\frac{1}{8}$ in.).

The class B audio amplifier provides full output from 350-mV signals. This input is compatible with most preamplifier designs. In the case of a stereo system, such monolithic preamplifiers can be used as *Motorola MC1303* or *National LM381*; in each case, each half of the preamplifier feeds a separate 15-W amplifier for a total of 30 W of IHF music power, operating on a single supply (V+ = 30 V). At rated 15-W output and 1 kHz, the harmonic distortion is listed at 0.8 percent. As given in its application note,[1] the use of external terminals is described for optional trimming of idle current and crossover characteristics. Provision for bass and treble boosts and response compensations are also shown when using the aforementioned preamplifiers. Provision is also made for short-circuit protection.

100-Watt Amplifier

The concept of a complete solid-state hybrid amplifier as a power component

[1] EAA0015 Application Note from Electronic Associates, Inc. (Precision Components Division), West Long Branch, N.J. 07764.

FIG. 7.9 Hybrid power amplifiers: (a) view of ceramic package of 15 W audio amplifier, EAA0015; *(Electronic Associates)* (b) view of 100-W (7-A) power amplifier, HC1000. *(RCA)*

has been pushed to 100 W by *RCA in its type HC1000*, shown in Fig. 7.9 (b). A quasi-complementary class B output circuit is utilized, incorporating hometaxial output transistors and built-in load-fault protection.

The compact package (weighing 100 g) has 10 heavy leads, appropriate for its capability of delivering a peak current of 7 A and providing 100-W rms power output into a 4-Ω load.

The features of this innovative design include direct coupling to the load and bandwidth to 30 kHz at 60-W output, while operating from a split supply of ± 37.5 V (it may also be operated from a single $+75$ V supply). As a rugged IC power component, this device offers the possibility of high-power applications, not only as an audio power amplifier, but also for such uses as servo amplifiers, deflection amplifiers, and as a power source in driven inverters.

8

CONSUMER/ COMMUNICATION CIRCUITS

8-1. IMPACT OF LINEAR INTEGRATED CIRCUITS ON COMMUNICATION EQUIPMENT

While monolithic integrated circuits have indirectly affected electronic usage for years, because of the pervasive use of the computers in which the digital ICs first realized mass use, it is in the growing acceptance of LICs in communication equipment that household use is most likely. Nearly everyone has one or more broadcast radio or television receiver at his or her disposal. Two-way radio and radio-paging systems are also rapidly increasing in both domestic and commercial applications.

Acceptance of LICs was a much slower process in communication equipment than in the case of digital-computer ICs. There is far less standardization in communication designs, and a more diverse group of equipment manufacturers is involved than in the case of computers. Another early roadblock was the inability of LICs to effectively perform frequency-selective filtering internally. While digital-computer circuits require fast switching speeds, they are still simply off–on switches, while LICs in communication systems must compete with individual discrete components specifically optimized for each part of the circuit. For example, a radio receiver may use a low-noise high-frequency RF amplifier transistor in its "front end" that has been optimized for resistance to cross-modulating signals, but this transistor would not make a good audio output amplifier. The same discrete-component receiver might use a low-frequency high-current pair of audio output transistors, made with a process radically different from that of the RF amplifier, to drive the loudspeaker. Monolithic circuits, however, are constrained to a common process. Thus communication LICs are often compromised in performance, compared to specially designed discrete counterparts.

A further problem, especially in television receivers, is the need to handle high power and high voltages in circuitry interfacing with the cathode-ray tube. While LICs have been developed to approach the power and voltage capability of discrete circuits, practical television receivers still use discrete components in such functions as a matter of economics. While modern LIC techniques permit certain frequency-selectivity functions to be performed monolithically, economics have continued to favor the use of conventional external tuning elements where needed. There is a trend for such tuning to be lumped into single, highly selective blocks when connected to the internally complex LIC, as opposed to the conventional "interstage tuning" of discrete circuitry, in which a number of "cans" are interspersed between discrete transistor stages. (Consolidation of tuning elements in communication LICs reduces the number of IC terminals needed to make connection between internal and external elements to a minimum.)

Integrated-Circuit Subsystems

Monolithic communication LICs are, however, coming into their own, as more and more separate functions are combined on a single IC chip. Drawing a parallel to early computer ICs, the replacement of a single function (whether it is a digital gate or a single IF amplifier stage in a radio receiver) was not any less expensive than performing the task with discrete transistors and associated components. Just as computer circuits became more complex, with two, then four, and now hundreds or even thousands of gates in a package (as in *large-scale integration*, or *LSI*), LIC communication circuits have begun to compete in cost with discrete components by building complete *subsystems* on a chip. An ultimate extension, within the reach of modern LIC technology, will be complete communication systems on a chip, such as radio receivers, transmitters, and the electronic portions of tape recorders. While the main reason for acceptance of communication LICs is lower overall cost, one-chip subsystems offer the kind of substantial size reductions that were predicted in the early days of ICs—two-way "wrist radios" and perhaps even "wrist televisions" are ideas that can only become practical by applying the LSI concept to communication LICs.

To cite one recent example[1] of a "radio-receiver" LIC, the *Ferranti ZN 414* is a 10-transistor TRF receiver in a 3-lead TO-18 package. While the manufacturer[2] suggests use with an audio amplifier, the IC can drive a sensitive earphone.

8-2. BROADCAST FM RECEIVER CIRCUITS

RF/IF Amplifiers

The earliest LIC to receive mass usage in consumer electronic equipment was the type *μA703* (Fig. 8.1), which is a simple differential or emitter-coupled amplifier intended to replace a single transistor or tube in the 10.7-MHz "IF strip" of a high-fidelity FM receiver. Typical of a class of devices known as RF/IF amplifiers, the *μA703* requires conventional *LC* tuned "IF cans" as both interstage coupling and to complete its dc biasing. The device includes a transistor constant-current source, which is biased by use of the monolithic current-mirror technique as previously described in the transistor arrays of Chapter 2.

Although the type 703 was not cost competitive with discrete transistor stages at the time of its introduction, it first found its way into higher-priced FM receivers, and later became universally used because of its superior FM limiter action. A conventional single-transistor FM-IF amplifier is required to produce a constant-amplitude "limited" output, even when the amplitude of the input signal varies. It does so by switching rapidly from cutoff (no conduction) to saturation (maximum conduction) when driven by small input signals. Saturation, unfortunately, is a mechanism that can vary with the strength of the incoming signal, since it comes out of saturation slowly because of stored charge in the transistor base. It thus presents a varying impedance to the tuned circuit driving the input, causing variations in tuned-circuit frequency and bandwidth with varying input-signal strength. The type 703, however, is a differential amplifier biased so that it cannot saturate, but merely shunts the current from the "mirror" constant-current source into either the input or output transistor of the pair. Thus excellent limiting action is obtained, compared to discrete transistor amplifiers, without any of the built-in discrete amplifier problems. A typical application is shown in Fig. 8.2.

[1] "State-of-Solid-State," by Lou Garner (periodic feature): *Radio Electronics*, Oct. 1973 issue.

[2] Ferranti Electric Inc., East Bethpage Rd., Plainview, N.Y. 11803.

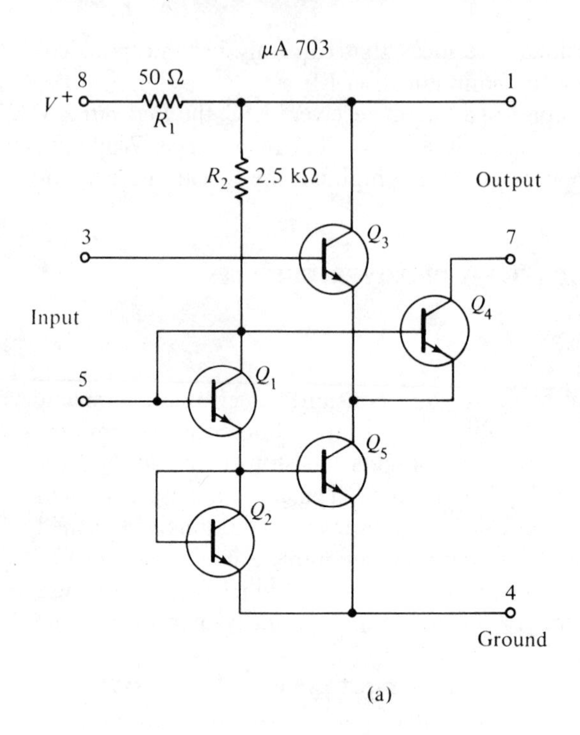

(a)

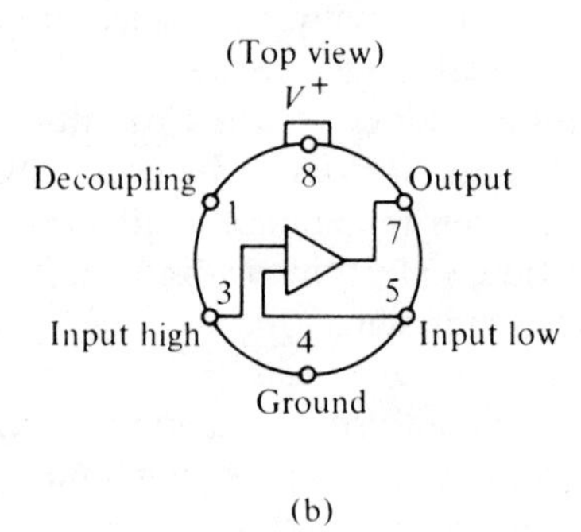

(b)

Note: Pin 4 connected to case.

FIG. 8.1 RF/IF amplifier (μA 703): (a) schematic diagram of monolithic amplifier, usable in limiting or nonlimiting applications to 150 MHz; (b) connection diagram. *(Fairchild Semiconductor)*

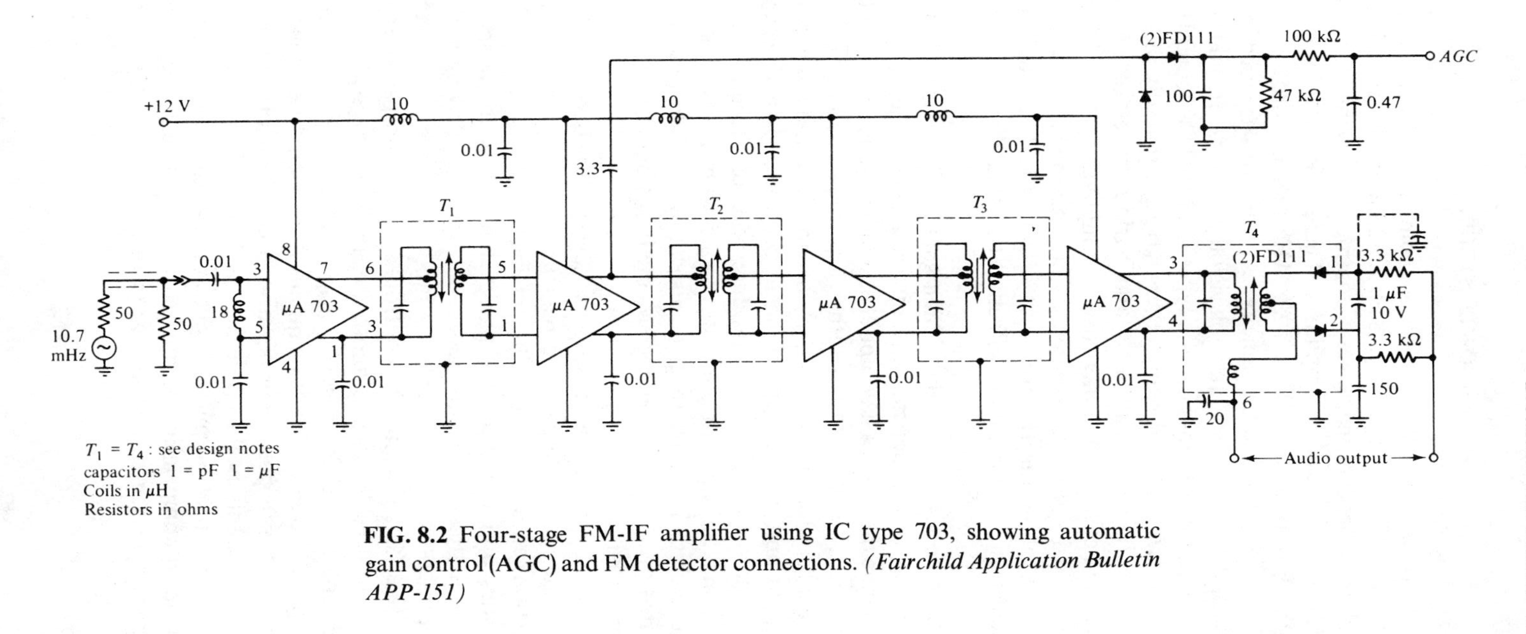

FIG. 8.2 Four-stage FM-IF amplifier using IC type 703, showing automatic gain control (AGC) and FM detector connections. (*Fairchild Application Bulletin APP-151*)

8-3. FM–IF STRIPS

Newer FM broadcast receivers use more complex "IF strip" LICs (such as the *RCA CA3043* shown in Fig. 8.3), containing several direct-coupled differential-amplifier limiting stages, a common dc biasing source, and often FM detectors and audio preamplifiers. There is essentially no difference between the FM broadcast IF at 10.7 MHz and the broadcast-television "sound IF" at 4.5 MHz, and the same types of IF strips are used interchangeably in both FM broadcast and television receivers. In addition to the CA3043, common "IF strips" include types *CA3014* (TV sound IF amplifier and detector), *MC1351*, and many others having essentially the same structure, but providing different gains and various types of FM detectors. Many designers of hi-fi FM tuners use the amplifier portions of such strips, adding their own specialized discrete-component detector circuits. Certain receivers may follow the IF strip with a "phase-locked-loop" (PLL) type of FM detector, a complex but effective IC type, discussed later in this chapter.

Practical application of these high-gain FM–IF strips requires very careful location of components and interconnections, as well as strategic shielding. While these strips have about as much voltage gain as an IC OP AMP (80 dB or more), the OP AMP has rather low frequency response, while the IF strip extends to the tens of MHz, where capacitive and inductive coupling between input and output is a significant problem. Most IF strips require only a few microvolts of input to produce volts of output. Thus there is a problem in making operating measurements with an oscilloscope probe, as enough pickup is added to give misleading results. Input and output are separated by a fraction of an inch on the IC, as opposed to several inches in a conventional "IF-can" type of strip.

Since the IC form of the IF strip has several direct-coupled gain stages, it requires a "lumped" external band-pass filter, as in Fig. 8.4; here it is usually a multi-section ceramic or crystal filter, giving very sharply shaped band-pass characteristics and relatively flat phase shift versus frequency response. Such characteristics are usually dependent upon the driving and load impedances presented to the filter; thus input resistance and capacitance of the monolithic IF strip must be controlled to give proper filter response.

The monolithic IF strip is another example of the widespread use of negative feedback in LICs. Since the strip is usually direct coupled and has very large gain, extremely small dc input offsets would be enough to drive the output to either its positive or negative limit, where it no longer amplifies signals. To avoid this problem, dc feedback is used, in which a negative-feedback resistor from output to input automatically adjusts the effective input voltage to compensate for any existing input offset conditions, thus guaranteeing that each stage will always operate in the center of its sym-

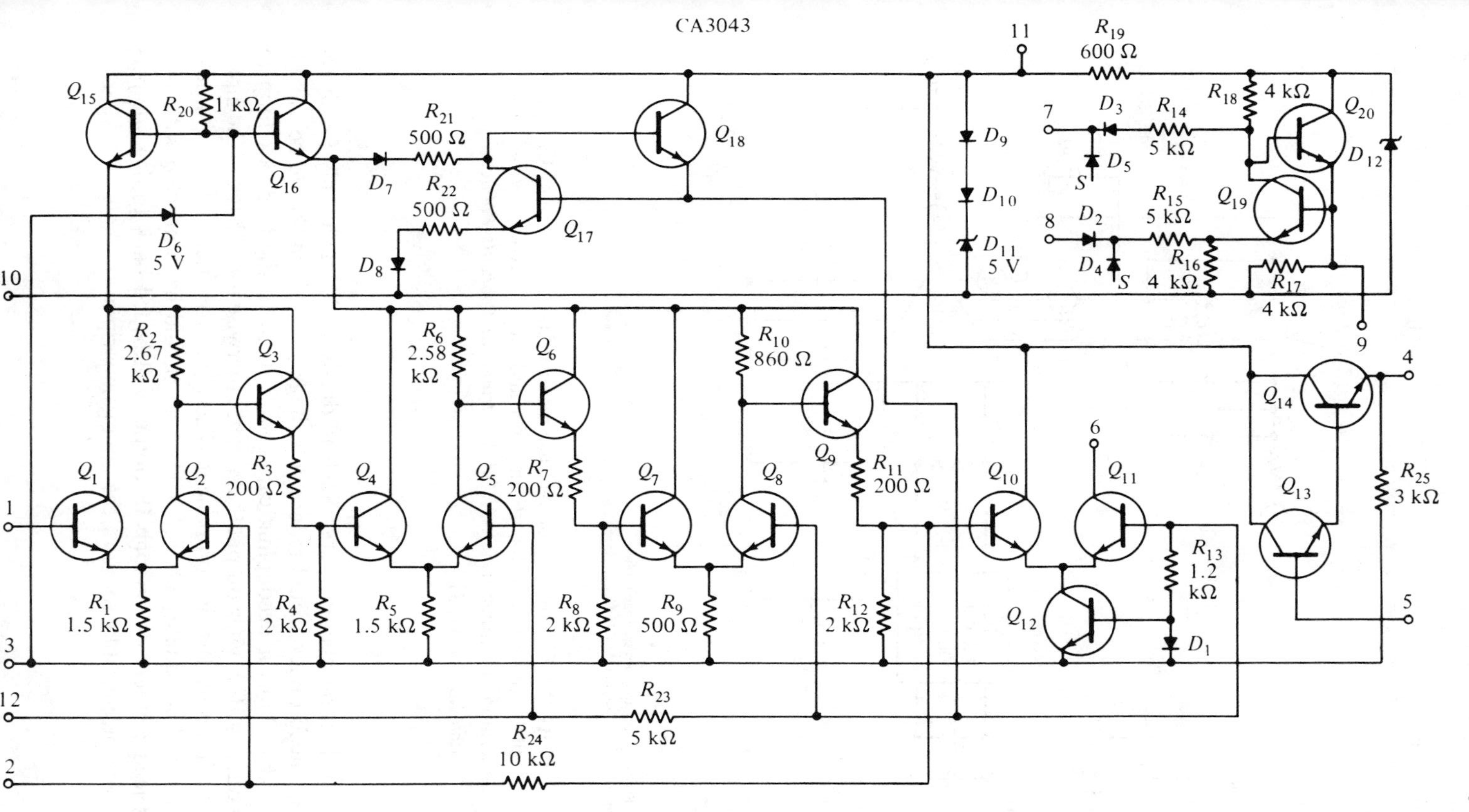

FIG. 8.3 Integrated circuit CA3043 IF amplifier and limiter; terminals 3, 10 and substrate (S) should be connected to the most negative point (see Fig. 8.4). *(RCA Electronic Components)*

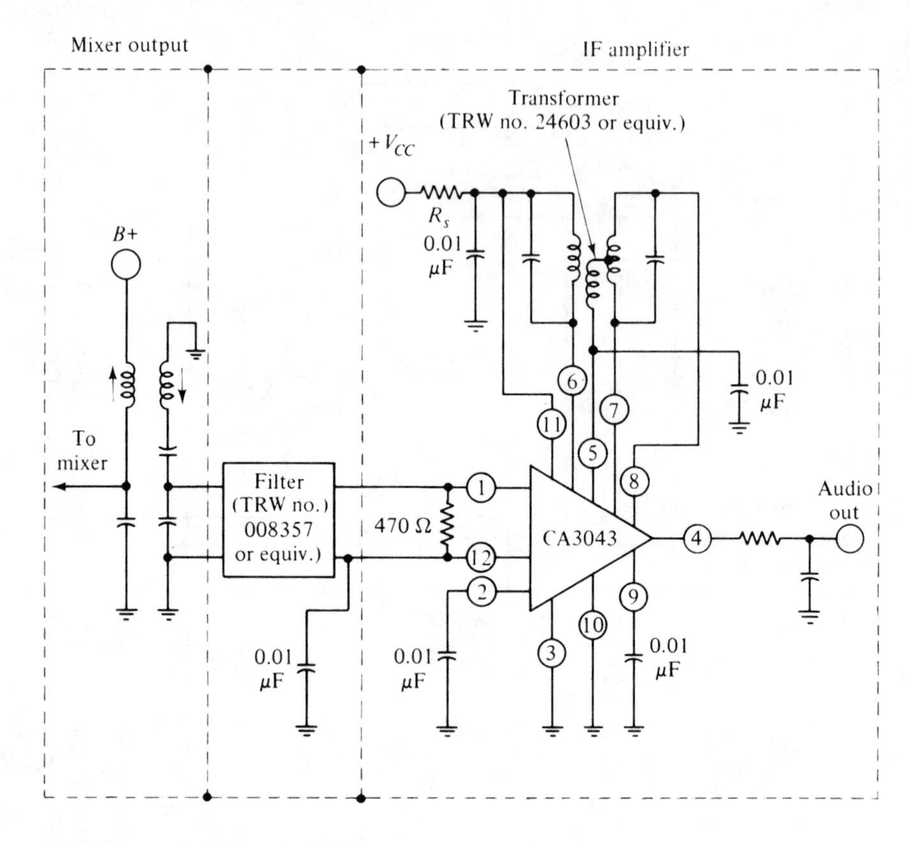

FIG. 8.4 FM receiver design, using conventional tuner, an LC filter, and a CA3043 as IF amplifier and limiter, feeding a discriminator transformer. *(RCA Electronic Components Publication No. ST-3773)*

metrical limiting characteristic. Such feedback is desirable only for direct current; ac feedback would reduce ac voltage gain, which should be as high as possible. Thus a decoupling capacitor is used in the feedback loop to short out any 10.7-MHz output that might otherwise travel back to the input, and leaving only the dc component. The capacitor used for this purpose must be of high quality at 10.7 MHz, and should be connected with very short leads; otherwise, enough IF output will be fed back to degrade performance, or, possibly, to allow the high-gain strip to oscillate.

8-4. FM STEREO MULTIPLEX DECODERS

The FM stereo decoder is an almost ideal candidate for monolithic construction; it must be relatively complex, and it demands accurate matching of components for best performance. Type *Motorola MC1305* is widely used in both home and automotive FM stereo receivers (Fig. 8.5). It consists of a number of separate functions, including amplifiers, summing networks, and the basic switching "multiplier" function used to derive the two output channels by locking onto an ultrasonic 19-kHz *pilot carrier*, which is transmitted along with FM-stereo signals. Certain versions of IC multiplex decoders also include a detector for this pilot carrier, which lights an external lamp if present, indicating the presence of a stereo-broadcast station.

The *MC1305* requires external tuned elements to separate the pilot carrier and filter out other unwanted carriers. The FM multiplex-decoder function may also be accomplished by use of the phase-locked loop (PLL) discussed elsewhere in this chapter, eliminating much of the external tuning requirement. A typical application of the *MC1305* is shown in Fig. 8.6.

Other IC Functions in FM Broadcast Receivers

While the FM–IF and the multiplex-decoder functions described previously are used primarily in specialized FM receivers, other functions, such as LIC audio preamplifiers and power amplifiers, are common to most communication circuits, and are discussed more generally in the previous chapter. Certain FM broadcast-receiver functions are not yet provided by LICs. Most significantly, the VHF "front end" must provide low noise and excellent rejection of cross-modulating nearby signals, and is a function that is best performed by specialized types of field-effect transistors not suited to present LIC process technology on a mass-produced scale. Such circuits may, however, become commonplace as IC technology advances.

8-5. BROADCAST AM RECEIVER CIRCUITS

The lack of mass usage of LICs in AM broadcast receivers thus far is testimony to the very low cost of building a satisfactory AM receiver with discrete components. Unlike the FM–IF strip, the IF function in an AM receiver does not require as large a gain, nor does it require limiting to remove the AM component of the signal. Instead, there is a requirement to linearly amplify the AM signal over a very wide variation in signal strength, yet maintain a constant output. This is accomplished by use of automatic gain control (or AGC). In this action, a dc voltage, proportional to the amount of signal reaching the detector, is fed back to one or more IF amplifiers,

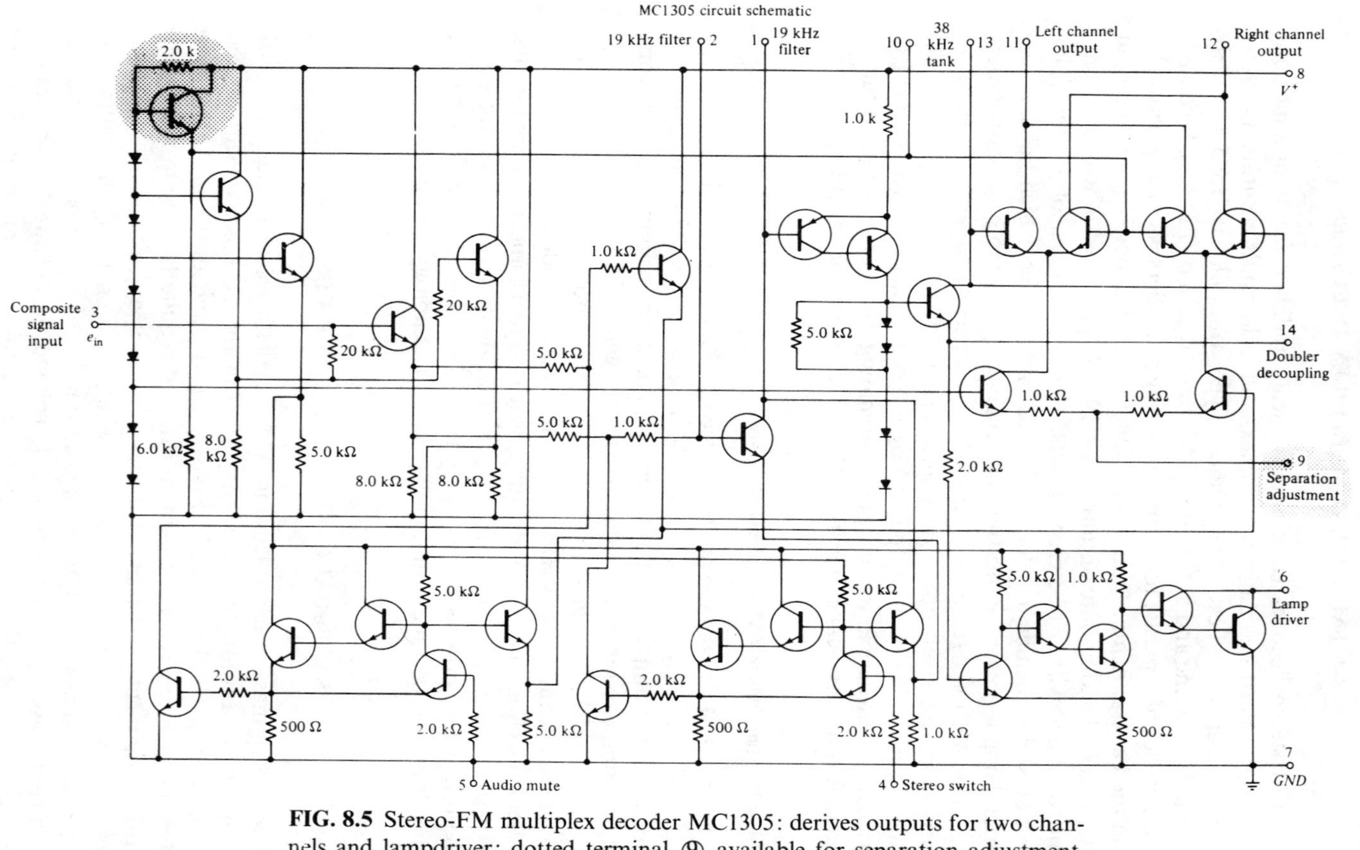

FIG. 8.5 Stereo-FM multiplex decoder MC1305: derives outputs for two channels and lampdriver; dotted terminal ⑨ available for separation adjustment. *(Motorola Semiconductor)*

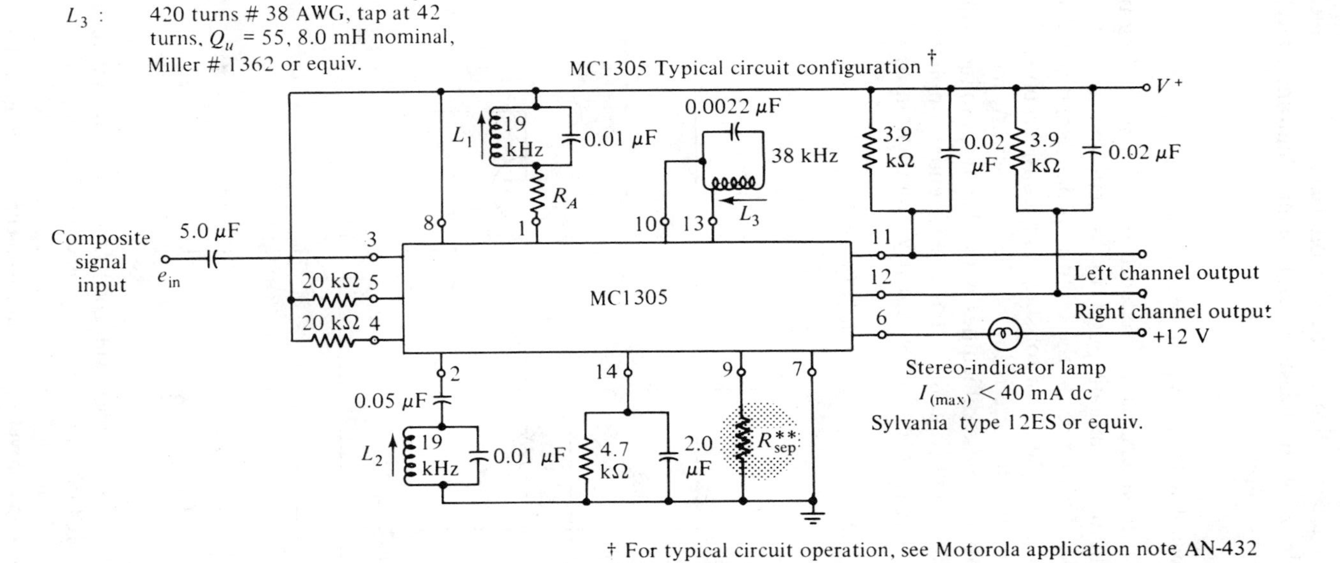

FIG. 8.6 Application circuit for demodulation of FM-stereo multiplex signals with MC1305. *(Motorola Semiconductor)*

whose gain can be controlled by the dc voltage. Thus, excessively strong signals reduce the IF gain and, as a result, gain is automatically adjusted over a wide range of input signals.

Automatic Gain Control in RF/IF Amplifiers

In discrete AGC amplifiers, gain is reduced by either increasing transistor current above the optimum-gain point, or by reducing transistor current to very low values. In both cases, the input impedance and signal-handling characteristics of the single transistor stage vary considerably as gain varies, upsetting the tuning characteristics of the receiver. A monolithic LIC designed to overcome these difficulties is the RF/IF amplifier with AGC, typified by the *MC1550* (Fig. 8.7). It is a differential amplifier; however, the current source (Q_1) is the high-frequency input amplifier, while one of

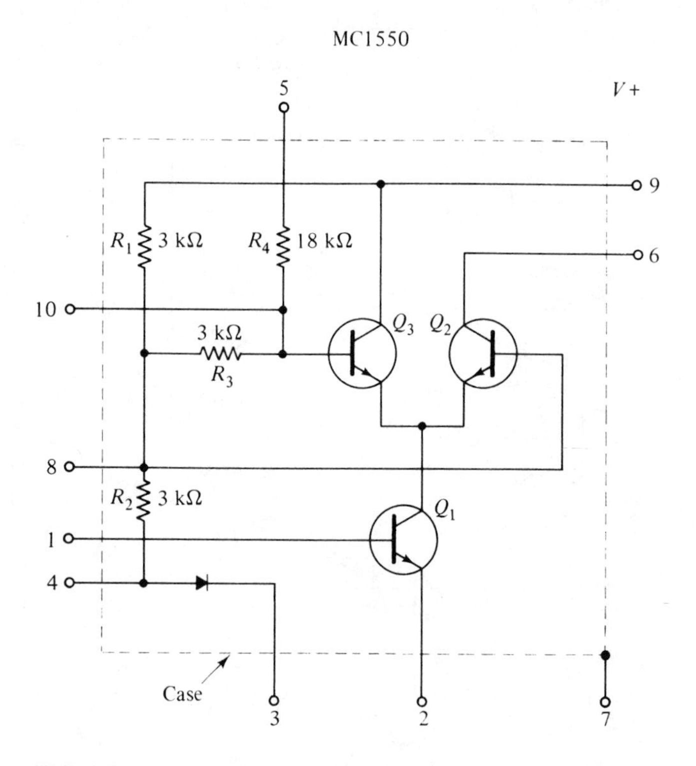

FIG. 8.7 RF/IF amplifier, with good AGC action for AM receivers (MC1550); with high-frequency input at terminal ① transistors Q_1 and Q_2 form a cascade circuit, while AGC voltage at terminal ⑤ provides automatic-gain-control action at the base of transistor Q_3. *(Motorola Semiconductor)*

the differential pair (Q_2) is a common-base amplifier connected to the common-emitter current source, thus forming a transistor analogy of the well-known "cascode" circuit. With negligible AGC voltage applied to its base (terminal 5), the transistor not connected to the output (Q_3) is biased fully off, and the circuit achieves maximum gain. With increased AGC voltage, the differential amplifier is balanced, and half the dc current from the current source is diverted into the nonoutput side (terminal 9), as well as half the output signal (a gain reduction of 2, or -6 dB.) As the nonoutput side is increased, more and more signal current is routed away from the output stage, until gain has been reduced nearly to zero. Because some leakages always exist, a practical AGC range of about 300:1 (or 50 dB) can be achieved with the monolithic AGC RF/IF amplifier. Because the current-source input stage is looking into two emitters whose parallel impedance remains constant as the AGC signal varies, the monolithic device gives a more constant input impedance than the discrete-component version. However, because it only replaces a single stage, and because the general quality required of AM broadcast receivers is low, this type of RF/IF amplifier has not received wide usage in consumer equipment.

AM-IF Strips with AGC

Following the same larger-scale integration trends as in FM–IF strips, devices such as the *National LM172/372* (Figs. 8.8 and 8.9) include enough gain to replace all stages in the conventional strip, and also provide self-contained AGC along with a built-in detector with characteristics suited to the AGC requirements. The *LM172/372* requires a single-lumped external bandpass filter, because all of its internal functions must be direct coupled. Gain variation is achieved by a pair of emitter followers, which act like diodes, one in series with the input, and another shunting the signal to RF ground. As their balance is varied, less signal reaches the output of the AGC section and more is shunted to ground, thereby giving an AGC operating range of about 1,000:1, or 60 dB. Gain is performed by a set of direct-coupled common-emitter amplifiers, which are forced to remain linear by use of a dc feedback loop (dc feedback loops are discussed in the previous section on FM–IF strips). The direct-coupled detector is far more complex than the simple diode detector used in the discrete AM receivers; it is a feedback amplifier with the detector diode enclosed by a negative-feedback loop, and this accurately sets its operating characteristics without the use of the usual tuning.[1]

[1] R. A. Hirschfeld, Application Note AN-15, National Semiconductor, 1968.

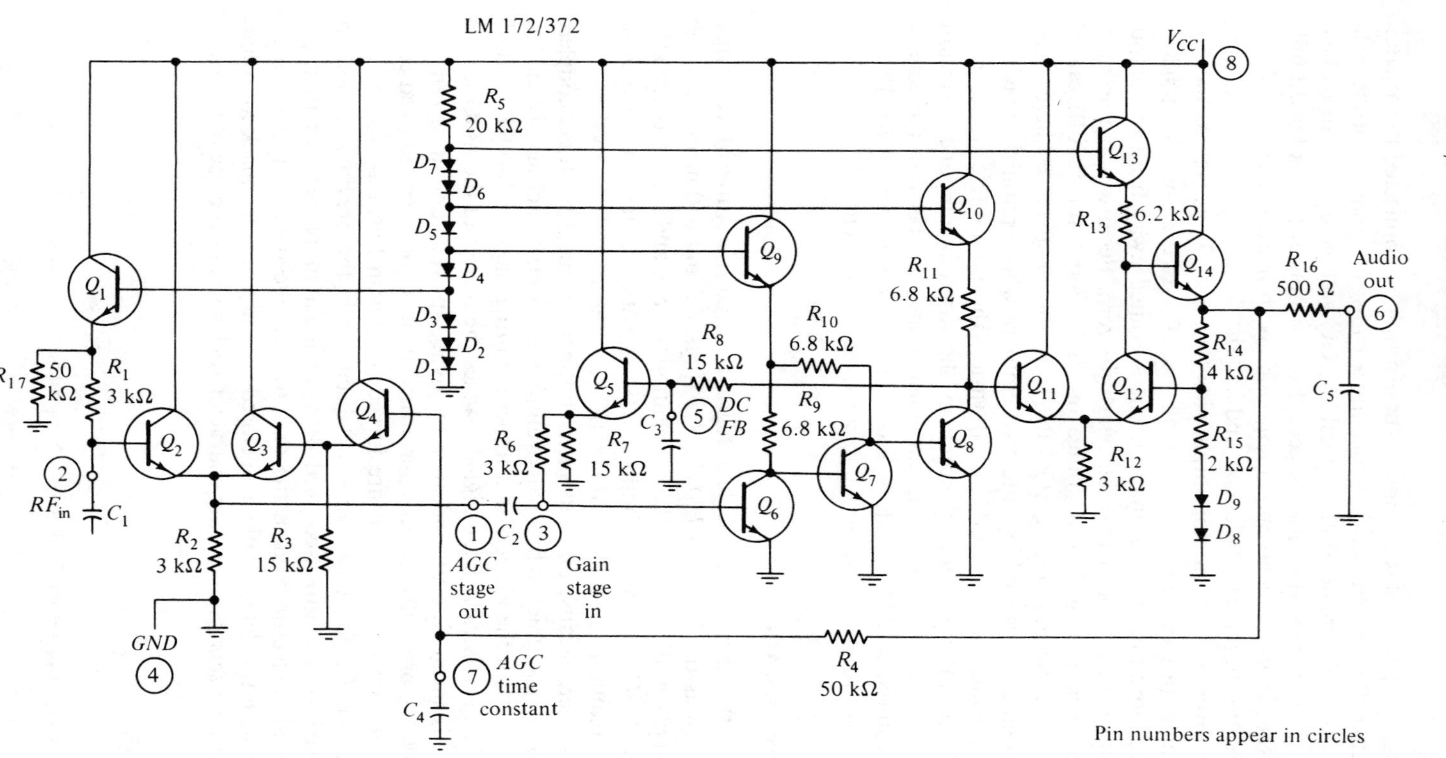

FIG. 8.8 AM–IF strip (LM172/372), a receiver subsystem, providing a high gain broadband amplifier (suitable for IF with an external filter) and, in addition, includes excellent AGC action and an active detector. (*National Semiconductor*)

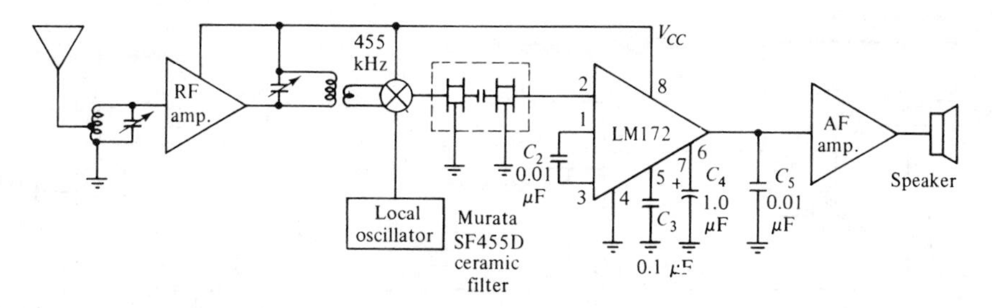

FIG. 8.9 Superheterodyne block diagram of an AM receiver circuit, using LM172/372, as an IF strip with AGC action and active detector. *(National Semiconductor)*

8-6. TELEVISION APPLICATIONS

While the critical high-power and high-voltage sections of television receivers are likely to remain in discrete form for many years, at least until monolithic LIC technology is able to provide such functions economically, many of the low-level functions, especially in the complex color-demodulation systems, are available in monolithic form. A typical television-receiver functional block diagram appears in Fig. 8.10. Four complex monolithic subsystems are used. A TV video-IF subsystem, the *RCA CA3068*, contains all these blocks; first, a video-IF/AGC generator for feedback to the tuner, circuitry for derivation of the 4.5-MHz sound IF from the composite signal, and a multiple-stage video IF detector and first video amplifier. A second IC, the *CA3065*, is a sound-IF, sound detector, and audio preamplifier. A third IC, the *CA3066*, provides chroma information through its chroma amplifier, band pass amplifier, burst amplifier, an automatic chroma control (ACC), and a crystal-controlled 3.58-MHz oscillator, which is locked to the color subcarrier of an incoming signal. The fourth IC, type *CA3067*, performs the critical color-demodulation function, splitting the red, green, and blue phospor-dot information channels for external amplification and distribution to the picture tube.

While a detailed examination of the circuits used in the block diagram of Fig. 8.10 is best left to applications literature available from the manufacturer, some observations can be made from the circuit schematics shown in Figs. 8.11, 8.12, and 8.13. First, it is apparent that a number of external components are still required, being primarily tuned and bypass elements, which are not easily integrated. Second, the complexity of the circuits and the limited number of pins have made these subsystems useful for only a single purpose, rather than being universally applicable types of LICs. To be economical in manufacture, such highly specialized circuits must have very large markets,

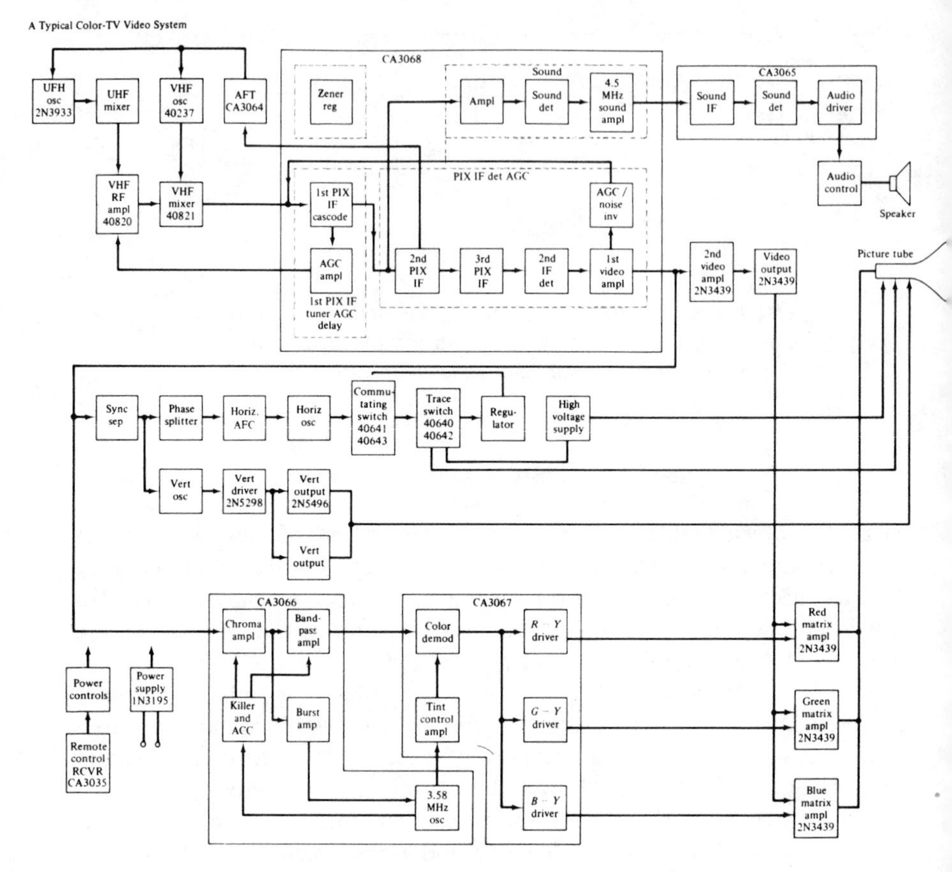

FIG. 8.10 Typical color TV receiver block diagram; shown are four TV subsystems, as follows: CA3068—video/IF system, CA3065—sound amplifier section, CA3066—Chroma amplifier section, CA3067—Chroma demodulator section. *(RCA Solid-State Division, File No. 467)*

and the television industry satisfies this criterion. Third, the circuit techniques used in these complex devices are basically the same as have been discussed with respect to single-function FM, AM, and general-purpose LICs and include current mirrors, differential AGC systems, multipliers, extensive direct coupling, and so on. In other words, complex TV circuits are simply logical extensions of existing LIC technology.

The degree to which these LICs can be called large-scale integration (or LSI) is limited primarily by the economics of packaging with large numbers of pins, by the difficulty in providing RF isolation, and by the higher power dissipation that even more consolidated circuits would require. Finally, it is

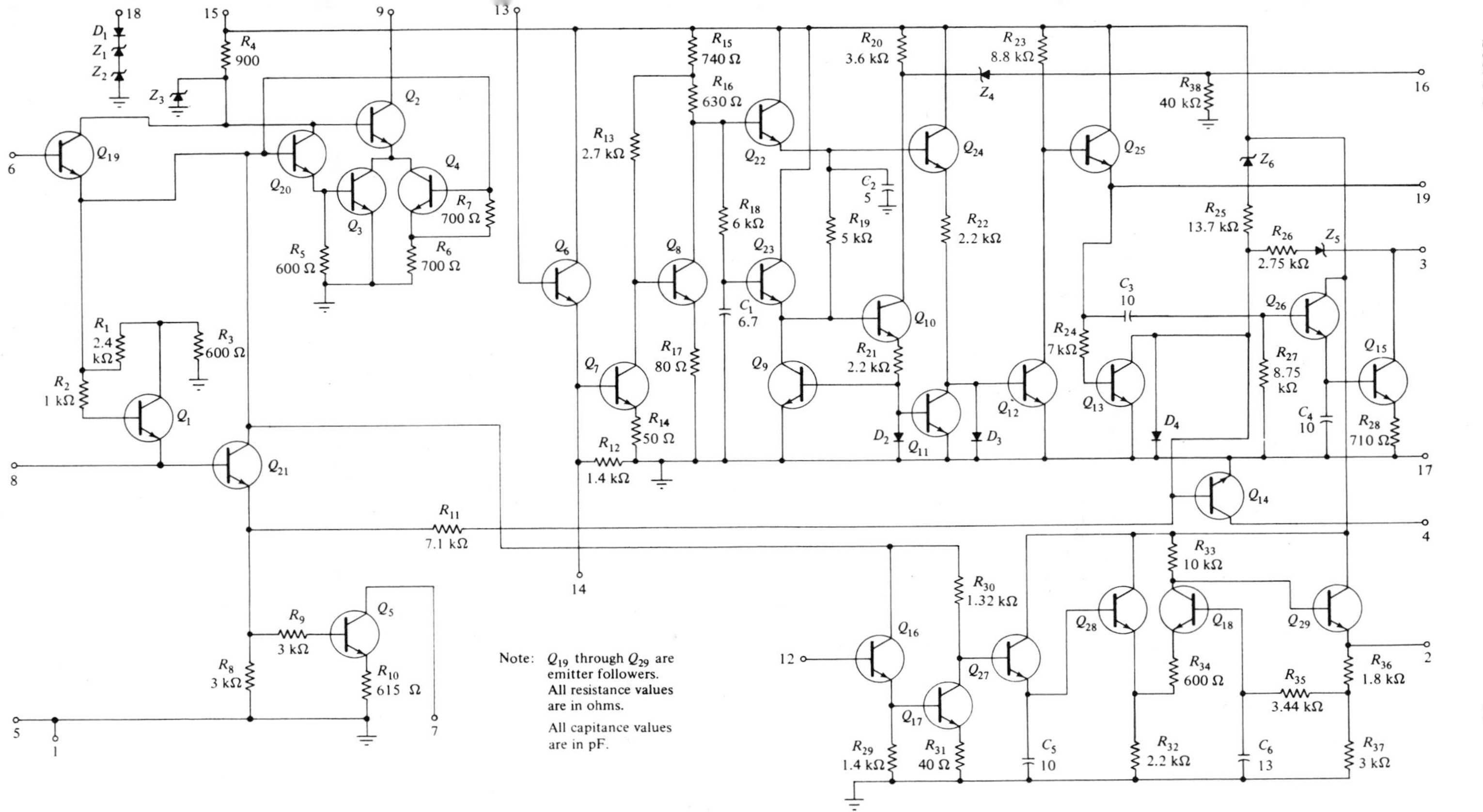

FIG. 8.11 Simplified schematic diagram of the CA3068 video/IF system. *(RCA Solid-State Division, File No. 467)*

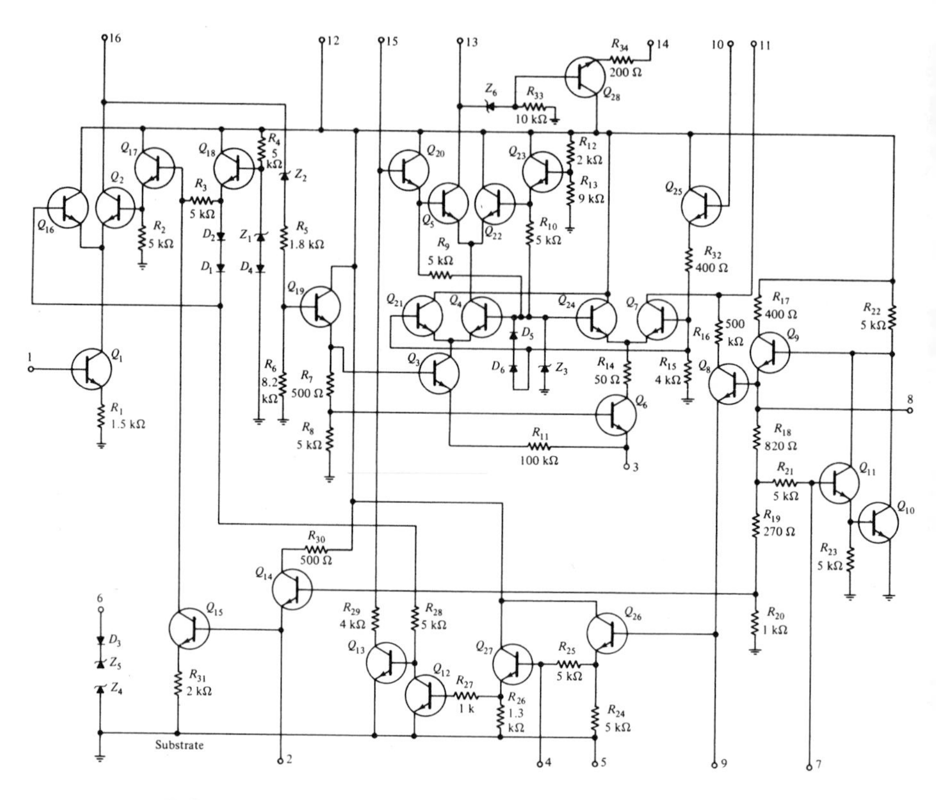

FIG. 8.12 Schematic diagram of the CA3066, Chroma amplifier system. *(RCA Solid-State Division, File No. 466)*

apparent that the reason for using such circuitry cannot be reduction in size, since the picture tube and associated power circuitry dominate the chassis of most TV receivers. Also, economic advantage is still a limited one, because quite a few external components are still required. Accordingly, the main advantage of LICs in TV receivers is modularity, that is, the use of modules to allow easy replacement in the field. By placing as many elements on a plug-in module as is practical, the television LIC allows immediate replacement of entire defective functions, as opposed to the replacement of only the active devices as in the past days of vacuum tubes, thus increasing the likelihood that a repair can be made without extensive troubleshooting.

General-Purpose Communication Applications

While the broadcast AM, FM, and TV receiver functions discussed are

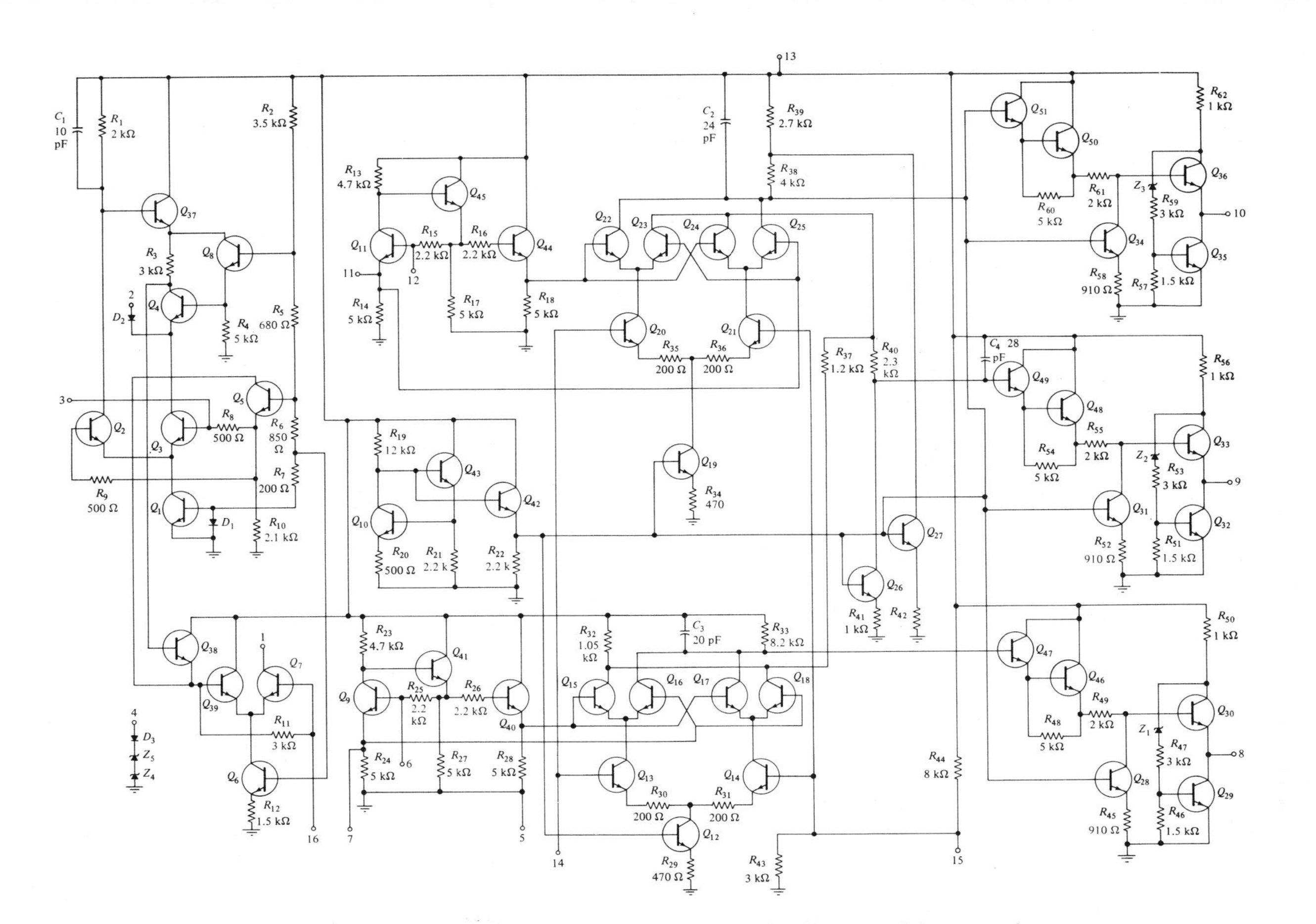

FIG. 8.13 Schematic diagram of the CA3067, Chroma-demodulator section. *(RCA Solid-State Division, File No. 466)*

responsible for the bulk of actual communication LIC usage, there remains a vast field of communication applications is served by a number of general-purpose functions. The applications include *two-way radio, navigation, telemetry, data transmission, remote controls, alarms,* and other general areas involving the transmission of signals from point to point.

Rather than examine each application separately, the following discussion will concentrate on several *general circuit functions,* which are common to many kinds of communication applications.

8-7. MIXERS, MODULATORS, AND MULTIPLIERS

A basic requirement in communication systems is to obtain an *output that is equal to the product of two input signals.* Theoretically, the process of signal multiplication in the time domain is the same as frequency addition and subtraction in the frequency domain—just two ways of looking at the same actual operation. An *RF mixer* can be considered as a multiplier:

$$V_{out} = V_{in_1} \times V_{in_2}.$$

Without dwelling on the mathematical theory, the result of mixing (or multiplying) two frequencies (f_1 and f_2) may be expressed as follows:

$$f_{output} = Af_1 + Bf_2 + C(f_1 + f_2) + D(f_1 - f_2).$$

That is, the output of such a mixer (or multiplier) contains not only components of both input frequencies (f_1 and f_2), but also the *sum and differences* of the input frequencies.

Similarly, an examination of various forms of amplitude modulation (AM) will show that the modulated output contains the same combinations of two input frequencies. If the inputs to an amplitude modulator are f_1 (RF carrier) and (f_2) (audio modulation), the output will contain both input frequencies, as well as their sum and differences, sometimes called "sidebands." In communication equipment, the undesired frequency components are filtered out, leaving only the carrier or carrier-and-sideband information frequencies to be transmitted. In all cases, including AM (carrier plus both sidebands), DSB (both sidebands with carrier filtered out), or SSB (single sideband, where only one sideband remains after filtering), *the necessary modulating process is one of multiplication of two inputs* (also called *mixing*).

Multiplier Types

A structure found in many communication LICs is the *tree-type multiplier,* typified by the *Motorola MC1596* (Fig. 8.14). It is a double-differential amplifier with its outputs "cross connected." Recalling the differential-amplifier AGC circuit (Section 8-5), it is a form of multiplier that produces an output equal to the product of an input signal and a dc voltage. In the AGC amplifier an increase (or decrease, if negatively connected) in dc AGC voltage causes an approximately proportional increase in the signal at the output. In the tree-type multiplier, two such differential circuits are balanced, so that the components of the two input frequencies are canceled at the output, leaving only the sum and difference frequencies. Such a structure is sometimes referred to as a *double-balanced mixer*, and is used in the generation of AM, double-sideband, and single-sideband modulation.

Product Detector: The same structure of double-differential amplifier, when driven by a beat-frequency oscillator (BFO), is used in the receiving-detector section of SSB receivers; the BFO frequency, when added to (or subtracted from) the SSB input, produces a component that is the original audio-modulating frequency. The multiplier nature of this type of SSB detector is shown by its name: *product detector.*

Phase Detector: If all output components except dc are filtered out, the tree multiplier becomes a *phase detector.* If two equal frequencies having a phase difference are used as inputs, a dc output is obtained proportional to the phaseshift. If phase shift is zero, the dc output is zero. This function is important in phase-locked loops, an increasingly important topic to be discussed later.

Quadrature FM Detector: If an *LC* phase-shift network is placed between the input signal and one multiplier input, with the input signal directly connected to the other multiplier input, a phase difference results that varies with input frequency. Thus the tree multiplier becomes a *quadrature FM detector*, which is used, for example, in such television sound IF detectors as the *Sprague ULN2111.* The advantage of such a detector is the requirement only for a simple external *LC* resonant circuit, rather than for a carefully constructed discriminator transformer.

The basic circuit of the *MC1596*, shown in Fig. 8.14, is not strictly a linear multiplier; a nonlinear distortion of the "upper" input signal results from the nonlinear diode characteristics of the transistor base–emitter junctions; the circuit is nonetheless useful in the functions listed previously, because the spurious frequency components resulting from the nonlinearity are usually filtered out. In *analog computation*, however, a precisely linear product of two inputs is desired. For a linear product, the tree structure has been compensated by using a pair of diodes to cancel the nonlinearity of the multiplier's diode characteristics; this is the type *MC1595, four-quadrant analog multiplier* (Fig. 8.15). The designation *four quadrant* refers to the capability of operating in all four quadrants of a graph, in which both positive and negative input #1 are shown on the vertical axis, and positive and negative

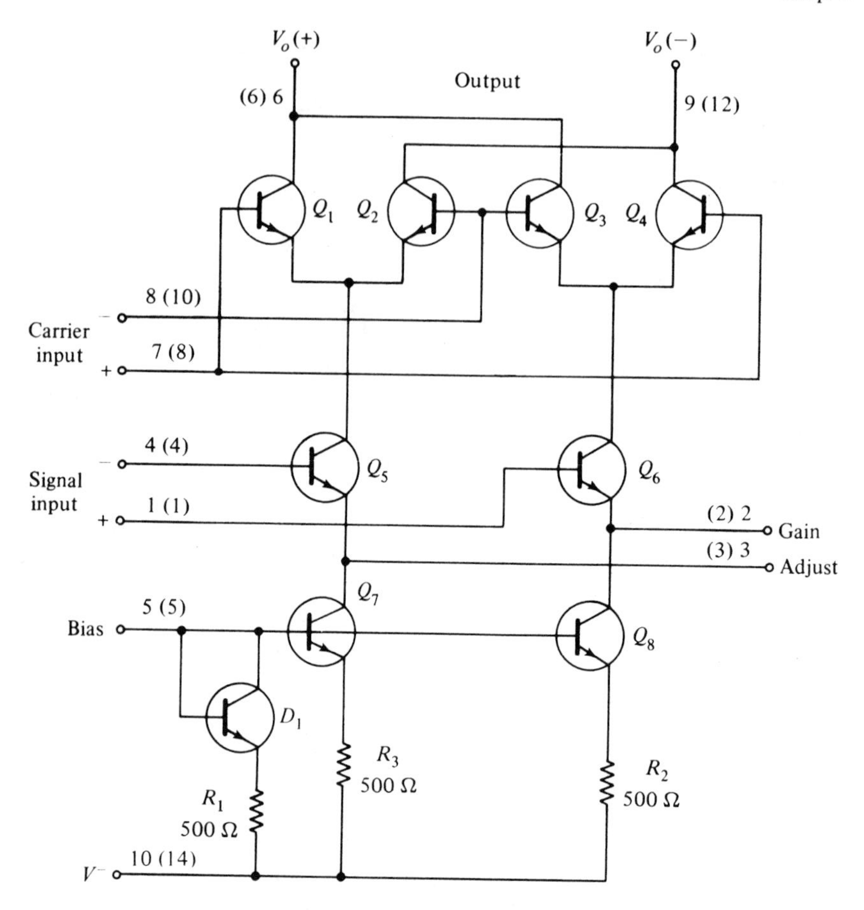

FIG. 8.14 Double-balanced mixer/modulator (MC1596): the balanced output is the product of the carrier and signal inputs producing their sum and difference frequencies, with internal carrier suppression, as in single-sideband (SSB) and other modulation applications. *(Motorola Semiconductor)*

input #2 are shown on the horizontal axis, thus dividing the possible output polarities into four quadrants.

8-8. PHASE-LOCKED LOOP

The basic idea of a phase-locked loop (PLL) has been known for many years and was used in radar systems and other complex modulation schemes. The

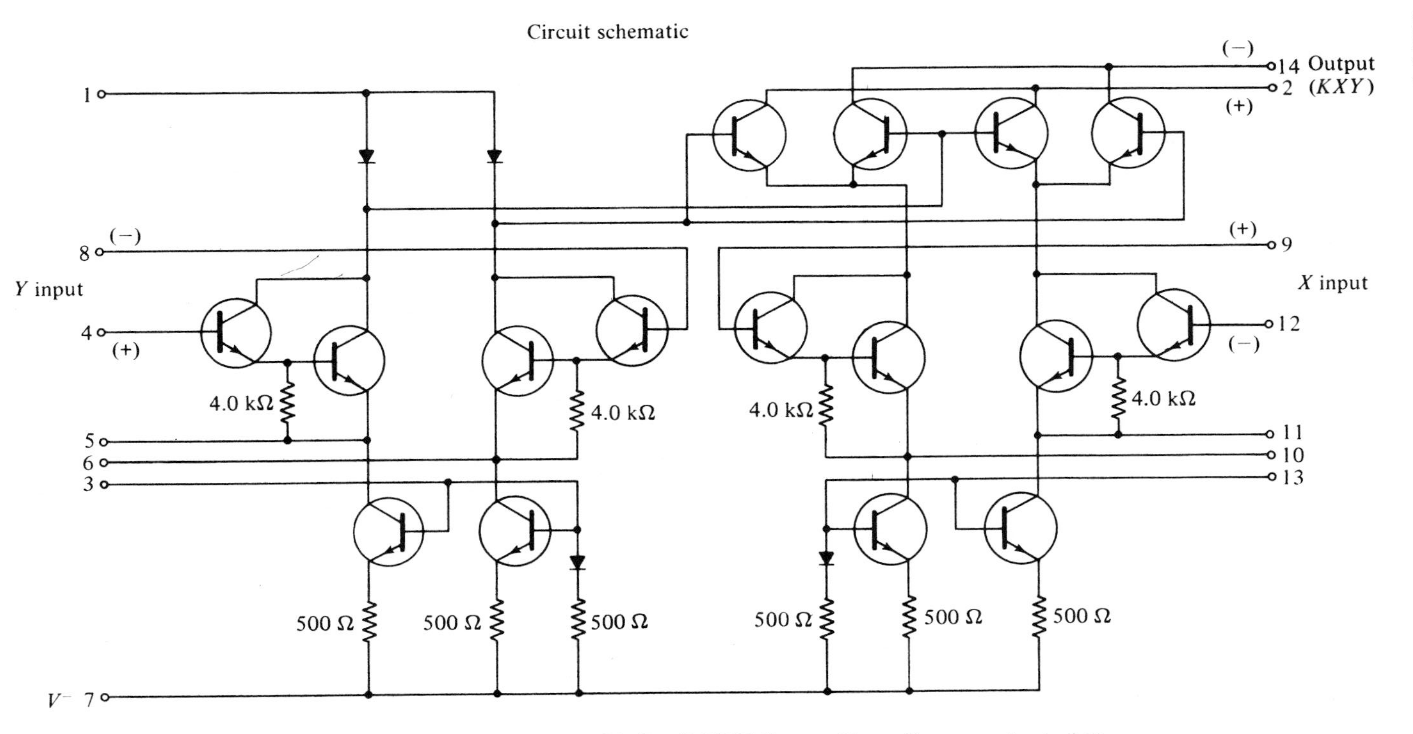

FIG. 8.15 Four-quadrant multiplier (MC1595): provides a linear product of X and Y inputs, for analog-computation applications, when used in conjunction with an OP AMP. (Typical applications are detailed on MC1595L data sheet.) *(Motorola Semiconductor)*

main elements are a *frequency reference*, a *phase comparator*, and a *voltage-controlled oscillator (VCO)* (Fig. 8.16). The phase comparator gives a dc output proportional to the phase difference between the reference frequency and the voltage-controlled oscillator (VCO). If the phase difference is zero, the dc voltage is zero. A negative feedback loop connects the phase-detector output, sometimes called the *error voltage*, and the control input of the VCO. Thus, if a phase difference does exist, the error voltage forces the voltage-controlled oscillator to change frequency (and thus phase) until it coincides with the reference.

Assuming that the reference oscillator is a very stable frequency source, the phase-locked loop can then be used as a *frequency synthesizer* by providing digital-counter-type frequency dividers in either the reference or VCO inputs to the phase comparator. By setting appropriate counter ratios, the VCO may then be forced to stabilize on one of a large number of possible frequencies, giving receivers and transmitters a large number of possible operating channels, which are as stable as the crystal-controlled channels, but which do not require a separate crystal for each channel.

Suppose, however, that the reference frequency is not a stable one, but is rather an unknown frequency that can vary over a wide range, such as the input frequency in an FM receiver. Since the negative-feedback loop tries at all times to force the voltage-controlled oscillator to equal the reference frequency, the VCO will "track" the unknown input frequency at a rate limited by the maximum rate of change of the dc error voltage of the VCO. Suppose further that there is a linear relationship between the dc VCO control (or error voltage) and the operating frequency of the VCO. Obviously, if the reference frequency varies, the error voltage will be a dc voltage that is proportional to the frequency variation. If the unknown input is, for example, a 10.7-MHz FM–IF signal that varies ±75 kHz, and if it takes a 1-V change in

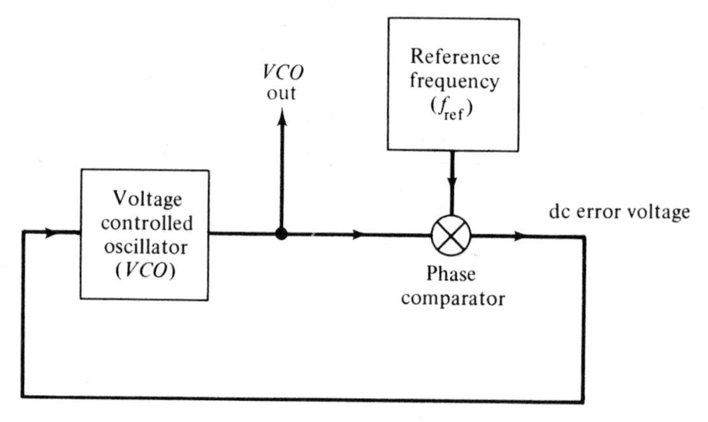

FIG. 8.16 Basic operation of phase-locked loop (PLL).

the error voltage to produce a 100-kHz change in the VCO frequency, then the error signal will be a demodulated replica of the frequency modulation of the input signal, giving ± 750 mV of audio output. FM demodulators using phase-locked loops offer the advantage of not requiring external tuned elements, since the VCO can be made with integrated *RC* elements. It is possible, however, for strong unwanted signals to "capture" the loop, which responds to the strongest signal within its lock-in range.

The phase-locked loop, such as the *Signetics SE565*, may be used to synchronously demodulate AM signals by locking on to the carrier, and using the locked VCO to "sample" incoming modulation on desired RF peaks. It may also be used to synchronously demodulate the multiplex subcarrier used in stereo FM signals.[1]

Because the dc VCO control (or error voltage) can be directly related to VCO frequency, the phase-locked loop may also be used to detect the presence of specific frequencies, as in *tone decoders*, such as the *SE567*. A dc comparator is used to sense whether the error voltage is above or below a limit corresponding to the desired frequency; dual dc comparators then sense maximum and minimum limits, giving a logic output when the signal is within a narrow band of frequencies.[2]

The main advantage of the phase-locked loop is its ability to perform *frequency-selective functions* without the use of inductors—a highly desirable asset in monolithic form. A receiver in which the PLL is the detector will only respond to signals within the lock-in range of the VCO. Once locked on to the incoming signal, the phase-locked detector has virtually no sensitivity to adjacent interfering signals, as long as they do not "capture" the loop.

In monolithic form the VCO usually takes the form of an *RC*-controlled sinusoidal or multivibrator oscillator, in which controlled monolithic current sources vary the charging time of the external timing capacitor. The phase comparator is usually of the "tree" multiplier variety discussed in Section 8-5.

[1] Signetics SE 567 data sheet.

[2] Signetics SE/NE 565 data sheet.

9

REGULATORS AND CONTROL CIRCUITS

9-1. POWER CONTROL WITH MONOLITHIC CIRCUITS

While the internal power-dissipation capability of a monolithic LIC is limited by the difficulty of removing substantial amounts of heat from a very small package, the precision and complexity of the circuit make it an effective control element for external power transistors, and also for thyristors (SCRs, or silicon-controlled rectifiers) and other devices that can do high-power work, but are themselves imprecise. Power-control LICs are expected to achieve widespread use in home appliances, lamp dimmers, motor controls, thermostats, safety circuit breakers, automotive applications, and much more. In all these control applications the *LIC performs timing, regulation, or overload protection*, or, at times, a combination of these functions. In low- and medium-power uses the LIC may contain the power-handling elements internally, but in most cases inexpensive external elements are used.

The use of external power components may be required for one of three reasons: power dissipation, maximum current handling, or maximum breakdown voltage limitations of the LIC fabrication process. The most economical, easily reproduced LICs are those operating at currents less than

1 A, voltages less than 40 V, and package dissipations less than 1 W. Each of these parameters can be increased by a factor of 10 with current technology, but trade offs must always be considered, especially when very inexpensive external power elements could be used instead.

9-2. VOLTAGE REGULATORS (LINEAR VERSUS SWITCHING TYPES)

Many linear, and nearly all digital, systems require a constant, precisely regulated power supply voltage, or several such voltages. Certain LICs provide such regulation internally for use in critical sections of their own circuitry. Certain systems use *local regulation*, with a separate voltage regulator for each section of the system, not only to provide constant voltages, but also under operating conditions to prevent variations in power consumption by one section from being transmitted back along the common supply lines to other sections.

Voltage regulators may be classified, on the one hand, by the method of achieving regulation, as in *linear regulators* or *switching regulators* and, on the other hand, by their means of connection, as in *series regulators* or *shunt regulators*. All voltage regulators contain three basic elements: a *precision voltage reference*, an *error amplifier*, and a *power-control element*. The type of LIC voltage regulators may consist only of the error amplifier, or may contain one or both of the other elements internally. Because of the limitations of design, most LIC voltage regulators are specifically designed for one supply polarity, either *positive or negative*, although universal, dual-polarity regulators are also available, such as the 723 type to be discussed later. Many LIC regulators also include *current limiting*, which may be of either the *constant-current* or *foldback-limiting* types.

Linear Versus Switching Regulators

The basic sections of a voltage regulator are connected in a negative-feedback loop, which, regardless of configuration, is connected to monitor the voltage at the load, and to increase the current to the load when the load demands an increase, so that the voltage across the load remains constant. There are two ways to vary the current to the load. In the *linear regulator* there is an active element whose impedance may be continuously varied by the error amplifier to suit the current demand. In the *switching regulator* a saturating, ideally lossless switch is turned on and off at a rate that delivers the desired average current in periodic pulses to the load.

The switching regulator is ideally a much more efficient circuit than the linear regulator, since the switching element theoretically dissipates no internal power in either the on state, with no voltage across it, or in the off state, with no current through it. The linear regulator must dissipate power in the control element continuously. Practically, however, few switching regulators are in use. First, there is a finite time in every switching cycle during which the switching element makes the transition between on and off; during this time, dissipation is substantial. Second, most loads cannot tolerate the periodic delivery of energy, even if smoothed by a capacitor. Third, the switching regulator is a troublesome source of transients and RF radiation. Thus, *most regulators using LICs are of the linear type.*

9-3. SERIES VERSUS SHUNT REGULATORS

In a series regulator the variable-impedance or switching-control element is placed between the raw, unregulated supply and the load. As the load demands more current, series impedance is decreased, so that the voltage divider formed by the series element and the load gives the desired output voltage.

In a shunt regulator the unregulated supply is either a high-impedance current source or an unregulated voltage source with a substantial fixed series output impedance. The control element is shunted across the load, and is increased in impedance as load current demand increases, so that the net current into the parallel combination of load and control element remains constant, and the voltage across the load remains constant.

A *series regulator is desirable in applications where load-demand current varies from zero to some fixed maximum;* dissipation in the control element is zero when load current is zero. A *shunt regulator is desirable when load current varies from a finite minimum value to a finite maximum value;* dissipation in the control element is low or zero when load current is at its maximum, but unnecessary power is then consumed in the fixed source resistance in the unregulated supply. *Most LIC regulators are designed for use as series regulators.*

9-4. CURRENT LIMITING

Although not a necessary part of a voltage regulator, the complex construction possible in monolithic form permits most LIC regulators to provide a form of current-limiting action. Usually, a small resistance is provided in series with the load; load current produces a small but predictable voltage across this *sensing resistor*, which is amplified and used to trigger a clamping circuit to defeat the negative-feedback loop that ordinarily controls the circuit. If load current continued to increase without this limiting action, the negative-feedback loop would attempt to provide the required large currents, ultimately destroying the control element, unless other resistances in the circuit or in the unregulated supply were large enough to prevent damage. *The current limiter, however, allows direct short circuits, in most cases, without permanent destruction.*

The simplest limiter is one in which a threshold is reached above which regulator output current, rather than voltage, is held constant. Thus a direct short circuit would have the same limited current through it as would a load slightly heavier than that permitted by the limiter. Heavy stress is placed upon *series regulators* in this kind of limiting, since they may have maximum voltage and constant, maximum current applied simultaneously, permitting a large but controlled power dissipation.

In the *foldback limiter*, when maximum load current is exceeded, the output voltage is folded back or shut down to a lower value, or even zero. Although this provides much safer operation, it does permit power to be shut off in case of fault from a load that may otherwise require continuous power. Such fault may be a very short, momentary one; nevertheless, the regulator will stay shut off until externally reset.

9-5. VOLTAGE-REGULATOR ELEMENTS: REFERENCE-VOLTAGE, ERROR-AMPLIFIER, AND CONTROL ELEMENT

The three main elements of a basic regulator (such as the widely used μA723) are shown in functional-equivalent form in Fig. 9.1 (a). Here the reference-voltage function (V_{ref}) is initially determined by the 6.2-V Zener diode; this is fed to one of the inputs of the error amplifier, while the other input receives a portion of the output, resulting in an amplified error that controls the output of the series-pass transistor.

Reference Voltage

This element is the standard against which the regulated output is compared. Accordingly, the accuracy of the regulated output is highly dependent upon the accuracy of the reference. Since monolithic processes are subject to wide variation, the absolute value of the reference is often hard to control, being established by forward or reverse diode breakdown voltages. Usually, a voltage divider external to the LIC is used to set the ratio between the output voltage and the reference, so that the actual initial value of the reference is less important than its stability with time and temperature. Since the temperature characteristics of various junction breakdowns are more predictable than their absolute voltage values, regulator references are often a series combination of forward and reverse (Zener) breakdowns calculated to give minimum temperature drift. In extremely precise regulators, a reference external to the LIC may be used.

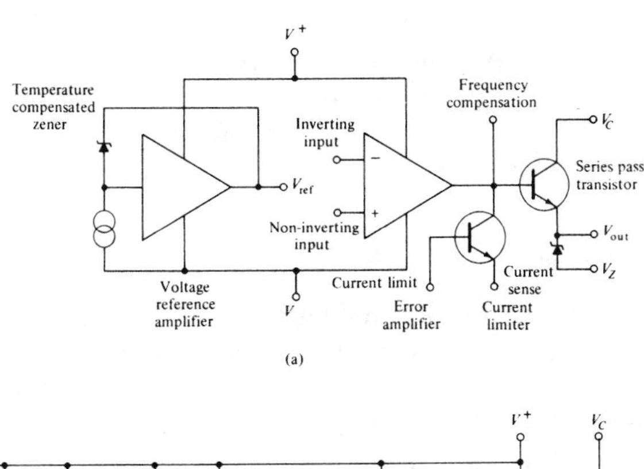

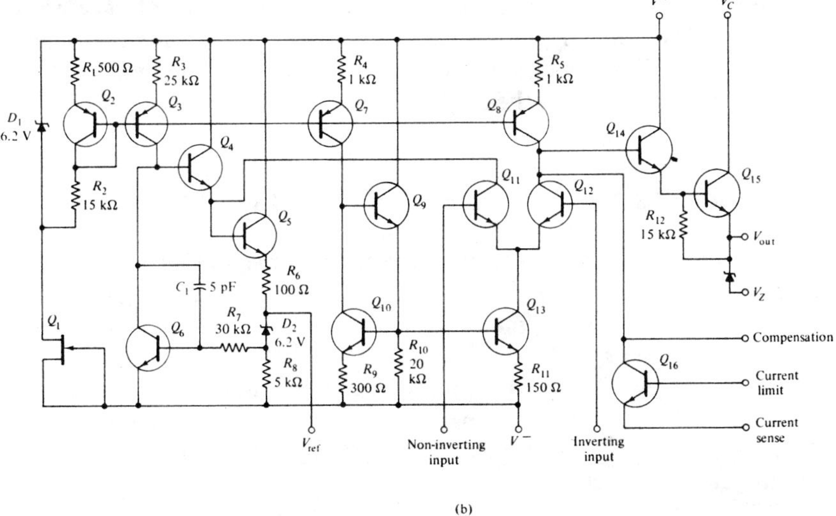

FIG. 9.1 Voltage regulator (μA723): (a) functional equivalent, with main elements—reference-voltage, error-amplifier, and control (series-pass) element; (b) schematic-equivalent circuit. *(Fairchild Semiconductor)*

Error Amplifier

This element compares the regulated output voltage against the reference voltage, and drives the control element, more or less, to compensate for load-induced variations in output voltage. It is desirable to have a large gain in the error amplifier, so that very small differences between desired and actual output voltages will cause large and immediate correction outputs. Very large gains, however, are often the source of high-frequency instability that is due to phase shifts in the feedback loop, which must be corrected by compensation capacitors (as is the case with operational amplifiers). The more rapidly the error amplifier can react, the better; otherwise, high-frequency variations in load impedance cannot be "followed" by a slow-acting error amplifier, and will produce unregulated outputs. It is also important that the error amplifier recover quickly from overload, and also from the initial transient resulting when the regulator is first turned on.

Control Element

This must be capable of handling both the maximum voltage and maximum current to be encountered, and of dissipating the power resulting from a worst-case combination of voltage and current. A fast-acting error amplifier may be worthless if the control element is not also capable of fast response. In the case of switching regulators, fast recovery from switching transients is also a must, as power is dissipated in such regulators during the switching transition period, as well as the on period.

9-6. REPRESENTATIVE EXAMPLE OF VOLTAGE REGULATOR

As an example of a general-purpose regulator, the Fairchild μA723 has become well known as an industry-standard type, featuring a highly flexible circuit that can be used in many different ways. The schematic-equivalent circuit is shown in Fig. 9.1 (b).

In general, the 723 type is capable of regulating either *positive or negative voltages, fixed or adjustable*, and in either *series or shunt* configuration. Its line regulation is typically 0.01 percent, and it will regulate typically within 0.03 percent against load changes (1 to 50 mA)

When output currents greater than 50 mA are required, it is designed for use with an external NPN or PNP power transistor. Terminals are shown on the schematic diagram for current limiting (CL) and current sensing (CS), and also provision is made for a sensing-current resistor (R_{SC}) and for a compensating capacitor (COMP). Connections to these terminals are made in various ways for the specific applications described in the next section.

9-7. APPLICATIONS OF THE 723-TYPE REGULATOR

A number of typical voltage-regulator applications are shown in the accompanying figures.

Basic Positive-Voltage Regulation

The first two illustrations, Fig. 9.2 and 9.3, show the connections for positive-

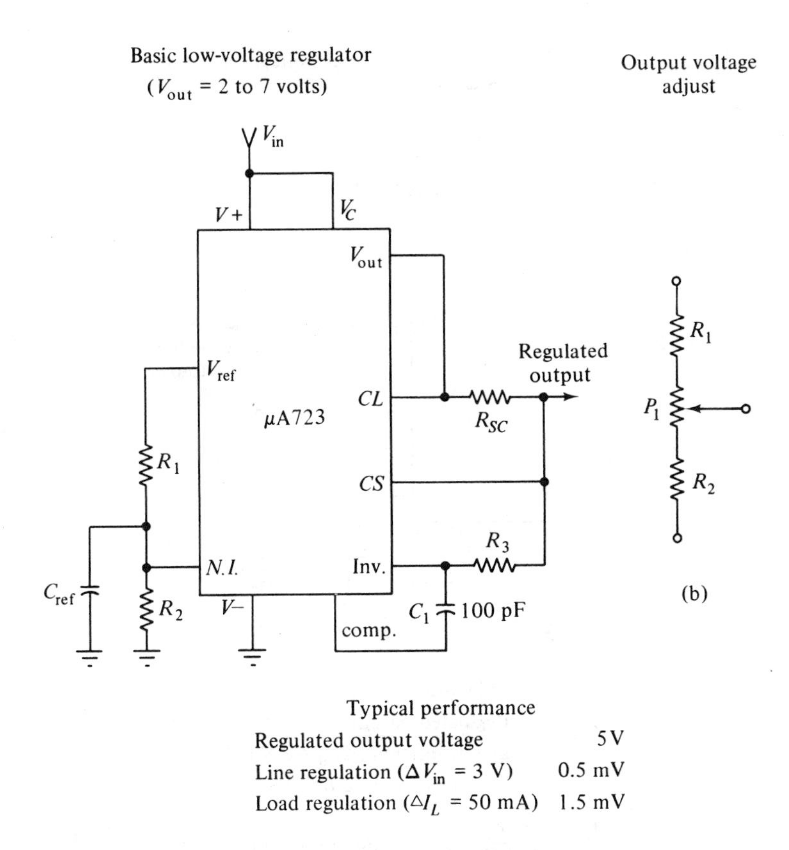

Note: $R_3 = \frac{R_1 R_2}{R_1 + R_2}$ for minimum temperature drift.

For outputs from +2 to +7 volts

$$V_{out} = V_{ref}\left[\frac{R_2}{R_1 + R_2}\right]$$

(a)

FIG. 9.2 Basic low-voltage regulator (to +7 volts): (a) for fixed output voltage; (b) for variable output voltage. *(Fairchild Semiconductor μA723 Data Sheet)*

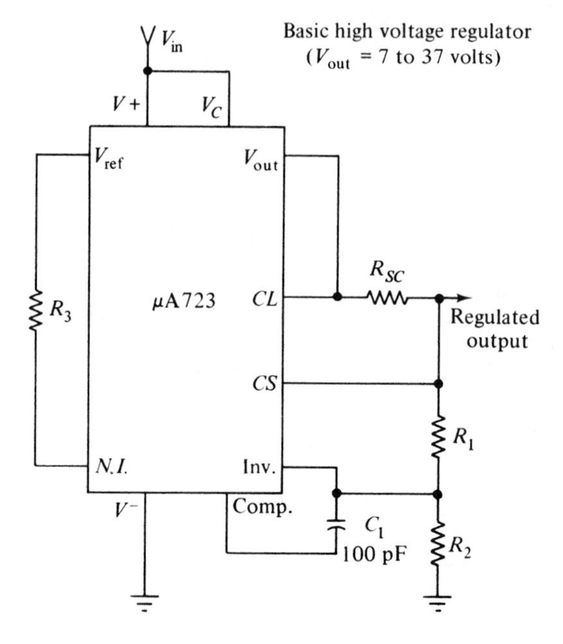

Typical performance

Regulated output voltage	15 V
Line regulation (ΔV_{in} = 3 V)	1.5 mV
Load regulation (ΔI_L = 50 mA)	4.5 mV

Note: $R_3 = \frac{R_1 R_2}{R_1 + R_2}$ for minimum temperature drift.

R_3 may be eliminated for minimum component count.

For outputs from +7 to +37 volts

$$V_{out} = V_{ref}\left[\frac{R_1 + R_2}{R_2}\right]$$

FIG. 9.3 Basic high-voltage regulator (to +37 volts): see text for note on specific voltage-divider values for fixed or variable output voltages. *(Fairchild Semiconductor μA 723 Data Sheet)*

voltage regulation. Figure 9.2 is for basic *low-voltage applications* (up to +7 V), which covers the familiar 5-V fixed output widely used in digital-logic circuits. The use of a *variable pot* (P_1) *for adjustable voltages* in this range is shown in part (b).

[*Note:* Specific values for R_1, P_1 and R_2 for this application (and the applications that follow) are given in the *μA723* data sheet.]

The connections for a basic *high-voltage regulator* are shown in Fig. 9.3. Here it will be noted that the junction of the voltage divider (R_1, R_2), which sets the desired output voltage, is now connected to the inverting (INV) terminal of the error amplifier, and thus determines output voltages higher than the reference, as shown in the formula on the diagram. Specific values for the voltage-divider resistors (R_1, R_2) are given in the data sheet up to +28 V fixed, and also for the variable pot (P_1) of Fig. 9.2 (b), when adjustable output is desired. This application covers the *+15-V output value* that is popularly used for the positive supply of OP AMPS. (Data for circuit connections and values up to +250 V are also given on the data sheet.)

Using External Pass Transistor

The use of an externally connected power transistor for output currents greater than 150 mA is shown in Fig. 9.4. Here an external NPN pass transistor is connected for positive-voltage regulation, and the performance data are given for a +15-V output, regulated for *load changes up to* 1 A. (The data sheet also gives connections and values for using a PNP pass transistor in a positive-voltage regulator.)

Negative-Voltage Regulator

The connections for a negative-voltage regulator, shown in Fig. 9.5, involve the use of an external PNP transistor with the unregulated input (V_{in}) applied to the collector of this transistor. Using the proper values for the voltage divider (R_1, R_2) as given on the data sheet, the performance data for the widely used *negative 15V supply* are shown on the figure, for both line and load regulation.

Current-Limiting Action

Connections for *current-limiting action of the foldback type* are given in Fig. 9.6. For the typical regulating performance indicated in the figure, the circuit will limit the short-circuit current to 20 mA.

Switching-Regulator Action

The connections for switching-regulator action of the 723 type regulator are

Positive voltage regulator
(external *NPN* pass transistor)

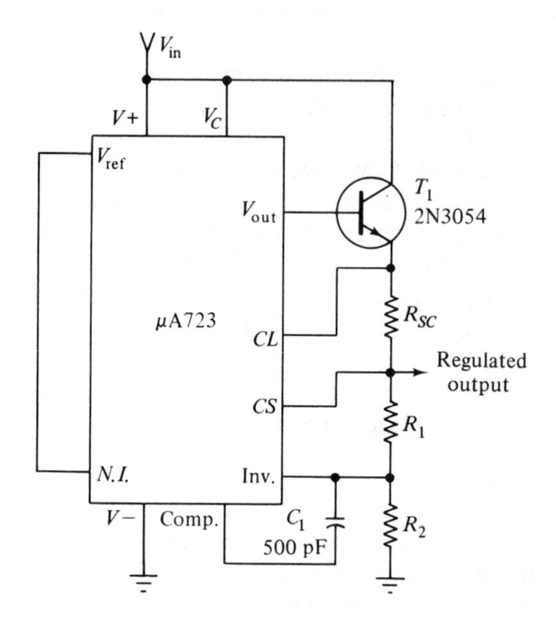

Typical performance

Regulated output voltage	+ 15 V
Line regulation (ΔV_{in} = 3 V)	1.5 mV
Load regulation (ΔI_L = 1 A)	15 mV

FIG. 9.4 Regulator with external pass transistor: used for output currents exceeding 150 mA (see text for note on specific values). *(Fairchild Semiconductor μA723 Data Sheet)*

shown in Fig. 9.7 (for positive output) and in Fig. 9.8 (for negative output). In this type of action, the regulator requires the use of two external transistors, and will handle relatively large changes in input voltage (V_{in}) and up to 2-A change in load current (I_L).

Shunt Regulator

The connections for applying the type 723 as a shunt regulator are shown in Fig. 9.9. Here the external transistor is connected to the V_Z terminal of the 723, and the regulated output is taken from the collector of this transistor, rather than from the ordinary V_{out} terminal of the device, when used as a series regular.

Additional applications, such as connections for remote control of the regulator and others, are given in the data sheet for the *μA723* type, along with the specific values of the voltage divider to be used for obtaining either fixed or variable output voltages.

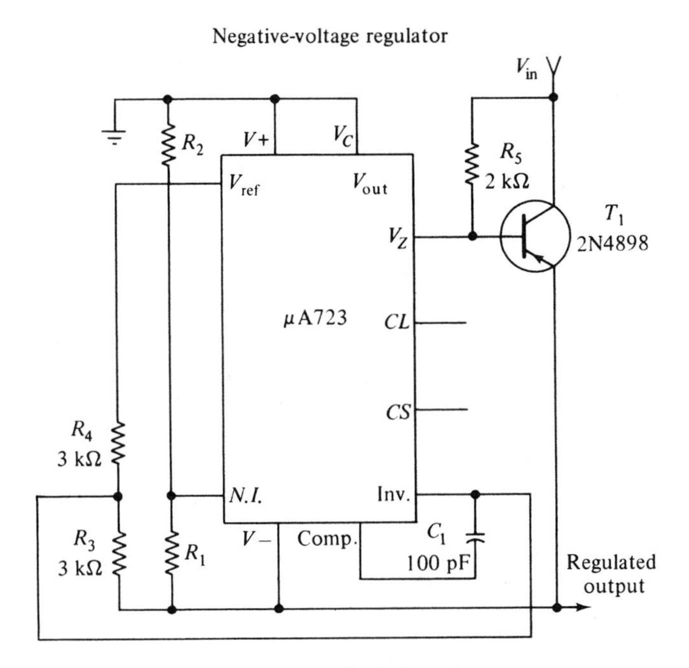

Typical performance

Regulated output voltage	−15 V
Line regulation (ΔV_{in} = 3 V)	1 mV
Load regulation (ΔI_L = 100 mA)	2 mV

For outputs from −6 to −250 volts:

$$V_{out} = \frac{V_{ref}}{2}\left[\frac{R_1 + R_2}{R_1}\right]$$

$$R_3 = R_4$$

FIG. 9.5 Negative-voltage regulator: specific values for voltage-divider resistors (R_1, R_2) for fixed output are given on data sheet; see Fig. 9.2(b) for adjustable output connections. *(Fairchild Semiconductor μA723 Data Sheet)*

9-8. OTHER REGULATOR TYPES

As representative of voltage regulators that also provide current regulation, mention may be made of the *Motorola* wide-range "floating type," *MC1566/1466*; also the *National LM109*, a local 5-V regulator used primarily for digital logic cards (but also capable of current regulation beyond 1 A in an alternative TO-3 power package); and the *LM376*, a positive-voltage regulator in a compact eight-pin mini-DIP package.

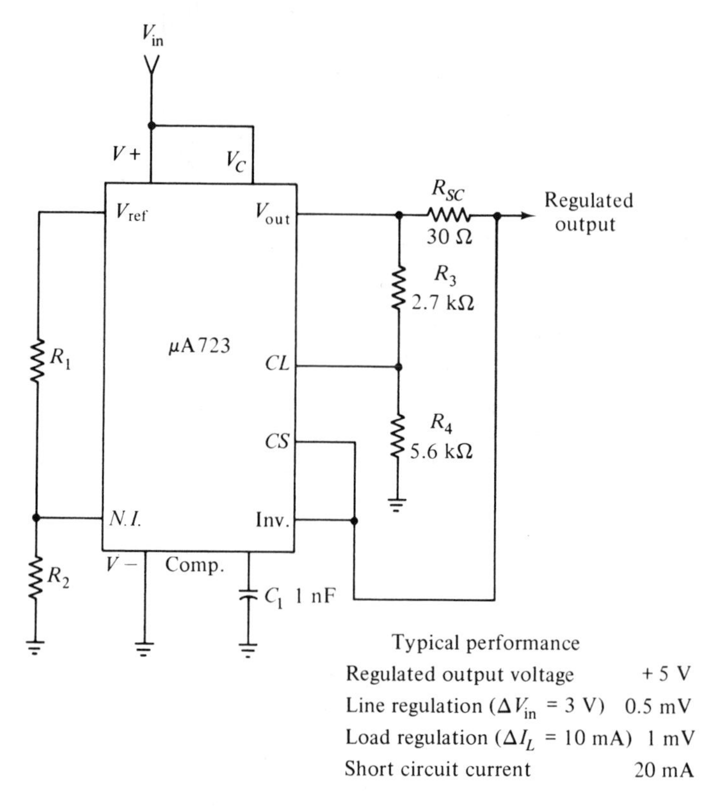

FIG. 9.6 Foldback current-limiting: with the values shown for the sensing-current resistor (R_{SC}) and at the current-limiting terminal (*CL*), the short-circuit current is limited to 20 mA. *(Fairchild Semiconductor μA723 Data Sheet)*

A voltage regulator that is internally set for dual-polarity output voltages (±15 V balanced to 1 percent) is the *MC1568/1468*. If adjustable voltages are desired, a single external adjustment can be used to change both outputs simultaneously from 14.5 to 20 V.

A versatile power-control type of LIC is the *CA3094A* programmable power-switch/amplifier. Not only can it function as a monolithic high-current- output device (as stated in Section 7-2), but is can also serve for programmable strobing and gating, making it suitable for control of temperature and motor speed. In the switching mode it can deliver 3 W average (or 10 W peak) to an external load. Its power dissipation (P_D up to $T_A = 55\,°C$) is 630 mW (without heat sink), and up to 1.6 W with heat sink.

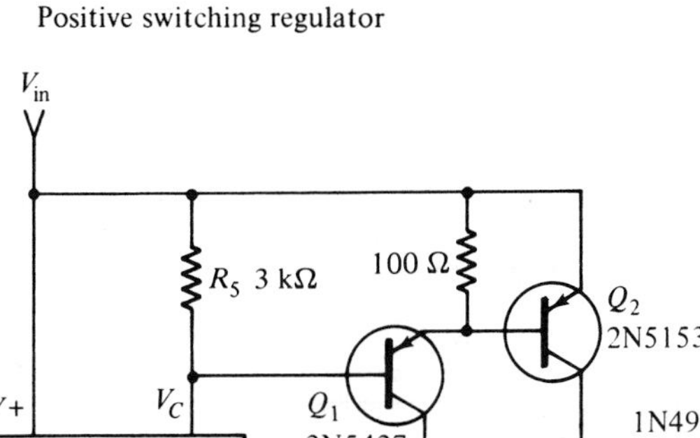

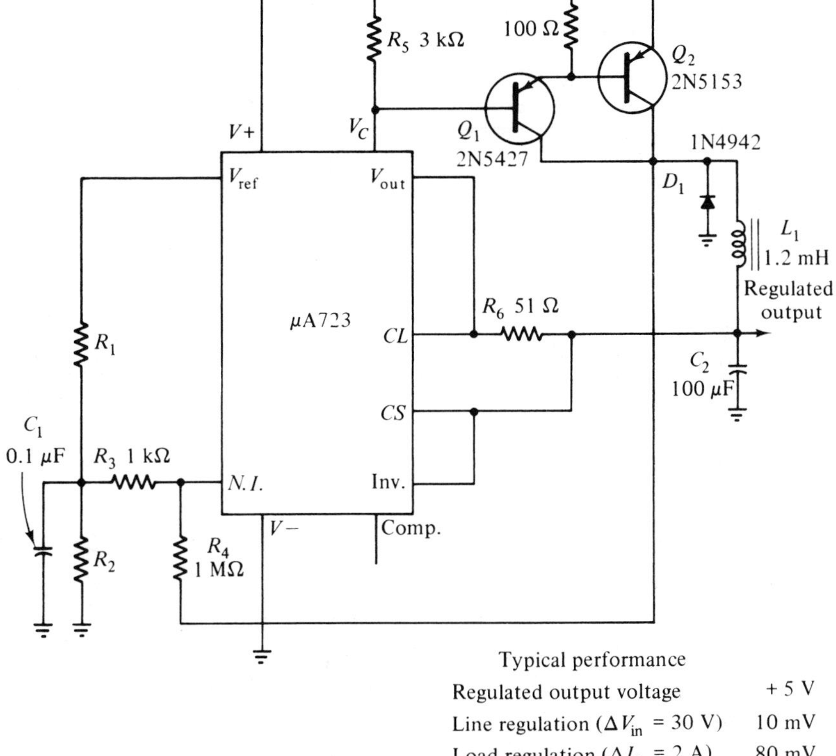

FIG. 9.7 Switching-regulator action for positive output voltage. *(Fairchild Semiconductor μA723 Data Sheet)*

A variety of power packages are offered for different power levels, as shown in the data manuals of the various manufacturers.

9-9. ZERO-VOLTAGE SWITCHES

In contrast to the control of dc power, which is the field of the voltage-regulator type of LIC, most industrial and household power is from the ac line, usually at 115 V ac, 60 Hz. The control requirements of such line-operated devices generally exceed both the voltage and current-handling capabilities of the previously described LICs.

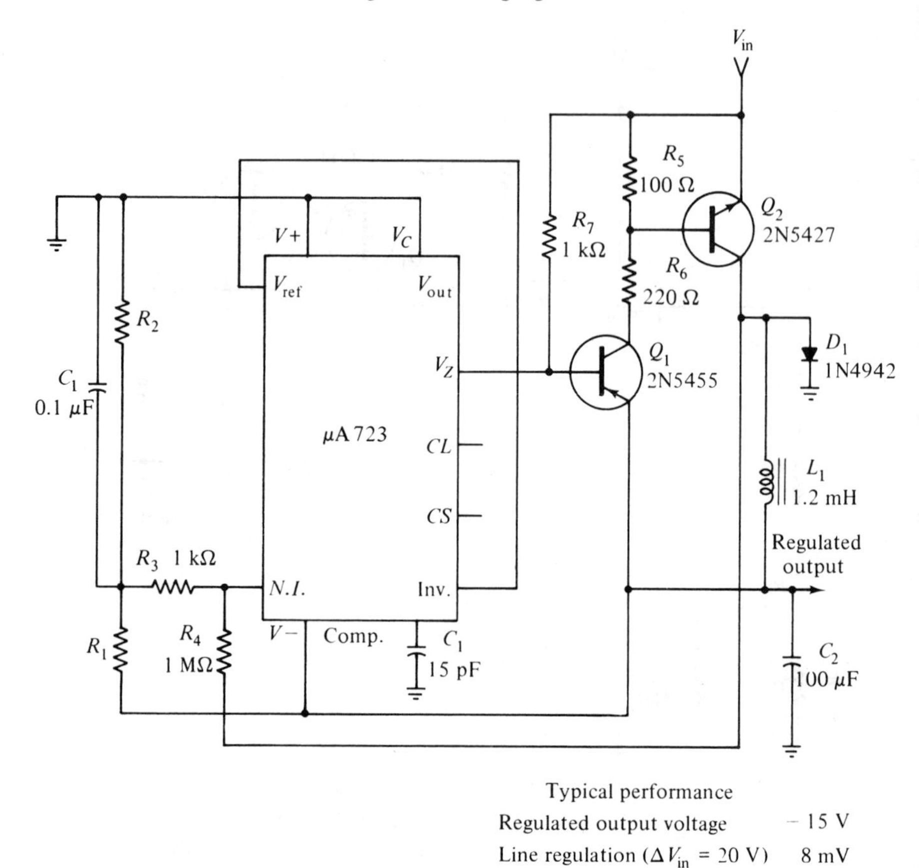

Typical performance	
Regulated output voltage	−15 V
Line regulation (ΔV_{in} = 20 V)	8 mV
Load regulation (ΔI_L = 2 A)	6 mV

FIG. 9.8 Switching-regulator action for negative output voltage. *(Fairchild Semiconductor μA723 Data Sheet)*

In controlling the high-power line-operated loads, such as heaters, valves, and motors, the older methods generally rely on such classic *electrical means* as relays, power rheostats and variable transformers. The newer *electronic methods* make use of the *silicon-controlled rectifier (SCR)* or various other triggered-device designations, such as *thyristor or triac.*

Controlling the SCR

The SCR, a four-layer semiconductor power device, is basically a latch that is either an open circuit or a low-resistance saturated two-terminal device. A

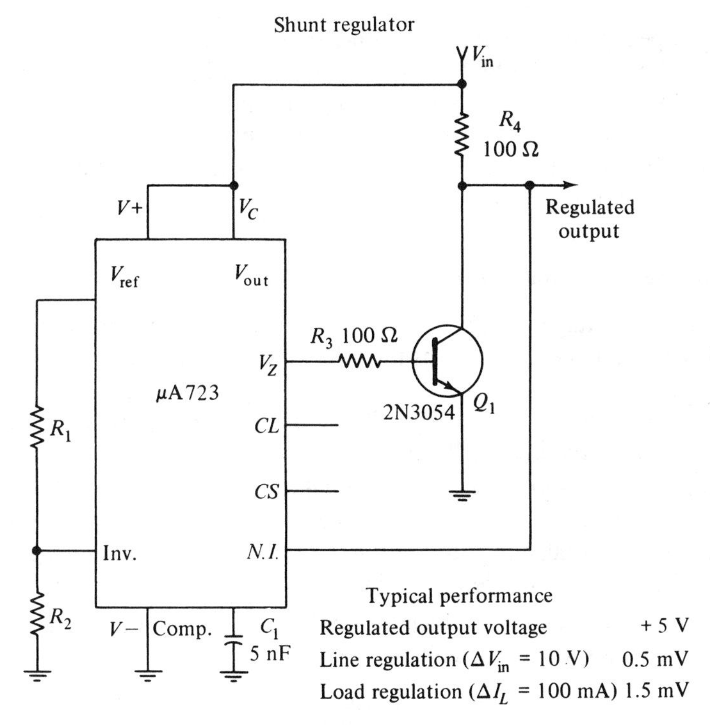

FIG. 9.9 Shunt-type regulators: for this type of regulation, the regulated output is taken from the external transistor rather than from the usual V_{out} terminal, when ordinarily used as a series type. *(Fairchild Semiconductor μA 723 Data Sheet)*

third terminal, the gate, is used as a trigger point, requiring relatively little drive current. Many SCR-operated appliances, such as variable speed drills, light dimmers, and the like, derive trigger current directly from the available ac power line through a variable resistor that sets the point on the ac wave form at which the SCR begins to conduct, with the latch turning off when ac supply polarity reverses on the next half-cycle. Simple SCR controls, therefore, permit controlled power to be delivered to the load during a portion of the ac cycle, varying from zero to the full cycle, depending on the control setting.

There are instances in which the triggering of the SCR must be more precisely controlled than is possible with a simple potentiometer, or in which a system control signal is used to drive the SCR rather than a manual control setting. Consider, for example, that a very large current load is to be switched, but that only a small SCR is available or economical. The small SCR can probably sustain fairly large pulses of current, but is likely to be damaged if the actual switching is accomplished while the ac-line wave form is at its maximum peak voltage, since this implies maximum power dissipation at the instant of turn on. If the SCR is turned precisely when the ac wave form is passing through its zero-crossing voltage, no power is dissipated until a later point in the cycle, when the switch is already closed. The *zero-crossing-switch LIC* is a sensitive detector that detects this critical timing, and drives the SCR trigger directly.

Representative Zero-Voltage Switch

A typical zero-voltage switch, the *CA3059*, is shown functionally in Fig. 9.10. The 14-pin DIP package contains the following basic functions:

1. *Limiter power supply*, which allows the LIC to derive its operating power from the high-voltage ac line.
2. *Differential on–off sensing amplifier*, which receives input on–off signals from an external NTC (negative-temperature coefficient) sensor or command signal, and gates ON in the rest of the circuit, in preparation for the next-arriving zero crossing.
3. *Zero-crossing detector*, which monitors the ac line voltage and gives an output pulse when that voltage is zero.

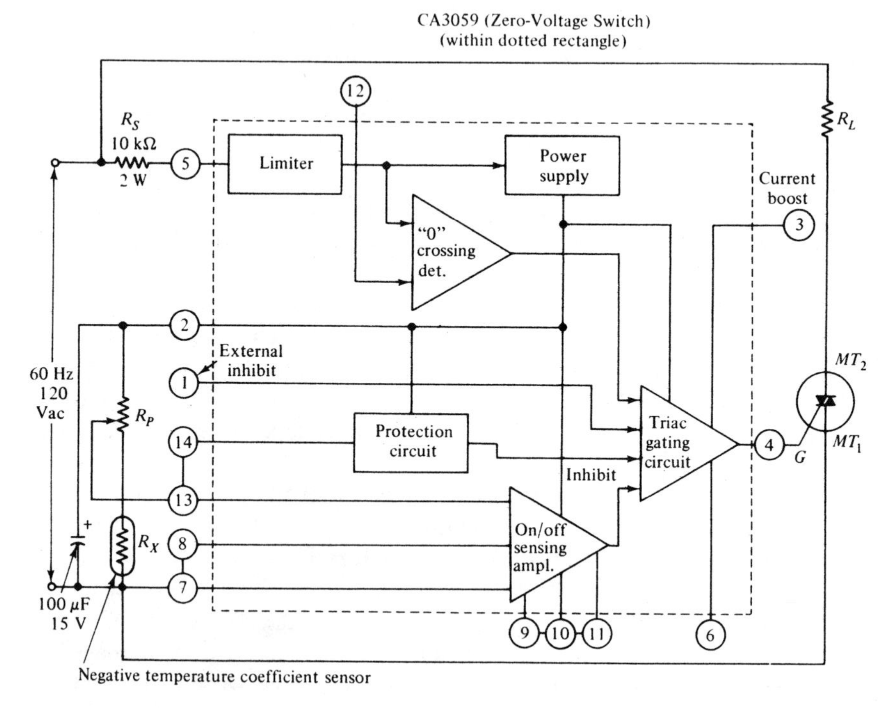

FIG. 9.10 Functional block diagram of the CA3059 zero-voltage switch. *(RCA Electronic Components)*

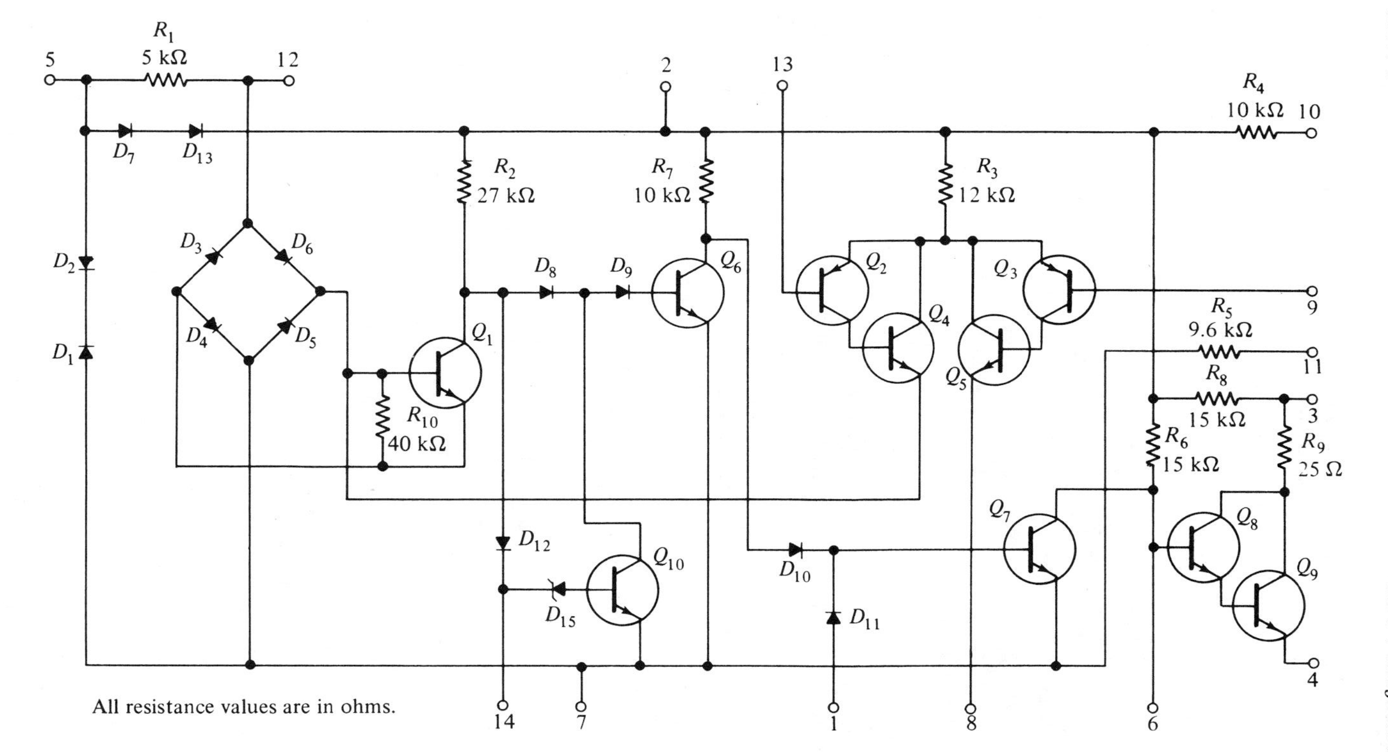

FIG. 9.11 Schematic diagram of the CA3059 zero-voltage switch. (*RCA Electronic Components*)

4. *The SCR gate driver*, which amplifies the output of the zero-crossing detector sufficiently to supply the moderate current required by the SCR's control-gate input.
5. *A protection circuit*, which senses opens, shorts, and other malfunctions, and removes drive from the SCR input.

As the block diagram of Fig. 9.10 and the internal schematic of Fig. 9.11 illustrate, the internal circuitry of the LIC is never subjected to the large ac line voltage, even though power is derived from the same source, because of an external series-dropping resistor and internal Zener-diode clamping. This is another example of the use of a complex, high-gain monolithic circuit to do critical work within the normal LIC process restrictions, while an external element (the SCR, also called a triac) inexpensively handles the larger voltage, current, and power dissipation required.[1]

Combination-Control IC: A very interesting and useful control IC is offered in the *RCA monolithic CA3097E device, designated as a Thyristor/Transistor Array*. In addition to the thyristor (or SCR) switching element, the 16-pin DIP package includes four other elements: a PNP/NPN transistor pair, a programmable unijunction transistor (PUT), a Zener diode and an uncommitted NPN transistor. The data sheet for this highly flexible device *(File No. 663)* includes a host of control-application schematics, including a pulse generator, one-shot timer, Schmitt trigger, and series and shunt regulators.

[1] For detailed application information, see the RCA ICAN-4158, "Applications of the CA3059 in Thyristor Circuits."

10

DIGITAL-INTERFACE CIRCUITS

10-1. DIFFERENTIAL VOLTAGE COMPARATORS

The integrated-circuit comparator is generally considered under the linear IC category, even though, strictly speaking, it belongs to both the linear and digital worlds. It appears almost universally in digital systems, which require that a logic signal be available only when a dc voltage somewhere in the system is more or less than a *critical threshold value*. It also finds use in many other instances when the level of an analog signal is to be a determining factor.

In its simplest form, it is provided with *two analog inputs* and *delivers a ZERO* or *ONE output, depending on which input is the larger*. It is thus a two-faced device, which is linear at its input and digital at its output. In this respect it can also be regarded as a basic, one-bit analog-to-digital converter, and so finds use in a wide variety of circuit systems. In addition to interfacing digital circuits, its analog applications include such uses as variable-threshold Schmitt triggers, discriminators, and other level detectors.

The dc voltage comparator is almost as widely used as the OP AMP and, in fact, is very similar to it with respect to the common property of high open-loop amplification. Each of these linear ICs, however, is specifically designed for its particular application, and the differences between them will become more apparent as we examine the properties of the voltage comparator more closely.

The comparator is specifically designed around the properties of *speed* and *accuracy* for differential-comparison purposes. This calls for a deliberate tradeoff of high gain with wide bandwidth (to ensure small propagation delays), and also an output that is compatible with logic levels. Because of the inherently opposing design requirements of high gain versus wide bandwidth (and the consequent tradeoffs between them), there are a number of different types of comparators available. Speed of operation is emphasized by the types designed for wider bandwidths (and consequently smaller propagation delays); other types favor accuracy, which depends on two main factors, one being high gain to reduce the "gain error" (the amount of difference signal required to cause the amplifier to switch), and the other being the minimizing of initial offset error with its attendant temperature drift. These properties are more successfully combined in the specially designed comparator than they are in general-purpose OP AMPS.

10-2. OPERATION OF COMPARATOR

Ideally, a dc comparator consists of an inverting (−) and non-inverting (+) input and an output, as shown in symbol form for the basic circuit in Fig. 10.1. Here an unknown voltage (E_{in}) is applied to the inverting terminal 3, while a precise reference voltage (V_{ref}) is applied to the non-inverting terminal 2. During the time that the unknown voltage is less than the reference, the output remains "high" (at a defined logic-compatible level, which is positive 3.1 V in this case). When the unknown voltage becomes only an infinitesimal amount more positive than the reference (1 or 2 mV more positive in this case), the output descends very rapidly to a "low" value (negative 0.5 V), signifying a ZERO, as opposed to the previous value for a ONE. Similarly, when the input connections are reversed (with V_{ref} at the inverting terminal), the output will also be reverse with a ONE output occurring whenever the unknown input (at the non-inverting terminal) slightly exceeds the reference.

10-3. COMPARATOR CHARACTERISTICS

The ideal dc comparator has infinite voltage gain and zero "offset" voltage and current errors at its input, and responds instantly to any change of condition at its input. These are similar to the requirements that apply to ideal OP AMPS; thus the internal construction of most dc comparators is very similar to that of an OP AMP. There are differences, however. The OP AMP must be

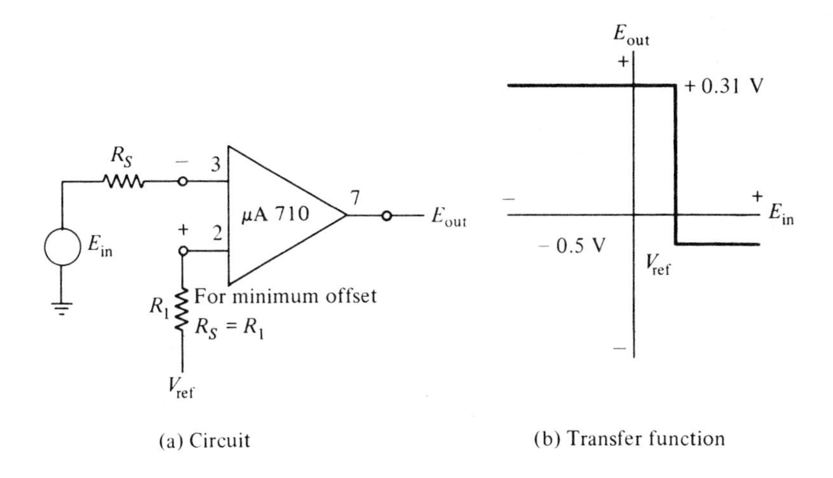

Note: Pin nos. apply to metal-can connections only.

FIG. 10.1 Basic comparator: (a) circuit in symbol form, compares level of input (E_{in}) with reference voltage (V_{ref}); (b) transfer function, showing ideal change of output from "high" (3.1 V for logical ONE) to "low" (negative 0.5 V for logical ZERO) whenever the input voltage exceeds the reference by as much as 1 or 2 millivolts. *(Fairchild Semiconductor)*

capable of linearly reproducing an input signal at its output, while this restriction does not apply to the comparator. The OP AMP must be capable of large output voltage swings, while the dc comparator must swing its output between two logic levels suited to the particular logic family it is designed to drive. Certain comparators are designed to produce substantial output currents, so that their "decision" may be used to directly drive a lamp or relay; this output current usually flows in only one direction, whereas high-current OP AMPS must both source and sink large currents. Since the OP AMP is generally used within a dc feedback loop, it is nearly always operating safely within its "linear" range; the dc comparator, on the other hand, is usually overdriven at its input, and must be constructed so that it can "recover" quickly from such overdrive, and, if possible, do so instantaneously, so that very large differences between input voltages do not cause internal damage. Both the OP AMP and the dc comparator must have good common-mode rejection (CMR); that is, they must respond only to the *difference* between input voltages, and not to their individual values. Finally, both types must have high input impedances, negligible input currents, and input errors that are negligibly affected as temperature varies.

The first commercially successful dc comparator, and probably the most widely used, is the type *μA710*, shown schematically in Fig. 10.2.[1] In addition to the basic dc comparison and "decision-making" function of the 710 type, the dual versions of this type (type 711) include a "strobing" input, which permits an external logic signal to determine when the output will respond to input differences. The "strobe" input may be used, for example, when a system must measure and perform logic based on an input voltage condition at a specific time, but not respond at other times, as in sampled measurement or control systems.

10-4. ADVANCED-PERFORMANCE COMPARATORS

In overcoming the inherently difficult task of producing a comparator that is fast, while simultaneously offering good input characteristics, precision, and versatility, advanced types of comparators have been produced to emphasize particular characteristics in the inherent *tradeoff between speed and precision.*

A limitation in the dual-comparator type *μA711* lies in the fact that while there are two sets of inputs there is only one output terminal (Fig. 10.3). This limitation is overcome in the dual types having two separate outputs, such as *Motorola MC1514*, which also has an improved output capability.

Added versatility to the single comparator type 710 is provided by the ability to operate from a *symmetrical* $\pm$*15-V supply*, and the *addition of a strobe terminal*, as offered by the *National LM106* type.

Multiple comparators are also available in *quad form*, as, for example, the *National LM339 and the Motorola MC3302/P.*

High-Speed Types

Other advanced *types emphasizing high speed* include the *Fairchild μA760* (response time of 16 ns) and the *Signetics SE259* (response time of 11 ns).

Precision Types

Among the advanced *precision comparators* is the widely used *National LM111*, considered the workhorse of precision comparators, just as the 710 is for high-speed comparators. (*Note:* the *LM111* is a single comparator—not to be confused with the *μA711*, which is a dual-comparator type.)

Although the 111 type is slower than the 710 (e.g., 200-ns response time versus 40 ns), the *input current* I_B (and resulting offset current) is *generally hundreds of times lower* (e.g., 100 nA I_B versus 20 μA).

[1] R. J. Widlar, *Operating of a Fast IC Comparator,* Fairchild Application Bulletin, APP-116.

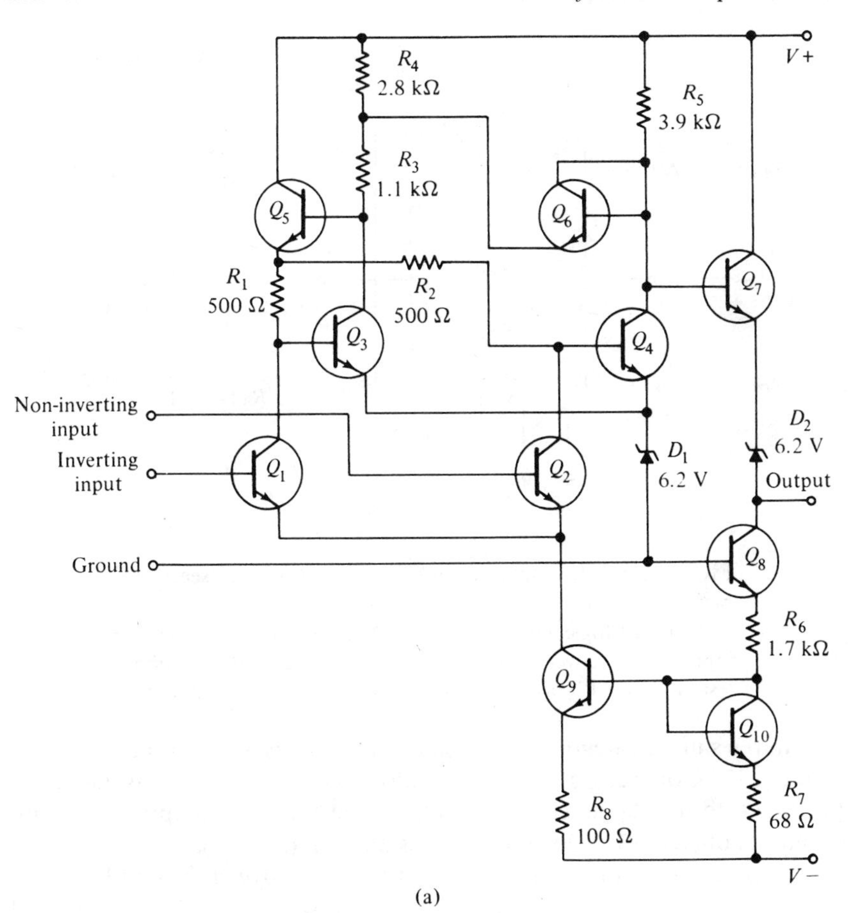

(a)

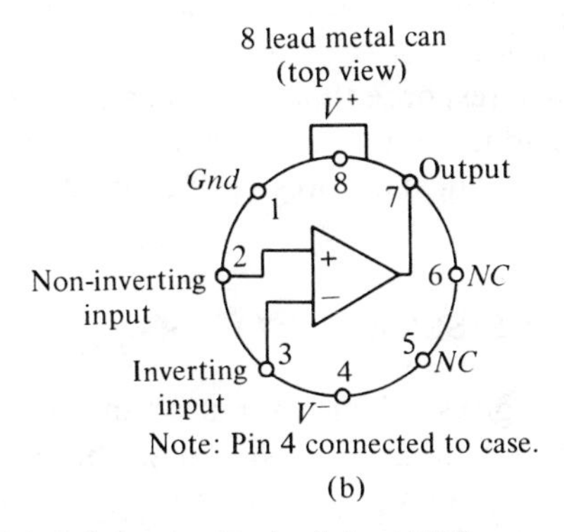

(b)

FIG. 10.2 Schematic circuit of the μA710 comparator: (a) input and reference voltages may be interchanged, depending on desired logic level at output; (b) connections for metal-can package with $V+ = 12$ V, $V- = -6$ V (also available in 14-lead dual-in-line package). *(Fairchild Semiconductor)*

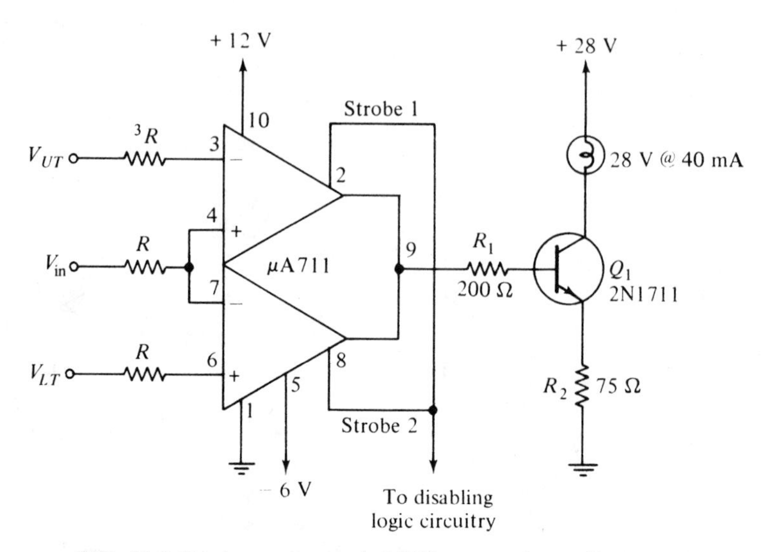

FIG. 10.3 Dual-comparator (μA711) arranged as a "sense amplifier"; shown connected to a lamp-indicator driver, it indicates a change of logic level (ZERO or ONE) to determine when the input (V_{in}) is between two precise reference thresholds (V_{UT} upper test limit and V_{LT} lower test limit). *(Fairchild Semiconductor)*

In the similar category of precision comparators, mention may be made of the *μA734*, offered as useful for analog-to-digital (A/D) conversion use (as discussed in a later section). Similarly, the *Intersil 8001* precision comparator is offered for its low-power consumption (30 mW).

Another precision comparator from *Precision Monolithics*, called *Mono CMP-02*, has a maximum offset specification of only 0.8 mV and typical drift of 1 μV/°C. While its response time of 160 ns is, as expected, greater than the fast types mentioned above, it is still better than other precision types, which typically have response times of 200 ns or more.

A trend toward including additional circuitry on the comparator chip is exemplified by the *μA750*, which features a self-contained voltage-reference capability.

10-5. SENSE AMPLIFIERS

While the dc comparator gives a logic output when an unknown input is above or below a reference voltage, it is often desirable to determine when an unknown input is *between* two precise reference thresholds. This can be accomplished by using two dc comparators, one for each threshold, and performing appropriate logic on their outputs so that the resultant logic output is ONE if the unknown is between the thresholds, and ZERO if outside the thresholds. Such a dual-threshold comparator is called a *sense amplifier.*

Sense amplifiers are LICs that are used in the heart of most digital computers. Magnetic memories in such computers, whether cores, plated wires, rotating drums, or magnetic tape, all give very small output signals (in the millivolt range) when interrogated. The interrogation process, in the case of core memories, requires a large current to pass through "address" wires running through the core in question; the few millivolts of response picked up in another wire are in the presence of a much larger, noisy "spike," resulting from the interrogation. The sense amplifier has good common-mode rejection, allowing it to respond differentially to the small memory content signal while ignoring the much larger interrogation signal. The sense amplifier is also "strobed" so that it transmits information to the computer only at times when such information is available and needed.

A typical sense-amplifier application of the *μA711* is shown in Fig. 10.3. It is essentially two type 710s on one chip, plus common output-logic circuitry. Other typical dual sense amplifiers are the *SN7524* and *SN7525 types from Texas Instruments.*

10-6. ANALOG-TO-DIGITAL AND DIGITAL-TO-ANALOG CONVERTERS

Information transmission systems sometimes require that a continuously varying analog signal be converted to digital form before transmission, to be reconverted after transmission back to analog form, or to be recorded in digital form for subsequent data analysis. Such conversion relies on the use of *dc comparators.* In the simplest analog-to-digital (A/D) converter, several comparators are used with an ascending set of reference voltages, as in Fig. 10.4. Those comparators having references below the instantaneous value of the input signal will give ONE outputs, while the others will give ZERO outputs. The resultant logic outputs can be fed into additional logic processing and transmitted. The accuracy of such a system depends on how many comparators are used, the spacing of their reference voltages, and how often the logic information is sampled and transmitted. The *μA734* comparator, for example, is offered for use in A/D converters with 12-bit accuracies and a 1-megabit conversion rate.

A more sophisticated A/D converter uses not only the dc comparator, but its counterpart, the digital-to-analog (D/A) converter as well. The input analog signal is compared by the dc comparator to the dc voltage created by the D/A converter. The D/A converter itself is driven by logic, which is fed by the output of the dc comparator in such a way that its dc value is "searched" until it is as close as possible to the value of the input voltage. The feedback loop then remains stable until either the input voltage changes,

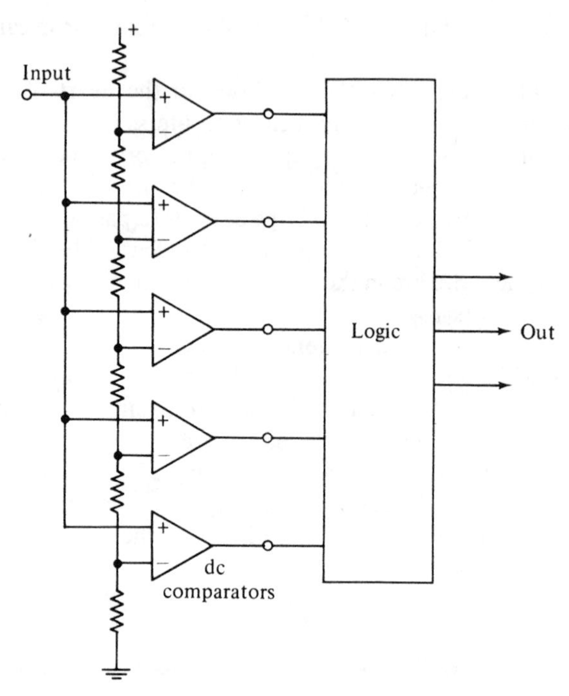

FIG. 10.4 Simple A/D converter: a series of comparators with an ascending set of reference-voltages to supply the corresponding ONE and ZERO outputs that make up the digital-logic output.

or it is commanded by a timing signal to begin in the search procedure anew. The basic blocks of such a system appear in Fig. 10.5.

Digital-to-Analog Conversion

The D/A converter is often a monolithic circuit that relies heavily on monolithic matching techniques, as in the *AIM/DAC100 (Precision Monolithics)*, Fig. 10.6. The input to the D/A converter is a combination of logic levels from the most significant bit (MSB, no. 1) to the least significant bit (LSB, no. 10). Each of these levels is used to determine whether a constant-current source of calibrated value is ON or OFF. By summing outputs of a number of such current sources, which are usually weighted with binary values to correspond to the thresholds of the original A/D converter, an analog-output current (AOC) is obtained, which is the *analog of the original input signal* before it was converted to digital form. There is, of course, inaccuracy, because the original conversion was only able to resolve the continuously varying input signal into certain discrete voltage steps, and because the D/A

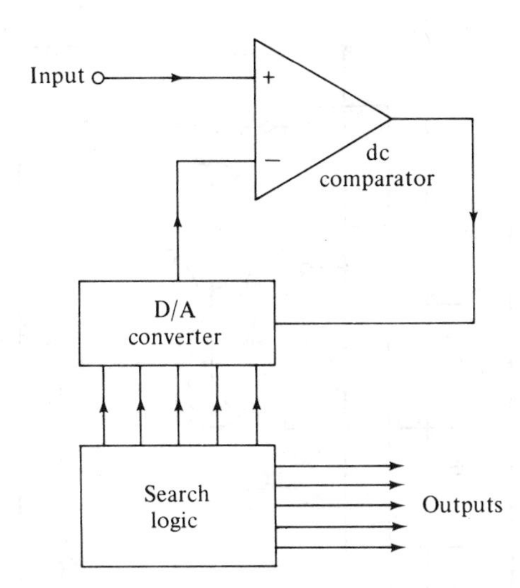

FIG. 10.5 Feedback A/D converter: contains a D/A converter in a feedback loop to produce digital outputs that correspond still more closely to the analog input.

converter can only produce a limited number of discrete output-current steps. The more steps there are, and the more often they are "updated," the better a replica of the original signal the D/A converter will produce. If the output of the converter is in the form of a current, it may be converted to an analog of the original voltage by running it through a calibrated resistance, or, in more sophisticated systems, by driving the current into the summing junction of an operational amplifier, thus reproducing a continuously varying voltage.

While the monolithic D/A converter uses inherent chip component matching to advantage, it is difficult to build very precise converters with resolution of better than about $\frac{1}{2}$ percent, because very wide ratios are not accurately controlled, even on the best monolithic chips.

Hybrid D/A Converter Modules

When the hybrid form of D/A converter is packaged to produce a complete module, greater accuracies can be provided. As an example, the *DAC372-12 (Hybrid Systems Corp.)* offers an accuracy to 0.0125 percent in a general-purpose version, providing an analog-voltage output. The size of the package (2 by 2 by 0.4 in.) is a fairly compact one in this hybrid form.

10-7. BORDERLINE LINEAR/DIGITAL INTEGRATED CIRCUITS

As may be seen from the numerous type numbers mentioned for comparators,

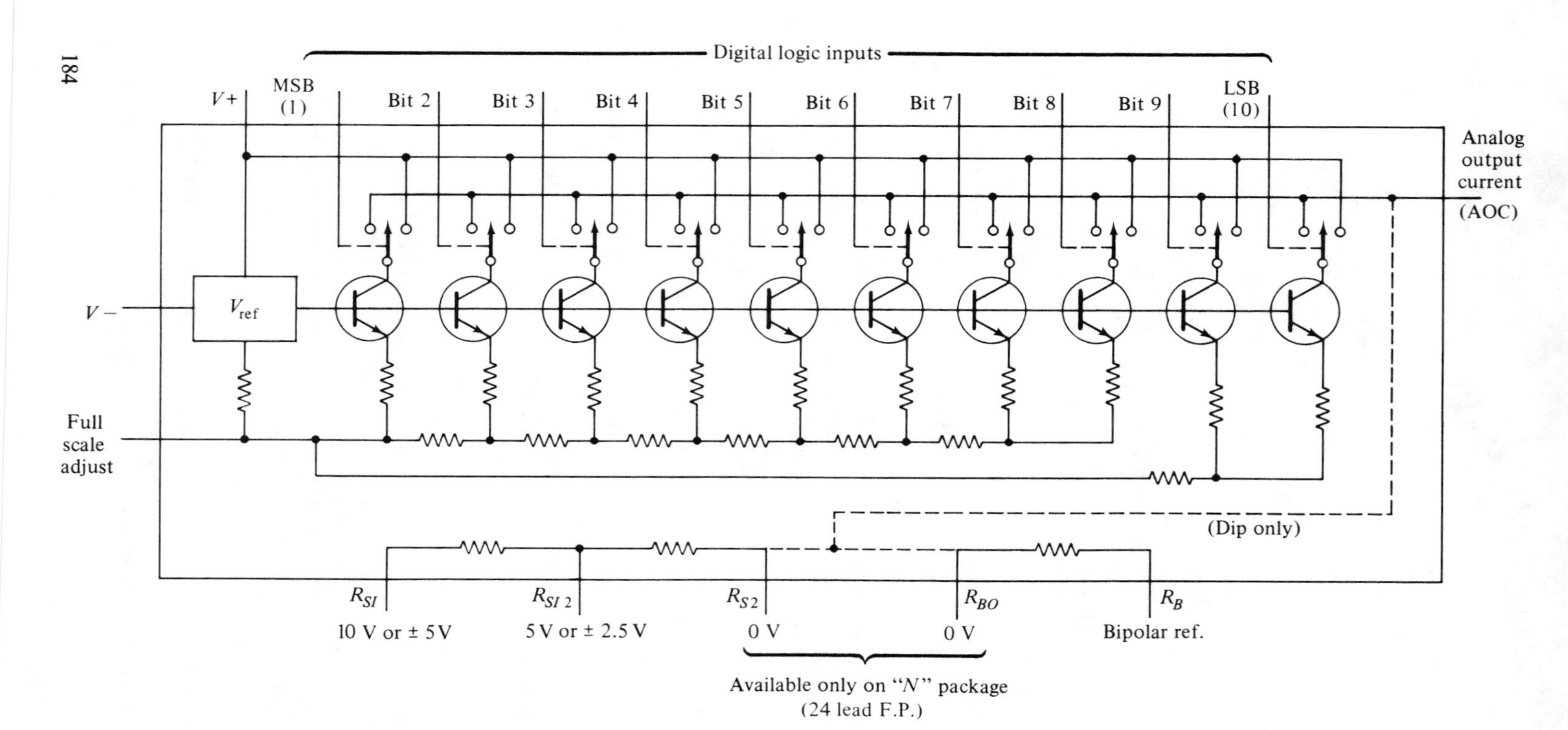

FIG. 10.6 Digital-to-analog (D/A) converter in monolithic form (AIM/DAC 100): the summing of the progressively higher constant-current levels from each digital bit produces an analog-output current (AOC) which is an analog of the original input signal to the A/D converter. *[Analog Integrated Microsystems (subsidiary of Precision Monolithics)]*

sense amplifiers, and A/D and D/A converters, there is a healthy competition for these kinds of LIC digital-interface circuits. This rivalry results in a multiplicity of types for the digital-interface units that approaches the amount of proliferation of the digital ICs for use in digital computers. Accordingly, within the limits of scope and size here, further specification details of these digital-interface circuits (*including line drivers and line receivers*) are best left to texts and manuals that concentrate on the digital ICs.

10-8. COMPLEMENTARY MOS TYPES

A good example of the overlapping trend between the linear and digital types is furnished by the complementary MOS types. In the digital field these FET types have made great inroads with their simplified circuitry and low power consumption, especially in switching applications. An extension to the LIC field of linear-circuit processing is offered by the *CD4007A*, an RCA "COS/MOS" IC, which provides a dual complementary pair, plus inverter, as the D/A switch and OP-AMP output stage for a digital-to-analog converter.[1]

In addition to the switching applications of the C/MOS (complementary MOS) type, there are many instances where it can be used beneficially as an alternate to the *linear bipolar IC*. Since it is possible to arrange the C/MOS circuit to perform *linear amplification* (within its smaller voltage-amplification limits), it may often be a preferred choice in special cases where its low power-consumption feature is an important factor.[2]

[1] "D/A Conversion, Using the CD4007A (COS/MOS IC)," RCA Application Note, ICAN-6080.

[2] See RCA brochure, "Linear Applications of COS/MOS."

11

PRECISION/ INSTRUMENTATION OPERATIONAL AMPLIFIERS

11-1. MAJOR HIGH-PERFORMANCE CHARACTERISTICS

Beyond the group of general-purpose OP AMPS previously discussed (Chapters 4 and 5), there is a large (and growing) list of advanced types—each one offering improved specifications of one kind or another. These represent the continuing advances in the innovative design and fabrication techniques of IC technology, along with the trend by designers toward the steady adoption of the improved LIC devices in place of many complicated circuit functions that were formerly feasible only in discrete-circuit form.

The improved devices provide *high-performance properties that are primarily aimed at overcoming potential error-producing characteristics of LICs* in three main areas, as follows:

1. *Decreasing input bias currents* (I_B) to obtain a higher input-impedance (Z_{in}) value (as with FET input), and to reduce errors associated with such input currents flowing through large resistors.
2. *Reducing drifts* generally associated with the temperature coefficients (TC) that cause changes in initially trimmable offsets, such as *input-*

voltage offset (V_{io}(TC) in μV/°C) and *input-current offset* (I_{io}(TC) in pA/°C).

3. *Reducing inherent noise voltages* that are large enough to be objectionable in the amplification of low-level signals.

There are other improved specifications that are particularly desirable for certain applications (such as requirements for *wide bandwidth* or for *micropower dissipation*), and these requirements will be considered separately as they appear in such applications.

11-2. DEVELOPMENT OF PRECISION OPERATIONAL AMPLIFIERS

Considering the continuing appearance of newer OP AMPS with ever-better refinements, it might be thought that such high-performance amplifiers are a fairly recent trend. Not so; these efforts at precision amplifiers go all the way back to the vacuum-tube era, when such OP AMP were satisfying a keen demand for improving the accuracy of the then-emerging *analog computers.* Similar efforts were subsequently applied to developing the discrete-transistor versions of these precision amplifiers to replace the bulky tube affairs with their mechanical choppers.

The present activity now centers around the vastly more reliable and efficient compact devices produced by IC techniques, with heavy accent on the ICs inherently better matching of differential-transistor pairs, with accompanying improvement, for example, in common-mode rejection capability.

Super-Beta Transistors in OP AMPS

The use of the super-beta (or super-gain) transistors is an outstanding example of an intensive development in the improvement of OP AMPS. This has resulted in a substantial reduction in the input bias current (I_B) by about three orders of magnitude, with a corresponding improvement in input impedance (Z_{in}) or (R_{in}). The exceedingly small bias currents used are shown in the characteristic curves of Fig. 11.1(a), along with a functional schematic of the super-gain voltage-follower operation in part (b).[1]

By means of advanced processing in the IC diffusion technology, the super-beta transistor can be fabricated with a *current gain exceeding 4,000*, but with its breakdown voltage at only about 4 V. However, when the IC chip is subjected to two separate emitter diffusions, it is possible to build the super-beta and the standard NPN transistors on the same chip. By this

[1] R. C. Dobkin, "New Developments in Monolithic Op Amps," *Electronics World*, July 1970.

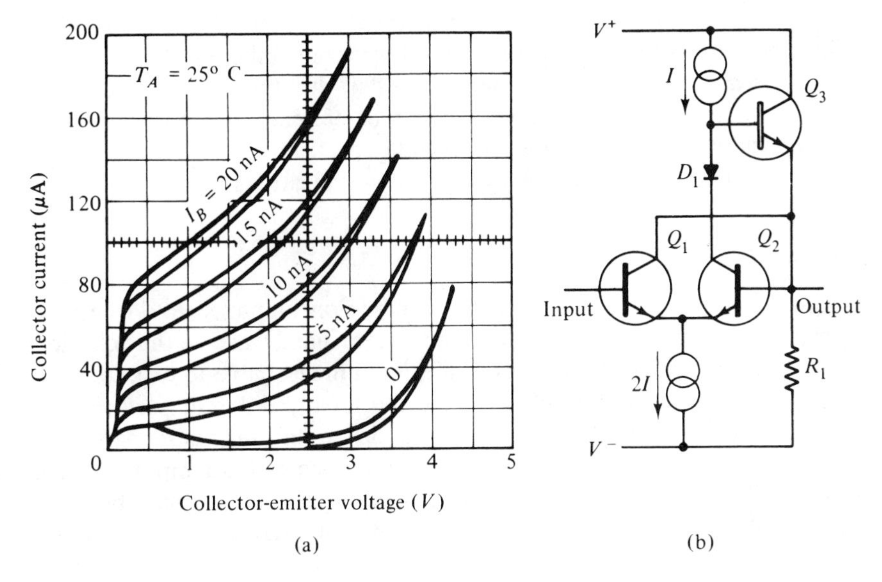

FIG. 11.1 The super-beta (or super-gain) transistor: (a) super-β transistor output curves; (b) functional diagram of voltage-follower circuit. Super-β transistors Q_1 and Q_2 are operated at very-low bias currents (in nano-amps), while buffer transistor Q_3 is the only one operated at relatively high voltage.

means (not available in discrete circuits), such IC circuits are designed to take advantage of high current gain, and yet can be operated at high voltages.

The super-beta-transistor types of OP AMP (exemplified by the National LM**108**/A, Texas Instruments SN52**771**, and Motorola MC1**556**) have input currents reduced to less than one tenth of the ordinary types, and offer a design alternative to the FET-input type (such as the Fairchild μA**740**). These examples offer high input-impedance alternatives that get around the FET limitations (particularly with respect to the bothersome leakage current of the junction FET, which doubles for every 10°C rise, as discussed later).

11-3. SELECTING OPERATIONAL AMPLIFIERS FOR HIGH-PERFORMANCE CHARACTERISTICS

It is well to note at this point that the emphasis in this chapter on small error-producing factors is not generally a major concern in a multitude of amplification applications, where the OP AMP is sufficiently close to an "ideal amplifier"; this emphasis, on the other hand, is primarily important in cases where *precise analog computation or highly accurate signal-con-*

ditioning applications are involved. It is in these areas that efforts at technical refinements are constantly being made to overcome inherent device limitations, such as temperature dependence, dc offset drift (not wholly susceptible to elimination by negative feedback), and ever-present noise, however small. This constant attention results in a wide spectrum of high-performance OP AMPS with a variety of specially tailored specifications.

Figure of Merit

If some standard figure of merit for a precision amplifier could ever be agreed upon, it would undoubtedly be a very handy aid in the selection process. But considering the facts of life, it is unreasonable to expect that a single figure should accurately evaluate the greatly different performance characteristics of amplifiers which are designed to emphasize particular properties. However, with this caution in mind, one such attempt to furnish a not too unreasonable comparison of overall performance may be mentioned.[1] This *comparison* includes a formula for arriving at an arbitrary figure of merit by including in the numerator such figures as voltage gain (V/V) and slew rate (V/μs) and dividing the numerator product by the product of input-voltage offset (V_{io}) voltage-offset drift (μV/°C), bias current (I_B), and supply current (I_{CC}). In practice, this notion of a single-figure of merit value would be valid only if proper weights for the pertinent parameters were assigned by each user for his particular requirements.

Chart of Typical High-Performance Op-Amp Characteristics

For practical initial selection purposes, we may divide those refined characteristics that are desired for a particular OP AMP application into two main categories, given in **Table 11-1**, Sec. 11-3, for quick comparison with general-purpose specifications. The main error categories are as follows:

1. Input bias current (I_B) and accompanying input impedance (Z_{in}).
2. *Input-voltage offset drift* (V_{io} in μV/°C).

(*Note: Noise voltage* specifications are most important in instrumentation amplifiers, and are discussed in Section 11-9.)

[1] Based on a performance-comparison chart compiled by J. Talley (of Texas Instruments): using the relative values described above, his figure of merit places the TI super-beta type SN52771 at 600, between the extremes of 0.4 (for the 741) up to a maximum beyond 1,000 (for the 108).

TABLE 11-1

Typical Precision Op-Amp Characteristic Values

	Low-Cost Types						*Premium Types*
	General-Purpose Type	*Precision Type*	*Super-beta Types*		*FET Input*	*Chopperless*	*High Speed (BW to 100 MHz)*
	μA741	μA777	LM108A	SN52771	AD503	AD508	AD507
Input bias current (I_B) (nA)	≈100	25	2	<25	0.01	20	2
Input impedance (Z_{in}) (MΩ)	>1	>2	>30	100	to 10^{12} Ω	4	40
Drift of offset voltage (V_{io} TC) (μV/°C)	—	15	<5	<15	25 ($I_{io}=2\times/10°C$)	1	15
Slew rate (V/μs)	0.5	0.5	0.2	2.5	6	0.12	35

Notes:
For micropower types (where values are adjustable), see Section 11-9.
For instrumentation type (*AD520*) (which is monolithic, but not strictly an op amp), see Sections 11-9 and 11-10.

Each main category is discussed in a separate section that follows, along with a representative example (or examples) of each type.[1]

11-4. LOW-BIAS-CURRENT (I_B)/HIGH-INPUT-IMPEDANCE (Z_{in}) TYPES

The value of input bias current (I_B) for this group of OP AMPS (and modular LICs) is given in nanoamperes, indicating a reduction by a factor of 10 or more compared to general-purpose OP AMPS. This reduction is achieved by designs using either super-beta transistors (as discussed in a previous section) or FET types in the input circuit (see **μA740** in Fig. 11.2). A comparison with the general-purpose OP AMPS of the 741 type can more easily be seen at a glance by listing in **Table 11-2** (Sec. 11-4) *representative rough values for comparable types.*

TABLE 11-2
Comparison of Input Impedance (Z_{in})

Type	I_B (nA)	Z_{in}
General-purpose type **741**	Approx. 100	Approx. 2 MΩ
Super-beta type MC**1556**	Approx. 10	Approx. 5 MΩ
Super-beta type LM**108**A	Approx. 1	Approx. 50 MΩ
FET-input type μA**740**	Approx. 0.1	Up to 10^{12} Ω
FET-input type AD**503**	Approx. 0.01	Up to 10^{12} Ω

The value for low bias current is naturally increasingly important in dealing with *high source resistances*, and also for improved accuracy in *integrators* and in *sample-and-hold circuits.*

Similarly, accuracy is improved by a low value of I_B to reduce input-current offsets. Here the *ultrahigh Z_{in} of the FET-input types*, while essential for some very high input-impedance sources, *may not be preferable to the super-beta transistors*, because the FET has the more rapid drift with temperature, as discussed in the next section.

11-5. LOW VOLTAGE-DRIFT TYPES

Where an application requires extreme stability against voltage drifts with

[1] For particular applications, very helpful guides to op-amp selection may be found in the product catalogs of many manufacturers, particularly in those with concentration on the precision-instrumentation types of LICs, such as Analog Devices, Teledyne/Philbrick, and Burr–Brown. (See Appendix IV for a more complete list of LIC manufacturers and their addresses.)

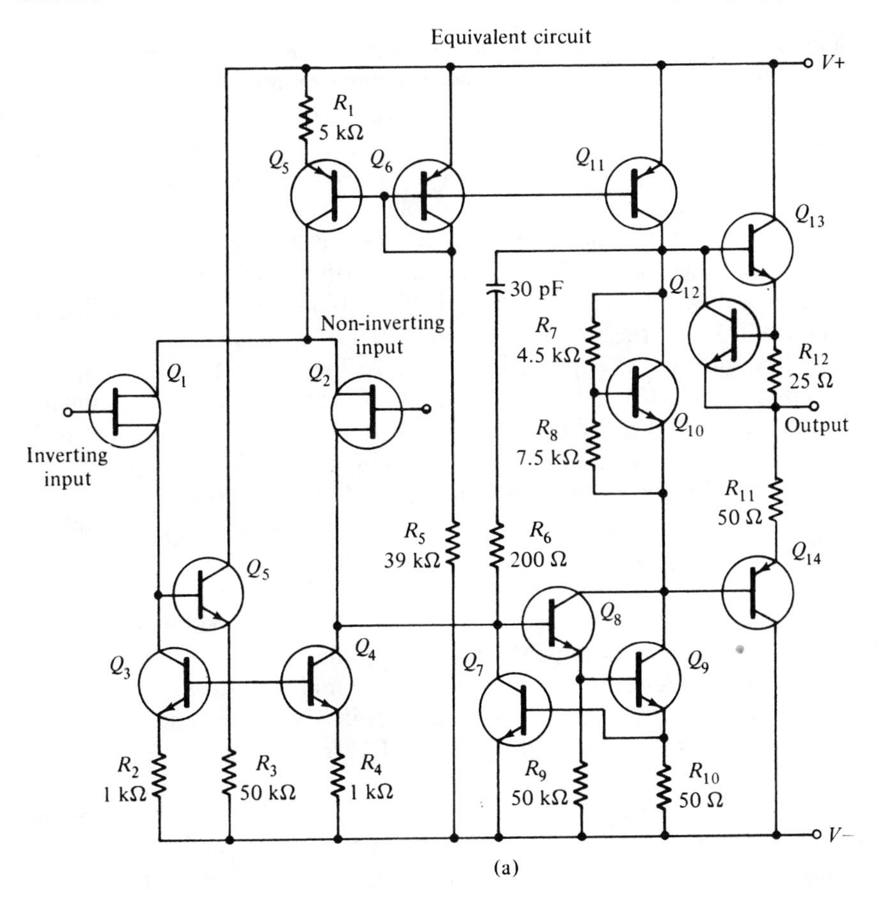

(a)

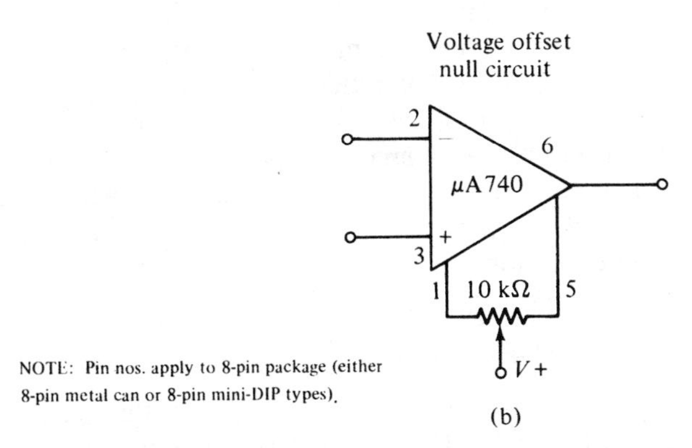

NOTE: Pin nos. apply to 8-pin package (either 8-pin metal can or 8-pin mini-DIP types).

(b)

FIG. 11.2 FET-input OP AMP: (a) schematic diagram of μA740 equivalent circuit, showing the FET differential pair in the input circuit; (b) pin connections are same as internally compensated 741 type. *(Fairchild Semiconductor μA740 Data Sheet)*

time or temperature, the traditional method has been to use *chopper-stabilized amplifiers*, where the "low-frequency" components most subject to drift have been chopped (or modulated), and are then processed in a practically driftless ac-coupled amplifier, which also handles the higher frequencies. This method is incorporated in *chopper-stabilized* OP AMPS, such as *Analog Devices type 234*, to achieve voltage drifts as low as 0.1 μV/°C and averaging about 1 μV/mo. These are generally available in *modular*, rather than monolithic, form, but there are noteworthy exceptions, such as *TI's recent model SN62/72088*, which is in the 14-pin DIP form. Another, in standard TO-99 form is the *Harris model HA2900*, which can also serve as a good example of the improved drift performance achieved in the chopper-stabilized models. Compared to the standard 741 types, the specification for this chopper-stabilized model of OP AMP offer better than an order-of-magnitude improvement in its low value for offset-voltage drift of 0.2 μV/°C; a figure that is associated with an extremely small input-offset current (I_{io}=0.05 *nA*) and an exceptionally large open-loop gain ($A_{VOL}=5\times10^8$ or greater than 170 dB).

"Chopperless" low-drift types have been designed in monolithic form, and these retain the advantage of the differential op amp with its high common-mode rejection ratio (CMRR)—an advantage that must be sacrificed in using the single-ended chopper-stabilized amplifier. Such "chopperless" types as the *AD 508*, a super-beta example, still offer a low value of voltage drift of 1 μV/°C and an input bias current (I_B) around 10 to 20 nA (as seen in Table 11-1).

While the value of I_B for the super-beta type *AD 508* results in an input impedance of only a few megohms (compared, say, with that offered by the FET-input type *AD 503* of up to 10^{12} Ω), it becomes important to realize the limitation imposed by the FET on its current-drift specification, which doubles for every 10°C rise in temperature. Hence, from the complete low-drift standpoint, the super-beta transistor may emerge as superior for its certified low-drift specification for a monolithic type.

Typical applications for the low-drift type of OP AMP include *accurate summing amplifiers* (as in servo loops), *precision comparison and regulation*, and generally as stable input amplifiers of low-level signals, where voltage drifts would become a significant part of the input signal, or would otherwise degrade the accuracy of operation.

Combination of Low-Drift and FET-Input Properties

An example of a high-performance monolithic amplifier that combines low input voltage-offset drift with FET-input performance is offered by the *Burr–Brown model BB3521* series. The overall performance in this type (and in similar high-performance types from other manufacturers) is achieved by advanced fabrication techniques, such as *active laser trimming.* The features include the expected high input impedance (from 10^{11} Ω up) of the input FET pair. This monolithic pair is carefully matched to keep the initial input offset voltage down to a small value (250 μV) and also to hold its drift down to a low level of 1 μV/°C.

When considering the ultrasmall input bias current (I_B) in FET types (10 pA in this case), we must also keep in mind the accompanying FET characteristic that calls for this input current to double for every 10°C rise. However, where temperature variations are acceptably low, this combination of the desirable properties of high input impedance and low voltage drift allows the use of the OP AMP for high-accuracy requirements (such as 0.01 percent buffer use), while still retaining a configuration similar to a general-purpose 741 type.

11-6. ADDITIONAL HIGH-PERFORMANCE FEATURES

The limited listing of typical high-performance specifications in **Table 11-1** was confined, for purposes of clarity, to just the few major characteristics of low input bias current (I_B) with corresponding high Z_{in}, and low voltage-offset drifts (μV/°C), as being two salient properties in characterizing precision types of OP AMPS. In addition to these, other high-performance features are desirable for particular purposes; such extra factors give rise to special types, such as the following:

1. *Wide-band (or high-speed) types* (discussed in Section 11-7).
2. *Low-power (or micropower) types* (discussed in Section 11-8).
3. *Instrumentation (or low-noise and high CMRR) types*, discussed in Section 11-9, summarizing most high-performance features.

A discussion of these additional types by no means exhausts the full range of advanced types in the active development of OP AMPS (and other LICs); it does, however, within space limitations, offer a sound basis for viewing the diverse performance capabilities of OP AMPS (including and beyond the general-purpose types), and should, hopefully, enable an intelligent interpretation of the maze of technical specifications given in manufacturers' data sheets. As a further aid, examples of these types are identified in Appendix II, in addition to a *cross-referenced* list of manufacturer's type and model numbers in Appendix III.

11-7. WIDE-BAND (HIGH-SLEW-RATE) TYPES

The property of wide frequency response is allied with high speed (i.e., fast rise and fall times) by the constant relation of their product; this is generally given in terms of the -3 dB high-frequency cutoff (f_c) and rise time (t_r) for good reproduction of a step input as a constant product:

$$(f_c)\,(t_r) = \text{constant} \qquad \text{(sometimes taken at 0.35)}$$

Unity-Gain Bandwidth

Consider first the bandwidth of an OP AMP: this extends from direct current to high-frequency cutoff (f_c). This ability to amplify a wide band of frequencies is generally specified in terms of the *−3-dB bandwidth of the* OP AMP at unity gain. This is typically the same value as the *gain–bandwidth product* ($G \times BW$), as exemplified by the value of 1 MHz for the general-purpose 741 type. Here the −3-dB bandwidth is 10 kHz at a gain of 100, giving a G × BW product of 1 MHz (100 × 10 kHz); similarly, when used with a greater amount of feedback for unity gain (at which gain it is still unconditionally stable because of its internal compensation), the unity-gain bandwidth is given as 1 MHz (1 × 1 MHz). Also, in the general-purpose class, the *extended bandwidth* 748 type is capable of extending the bandwidth at a gain of 100 to a frequency of 100 kHz (by the use of a 3 pF external compensation capacitor), giving a G × BW product 10 times greater to 10 mHz (100 × 100 kHz). However, for cases requiring still larger bandwidths, we must turn to the high-performance wide-bandwidth types, such as, for example, the *AD507*, which provides a gain–bandwidth product out to 100 MHz in an OP AMP configuration, as shown in Fig. 11.3 (a). It will be seen in the figure that some external compensation (20 pF at terminal 8) is required to go down to unity gain, although this is not required for closed-loop gains greater than 10.

Slew Rate

When a large-amplitude signal is fed into an OP AMP, the full output appears after the lapse of a small amount of time, owing to the presence, generally, of internal capacitance. This limitation on the *rate at which the output can change (or slew)* in response to an input step is expressed as the *slew rate in volts per microsecond* (V/μs). Since the same limiting factor of capacitance (internal or distributed) is involved in both bandwidth and slew rate, it can be expected that the high-speed amplifier types will also be found in the wide-band group.

In the case of the example of the wide-band amplifier (type *AD507*), the slew rate is specified as 35 V/μs, as compared to the value of 0.5 V/μs for the general-purpose 741 type, an improvement of 70 times.

This improvement in bandwidth and slew rate is obtained, as can be

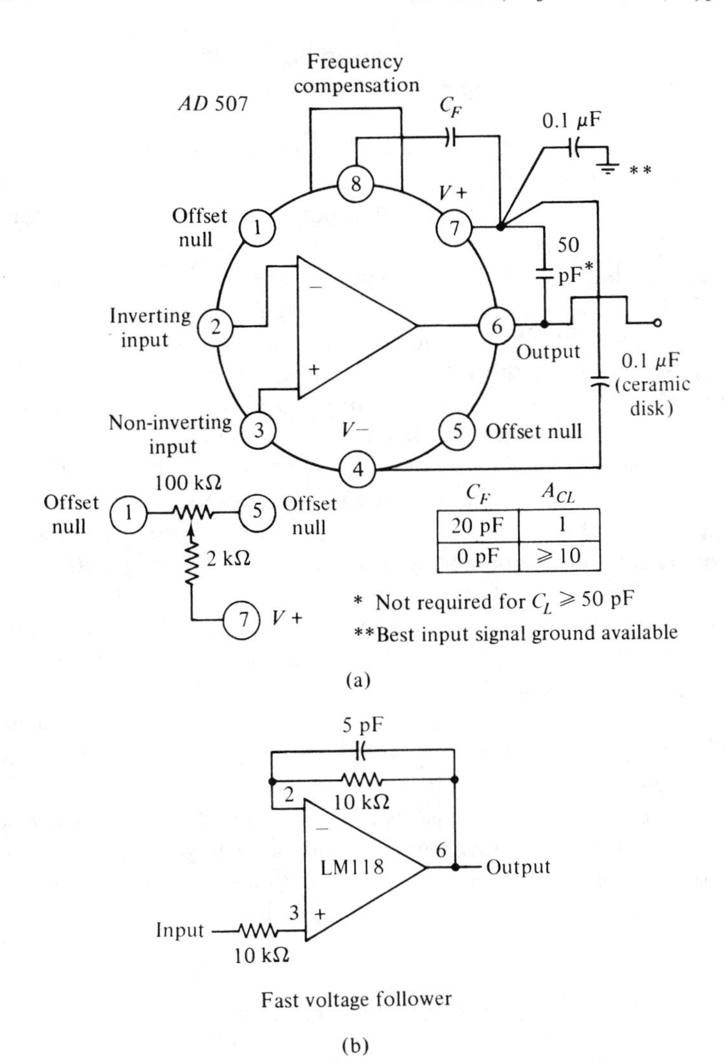

C_F	A_{CL}
20 pF	1
0 pF	⩾10

NOTE: Pin nos. apply to 8-pin package (either 8-pin metal can or 8-pin mini-DIP types).

FIG. 11.3 Wide-band and high-speed monolithic OP AMPS: (a) wide-band (100 MHz) type AD507 *(Analog Devices)* connection diagram; (b) high-speed voltage-follower, with feed-forward compensation circuit of the LM118 type *(National Semiconductor)* to boost slew rate beyond 100 V/μs.

expected, by various tradeoffs of other high-performance characteristics. When compared, for example, with the precision super-beta type *AD508*, which is optimized for low-voltage-drift properties, the offset-voltage-drift is only 1 μV/°C, contrasted with a drift of around 15 μV/°C for the *AD507*.

With regard to the *correspondence between wide bandwidth and slew rate*, it should be noted that the correspondence is not strictly a constant, depending on the particular fabrication techniques employed. For example, the precision high-speed *LM118* type offers a small-signal bandwidth of only 15 MHz (much below the 100-MHz figure of the *AD507*), while still providing a guaranteed slew rate of 50 V/μs (as opposed to 35 V/μs for the *AD507*). Moreover, in the *LM118*, which is internally compensated down to unity gain, it is possible to use external feed-forward compensation for inverting applications that will boost the slew rate to over 150 V/μs, and almost double the bandwidth. A fast voltage-follower circuit for the *LM118* is shown in Fig. 11.3(b), where a feed-forward compensation of only 5 pF is required for obtaining a generally ample high-speed action; similarly, a simple capacitor can be added to reduce the *0.1 percent setting time to under 1 μs*. Used in such high-speed applications as oscillators, active filters, and sample-and-hold circuits, the *LM118* represents an order of magnitude better performance than the original wide-band *μA709* type, a standard long used in industry as the first popular OP AMP. A similar improvement over the original *μA709* type is seen in the *μA715*, which offers a bipolar combination that can be frequency compensated for unity-gain bandwidth of 65 MHz, and a slew rate of 100 V/μs.

Where even greater emphasis is placed on *high slew rate and nanosecond settling times*, as in A/D converters, the tendency is to go to modules, such as, for example, *Data Devices model WB-23*, with a slew rate of 1,000 V/μs, or an FET-input type, *Intech model A-132*, which offers a typical slew rate of 1,500 V/μs.[1]

11-8. MICROPOWER TYPES

Low-drain performance characteristics for *battery-powered applications* are provided by the micropower types of OP AMPS. A representative example, *National type LH0001A*, offers a *no-load power dissipation of less than* $\frac{1}{2}$mW at supply voltages of ±5 V. With this small amount of dc input power, there is still an ample open-loop gain of more than 30,000 (or around 90 dB), with a typical low noise-voltage value of 3 μV rms. These specifications allow for quite satisfactory operation in many low-power instrument and transducer amplifiers. A premium version of the micropower OP AMP, the *CA3078A (from RCA)*, is tailored for operation from a single supply, using

[1] N. Sclater, "On Op Amps," *Electro-Procurement*, Dec. 1972.

a 1.5-V battery (size AA cell). Operation in this fashion, as an inverting 20-dB (gain of 10) amplifier, is shown in Fig. 11.4; the total power consumption for this circuit (and also for the noninverting version) is approximately 675 nW (or less than $\frac{3}{4}$ μW). The output voltage swing is 300 mV (or about 107 mV rms).

The operating point of the *CA3078A* in the circuit of Fig. 11.4 is obtained by an external resistor ($R_{set} = 30$ MΩ), connected from the positive supply to terminal 5. Other operating points may be adjusted by the external bias-setting resistor (R_{set}) to change the microamperes of quiescent current (I_Q) over a wide range of 3 orders of magnitude (from 1 to 1,000 μA). This provides for a resulting change in open-loop gain (A_{vol}) of roughly from about 60 to 90 dB (or from about 1,000 to 30,000 times). Such amplifiers, having the ability of changing bias current (and so gain or transconductance) by an externally adjusted resistor, have been designated by RCA as *operational transconductance amplifiers (OTA).*

In the OTA category there is type *CA3080A*, where the amplifier-bias input terminal allows variation of the amplifier bias current (I_{ABC}) from 0.1 to 1,000 μA, to produce corresponding changes in the transconductance (g_m) from 2 up to 20,000 μ℧. Although not strictly a micropower amplifier, the device power dissipation, at a supply voltage of ±3 V, can be made to change from just over 1 to 10,000 μW. A similar scheme for controlling micropower operation is called *programmable* OP AMP, in *type μA776*. Here

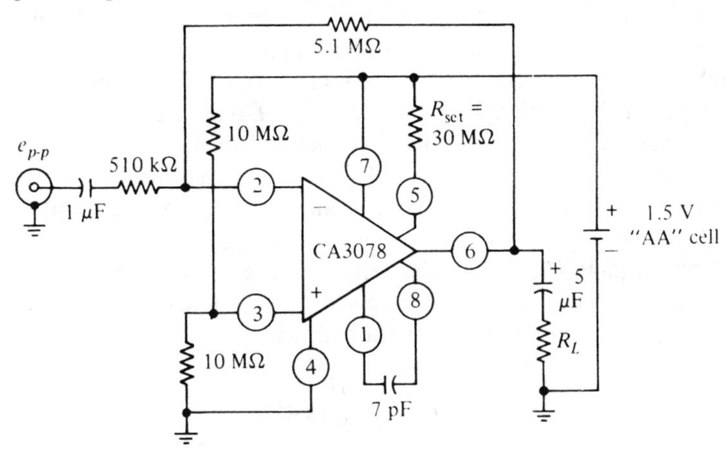

NOTE: Pin nos. apply to 8-pin package (either 8-pin metal can or 8-pin mini-DIP types).

FIG. 11.4 Micropower OP AMP circuit: the CA3078 type, operating with a single supply of 1.5 volts, consumes less than 1 microwatt of quiescent power; in this 20 dB inverting amplifier circuit, the bias-setting resistor (R_{set}) is externally adjustable. *(RCA Solid-State Division Data Book SSD-201)*

the quiescent current is again controlled by an external resistor (R_{set}). By selecting values for this resistor to produce an I_{set} current variation, the input bias current can be changed from about 0.1 to 20 nA. As a result, with a supply voltage of ± 3 V and $I_{set} = 1.5$ μA, the resulting supply current of 15 μA produces an input power of 90 μW (15 μA × 6 V). With $R_{set} = 22$ MΩ and a power supply of ± 1.2 V, the power dissipation (P_d) for a 20 dB gain is reduced to 600 nW (or 0.6 μW).

11-9. THE INSTRUMENTATION AMPLIFIER

The amplifier requirements for instrumentation purposes are among the most severe for any group of linear IC's. In addition to all the properties needed for precision amplifiers (as previously discussed), there is also the added consideration of handling the changing conditions produced by small tranducer variations on accurate operation. This requires the instrumentation amplifier to preserve its initial high accuracy against expected variations of its input sensor or transducer while rejecting any undesired voltages. The strategies for meeting these requirements can also serve as a practical summary of the refined OP AMP characteristics that are significant in applications demanding very stable and highly accurate amplifiers.

The situation can be clarified by examining the instrumentation amplifier in its frequent use for amplifying the output of a transducer-bridge configuration,[1] shown in functional-equivalent form (as if it were a single OP AMP) in Fig. 11.5. Here the signal of interest (e_d) is the difference between the voltages at the corners of the bridge ($E_1 - E_2$), as a result of one of the four resistors, acting as a sensor, changing its resistance. If, for example, the resistors (R) are originally 500 Ω, the bridge has originally a source resistance (R_s) also equal to 500 Ω, with an output signal (e_d) equal to zero, but with a superimposed common-mode voltage (E_{cm}) roughly equal to half the supply voltage ($E_R/2$). As the sensor changes its resistance ($R + \Delta R$), the bridge output becomes the difference signal (e_d), which is given in both its exact and approximate forms as the following[2]:

Exact:

$$e_d = E_R \frac{\Delta R}{4R + 2\Delta R};$$

Approximate (for small deviations):

$$e_d = E_R \frac{\Delta R}{4R}, \qquad \text{where } 2\Delta R \ll 4R.$$

[1] Analog Devices: *Product Guide*.

[2] The derivations of the unbalanced-bridge output are given in S. Prensky, *Electronic Instrumentation*, 2nd ed. Prentice-Hall, Inc., Englewood Cliffs, N.J., 1971.

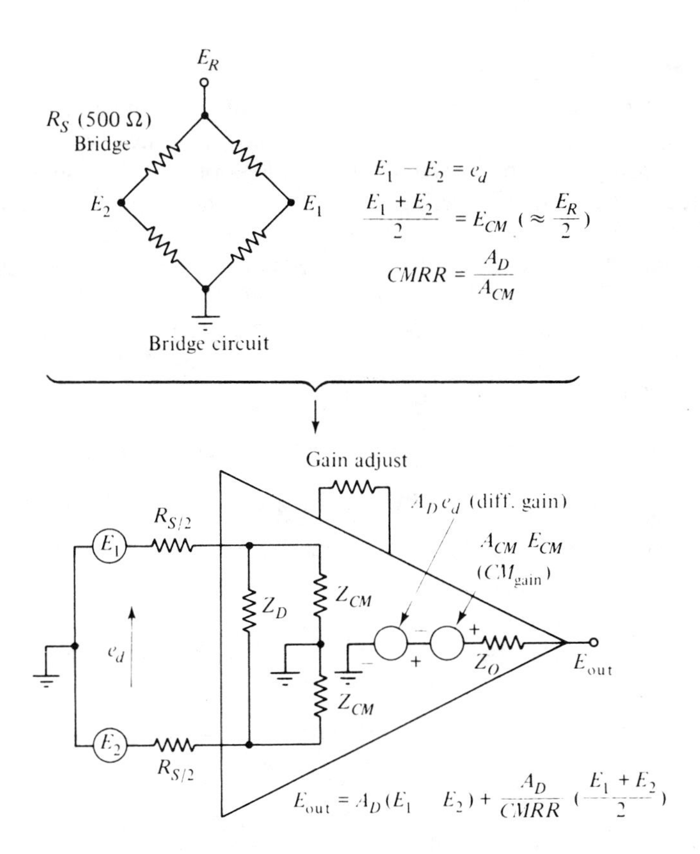

FIG. 11.5 Functional diagram of instrumentation amplifier; the equivalent input from the balanced bridge at the left is shown connected as a differential input to the amplifier equivalent at the right. *(Analog Devices)*

This difference signal (e_d) is applied as a double-ended input to the operational amplifier, by using both inputs of the amplifier, which then acts as a *differential amplifier*; this type of amplifier should (and initially does) have a satisfactorily high common-mode rejection ratio (CMRR), usually well beyond 80 dB (or 10,000:1).

As shown in the right-hand portion of Fig. 11.5, each input originally has an equal source resistance ($R_s/2$), under which condition the output (E_{out}) contains a practically negligible amount of amplified common-mode interference. Before examining what happens as the bridge is unbalanced, it is instructive to see how this differential configuration of the amplifier accomplishes the highly desirable rejection of common-mode interference.

Advantages of Differential Input

The beauty of the differential input of an OP AMP lies in the desirable feature of amplifying a small difference between two large, but nearly equal, signals. This is accomplished by *symmetrical circuitry*, obtained by using both the inverting (−) and non-inverting (+) inputs of the amplifier, as shown in Fig. 11.6. A good insight into the working of this symmetrical circuit[1,2] can be gained by regarding it as a combination of a unity-gain inverter [with e_1 applied to the inverting (−) terminal] and a voltage follower [where e_2 is applied to the non-inverting (+) terminal]. When all four resistors have the same value (R), the output of the circuit (e_o) is unity times the input difference:

$$e_o = 1(e_2 - e_1), \qquad \text{with generally negligible } E_{CM}.$$

An analysis of the circuit by superposition is instructive to show how

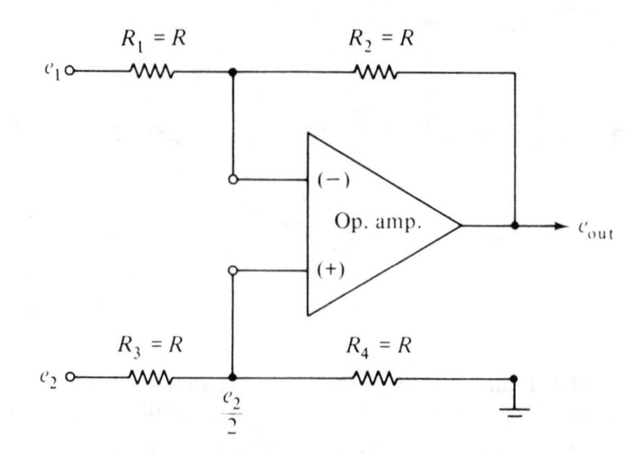

FIG. 11.6 Symmetrical arrangement of instrumentation amplifier, using differential (double-ended) input connection; the output (e_o) is an accurate representation of only the difference between the input voltages ($e_2 - e_1$) with excellent, common-mode rejection ratio (CMRR).

[1] A delightful analysis of this circuit is given (even including humor) in J. I. Smith, *Modern Operational Circuit Design*, John Wiley & Sons, Inc. (Interscience Division), New York, 1971, Chapter 5.

[2] This circuit and its variations are also well analyzed in G. E. Tobey, L. P. Huelsman, and G. G. Graeme, *Operational Amplifiers*, McGraw-Hill Book Company, New York, 1971.

the symmetrical circuit produces an output directly proportional to the input difference. With all four resistors equal to R, we have the following two conditions for superposition:

First, for $e_2=0$, the output is for an inverting amplifier, which equals

$$-e_1\frac{R}{R}=-e_1.$$

Second, for $e_1=0$, the output is for a non-inverting voltage follower with a gain of 2 [or $1+(R/R)$], acting on $e_2/2$ (or half the input signal), making this output equal to

$$2\frac{e_2}{2}=+e_2.$$

Then, by superposition, adding the two outputs,

$$e_o(\text{overall})=e_2-e_1.$$

The action of the common-mode voltage on the symmetrical circuit is shown on the right side of Fig. 11.5 to simulate the same undesired signal $(E_1+E_2)/2$ on both inputs. Under this condition, the high value of CMRR allows the common-mode voltage to swing within fairly wide limits, while the output magnifies only the difference E_1-E_2 by the differential gain (A_d). Thus, even if the differential gain is only unity, the symmetrical circuit still has a great advantage over a single-ended configuration; in rejecting the unwanted common-mode voltages, it retains the important property of high CMRR. Moreover, this concept is very useful in *computation circuits* serving as *subtracter* and also as *adder–subtracter* functions.

While the symmetrical concept is highly attractive in theory, the simple circuit of Fig. 11.6 *does not work out* to be equally good for a practical instrumentation amplifier. The necessity of working with four accurately matched resistors makes the calibration for a given gain quite difficult. For a gain (K) greater than 1, it is necessary to keep resistance ratios carefully matched to satisfy the relation

$$\frac{R_2}{R_1}=\frac{R_4}{R_3}$$

Bridge Amplifiers

An additional problem presents itself when the input voltages e_1 and e_2 are derived from an unbalanced bridge. This arises from the relation previously given for the output of a bridge which is unbalanced by a resistance change in the active transducer that forms one arm of the bridge, $R+\Delta R$.

This output is linear for only small fractional deviations ($\Delta R/R = \delta$), as seen in the formulas expressed in terms of

$$\text{exact} \quad e_{\text{out}} = \frac{E_R}{2}\frac{\delta}{2+\delta},$$

$$\text{approx.}\ e_{\text{out}} \approx \frac{V}{4}\,\delta.$$

As a rule of thumb, for a given deviation, say $\delta = 5$ percent, the discrepancy in the linear output given by the approximate formula is about half (or error $= 2.5$ percent).

A circuit that overcomes this limitation on wide deviations is shown in Fig. 11.7, where the output remains linear with the deviation δ. This circuit [1] is drawn to show its relation to the traditional bridge, and its output, a *linear readout* of the resistance deviation δ, is given as

$$e_{\text{out}} = \frac{E_R}{2}\,\delta,$$

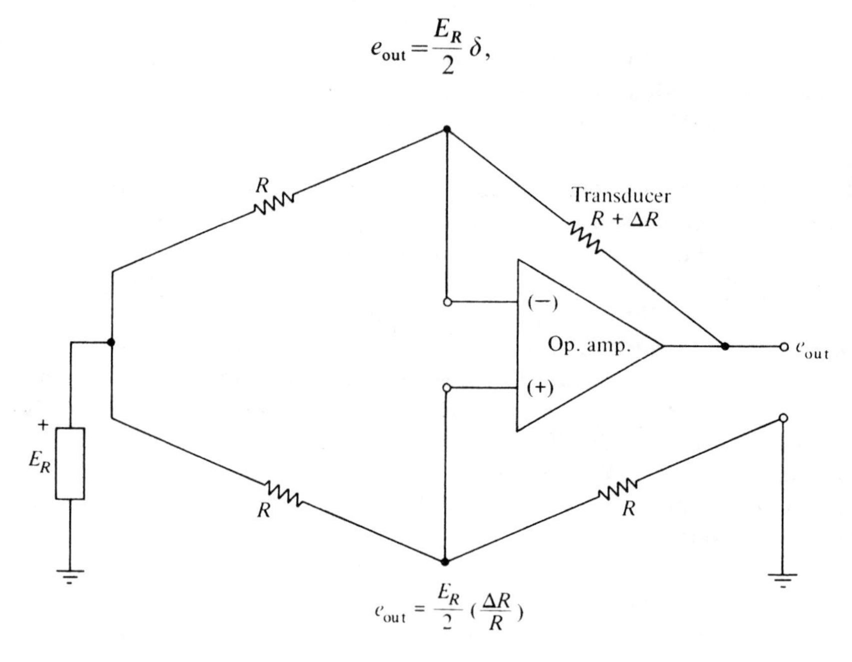

FIG. 11.7 *Linear* bridge output circuit, where the differential input connection to the OP AMP produces a bridge output directly proportional to the fractional deviation ($\delta = \Delta R/R$) of the transducer element ($R + \Delta R$).

[1] Smith, *op. cit.*

which is an exact form, as compared to the preceding similar-looking formula, which is only approximate for the traditional bridge.

Another trouble with the simple version of the symmetric difference amplifier of Fig. 11.6 is that it *lacks the high-resistance inputs* that are desirable in an instrumentation amplifier. One possible method of overcoming this limitation would be the use of two additional OP AMPS to serve as high-input-impedance voltage followers feeding the third OP AMP connected as a difference amplifier. This basic idea is shown in Fig. 11.8(a), with the resistance ratio for gain (K) obtained by the ratio KR/R.

This theoretical method still leaves us with the difficulty of tailoring the gain (preferably variable) by the awkward need for changing the exactly matching ratios of four transistors, with the added chances of changing the CMRR value as the gain is changed.

Variable Gain

The methods for obtaining variable gain in an instrumentation amplifier are necessarily different from the simple arrangements that can be employed in a general-purpose OP AMP. For general use, the gain can ordinarily be changed by simply changing the feedback resistor (R_f) to change the gain-determining ratio R_f/R_i. However, it is well to note that this easy method has a severe limitation in some cases, owing to the fact that the larger values of R_f produce correspondingly larger values of voltage offset, caused by the increased drop (I_bR_f) as R_f increases. This becomes more evident in dealing with a circuit where the signal source, for example, requires a fairly large input resistor R_1, of, say, 10 kΩ, and we wish to increase the gain to 100. This would require making R_f equal to 1 MΩ, which might easily cause an unacceptably high voltage offset.

A useful method for avoiding undesirably high values of R_f is shown in Fig. 11.8(b), where the feedback connection for R_f is taken from a voltage divider across the output. In the example, where we have set R_1 equal to 10 kΩ, we can use a lower value of 100 kΩ for R_f and still obtain a gain of 100 by connecting to the tap on the voltage divider, which supplies only one tenth the output to R_f. In this case, the gain is obtained by the R_f/R_1 ratio of 10 multiplied by the reciprocal of the voltage-dividing ratio $(R_2+R_3)/R_3$, also 10, to give the desired gain of 100.

This method illustrates a *general rule of thumb that a voltage-dividing network in the feedback path produces the opposite effect of such a network in the forward path.* This relation can be intuitively verified by regarding a smaller feedback voltage as equivalent to a larger resistor in determining the amount of feedback current.

With regard to the instrumentation amplifier, we find that neither of the preceding gain variations is feasible because of the necessity of maintaining

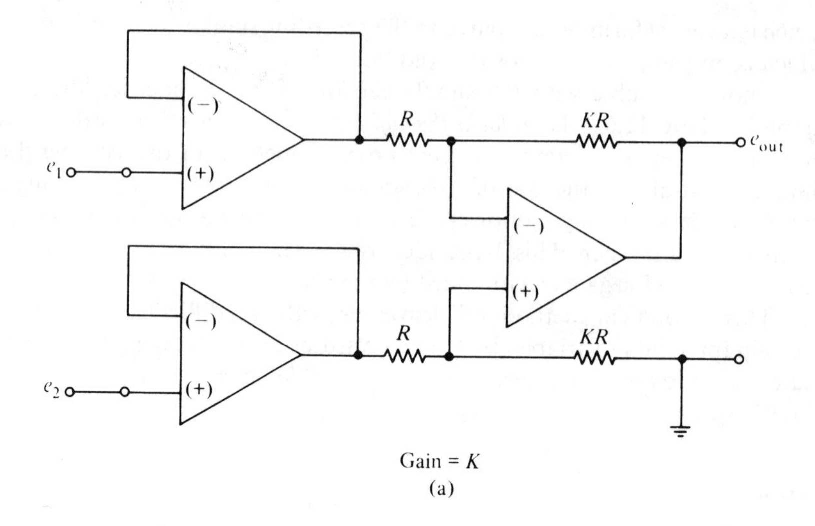

Gain = K
(a)

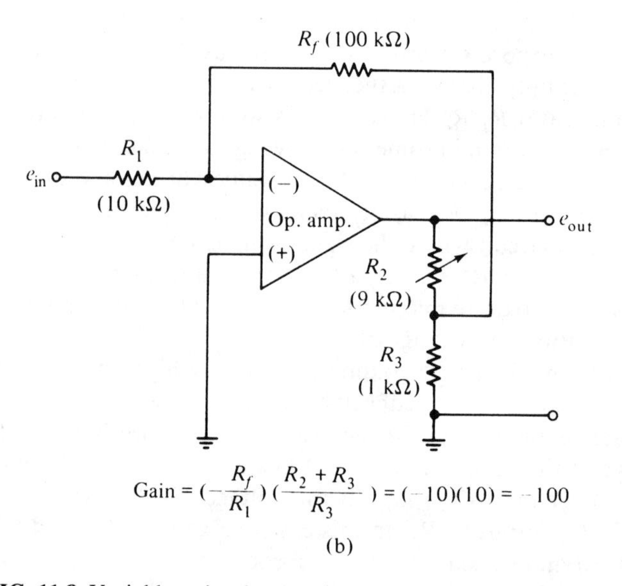

$$\text{Gain} = \left(-\frac{R_f}{R_1}\right)\left(\frac{R_2 + R_3}{R_3}\right) = (-10)(10) = -100$$
(b)

FIG. 11.8 Variable gain circuits: (a) symmetrical arrangement for an instrumentation amplifier vs. (b) an inverting OP AMP circuit; the use of three amplifiers in (a) is one method of preserving high Z_{in} and CMRR, despite changes in gain.

accurate resistance ratios to preserve the ability of the amplifier to reject common-mode voltages.

Practical Approach to Instrumentation Amplifier

A different approach[1] to accomplish practically all the desired objectives for an instrumentation amplifier, and to largely overcome the difficulty with variable gain, requires the use of only two-amplifier OP AMPS, as shown in Fig. 11.9(a). It permits the gain variation to be set by an external GAIN ADJUST (R_3) separate from the accurately matched resistors making up R_1 and R_2 that are required for the symmetric arrangement. This arrangement possesses the very desirable feature that the CMRR value is completely independent of the setting of the gain adjust R_3. As a result, with the use of precision ratios for R_2/R_1, the gain can be set and easily changed without degrading the amplifier performance.

The gain formula, along with the graph of increased gain as the resistance of R_3 is changed to lower values, is shown in the plot of Fig. 11.9 (b). The curve is seen to be quite nonlinear. In the example, using a 9-kΩ rheostat at its full value for gain adjust R_3, the gain is seen to increase quite regularly as the resistance is lowered for a gain increase of 5 to 10, but there is a sharp nonlinear increase (up to a gain of 40) as the resistance is further decreased to a minimum value of 500 Ω. Accordingly, to obtain a wide range of gain variation, a tapered (or vernier type) of variable resistor would be called for.

It can be seen from the previous discussion that many strict characteristics are required for an instrumentation amplifier, and it should therefore be no surprise that the resulting form of this type of amplifier would tend to be more advanced than a single op amp, even of the precision variety. This, in fact, is the proud claim of many of the instrumentation amplifiers that have appeared in *module form*. It is also true in the case of the *monolithic form* of the representative example discussed in the next section, where the manufacturer's initial description emphasizes the fact that it is not an OP AMP, and that, although it can be made to operate in a similar fashion, it has measurement properties beyond those found in any single OP AMP.

11-10. REPRESENTATIVE INSTRUMENTATION LINEAR-INTEGRATED-CIRCUIT AMPLIFIER (MONOLITHIC)

In monolithic integrated-circuit form, *Analog Devices AD520* represents an instrumentation amplifier that includes two OP AMPS and associated circuitry in its internal closed-loop gain block, all in a 14-lead DIP package.

[1] Smith, *op. cit.*

The circuit configuration is shown in Fig. 11.10 (a), accompanied by the pin configuration in part (b). Here it will be noted that provision is made for connecting an external resistor (R_{gain}) and another for R_{scale}, with the closed-loop gain thus determined for any desired value from 1 to 1,000 by the ratio

$$A_{VCL} = \frac{R_{scale}}{R_{gain}}.$$

A most important result for this arrangement is the fact that the *gain adjustment does not degrade either the input impedance* (Z_{in}) *or the rejection ability* (CMRR), two critical factors in an instrumentation amplifier. Thus the *AD520* performs like a modular (or hybrid) instrumentation amplifier, even though its monolithic form gives it the appearance of an OP AMP.

As an instrumentation amplifier, it should not be confused with the class of OP AMPS that also provides the feature of variable gain setting by an external resistor (such as the operational-transconductance-amplifier mentioned previously). As examples of the difference, in the *AD520*, both inputs are at high input impedance (2,000 MΩ), and it also bears repeating that *the* Z_{in} *and CMRR remain high at all gain settings.* (The obstacles in the way of achieving these properties with any single OP AMP were described in the previous section.)

Additional features, also shown in Fig. 11.10, include the optional use of the REF and SENSE terminals. The *output reference terminal* (REF) allows the output offset to be adjusted independently of gain setting, and the high-impedance *sense terminal* (SENSE) allows the circuit's feedback to be derived from either the output terminal or an arbitrary external point.[1]

Using the REFERENCE Terminal

An application where the amplifier output is connected to a recorder is shown in Fig. 11.11 (a). In many cases the input (V_{in}) is from a transducer bridge, and it is desirable to position the pen of the recorder at a desired (or set point) part of the chart; this is accomplished by connecting the arm of an external voltage divider to the REF terminal. (Ordinarily, the REF terminal is grounded.)

Using the SENSE Terminal

In ordinary use, the SENSE terminal is connected to the output, and it thus provides the internal feedback link. In Fig. 11.11 (b), this terminal is used to

[1] These and other instrumentation considerations are comprehensively discussed in the eight-page data sheet for the Analog Devices AD520 J/K/S (see Appendix IV for addresses).

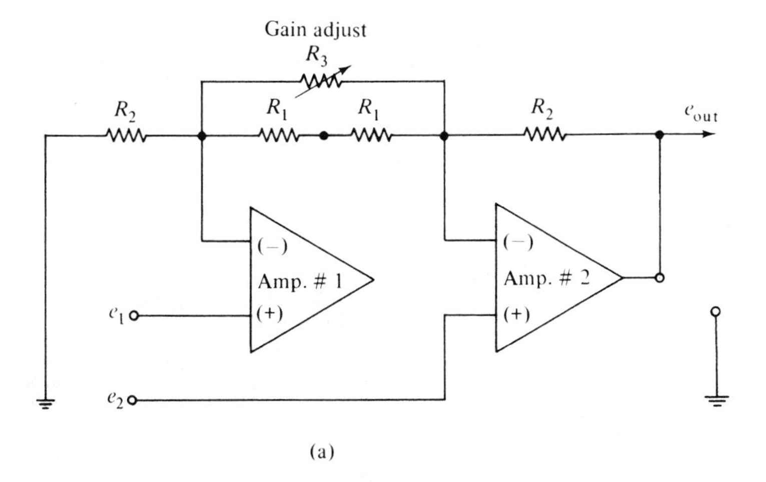

(a)

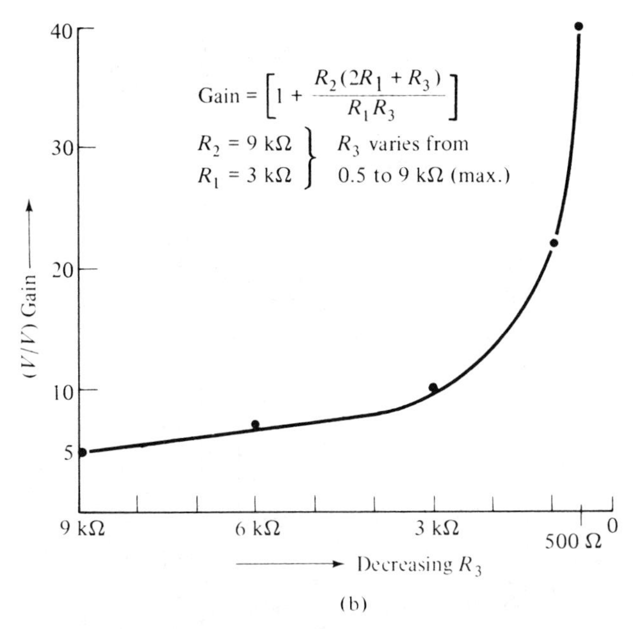

(b)

FIG. 11.9 Condensed version of instrumentation amplifier circuit, using only two amplifiers: (a) the gain adjustment of external resistor R_3 does not affect the accuracy or CMRR, which depends on close-matching of external resistors R_1 and R_2; (b) gain variation.

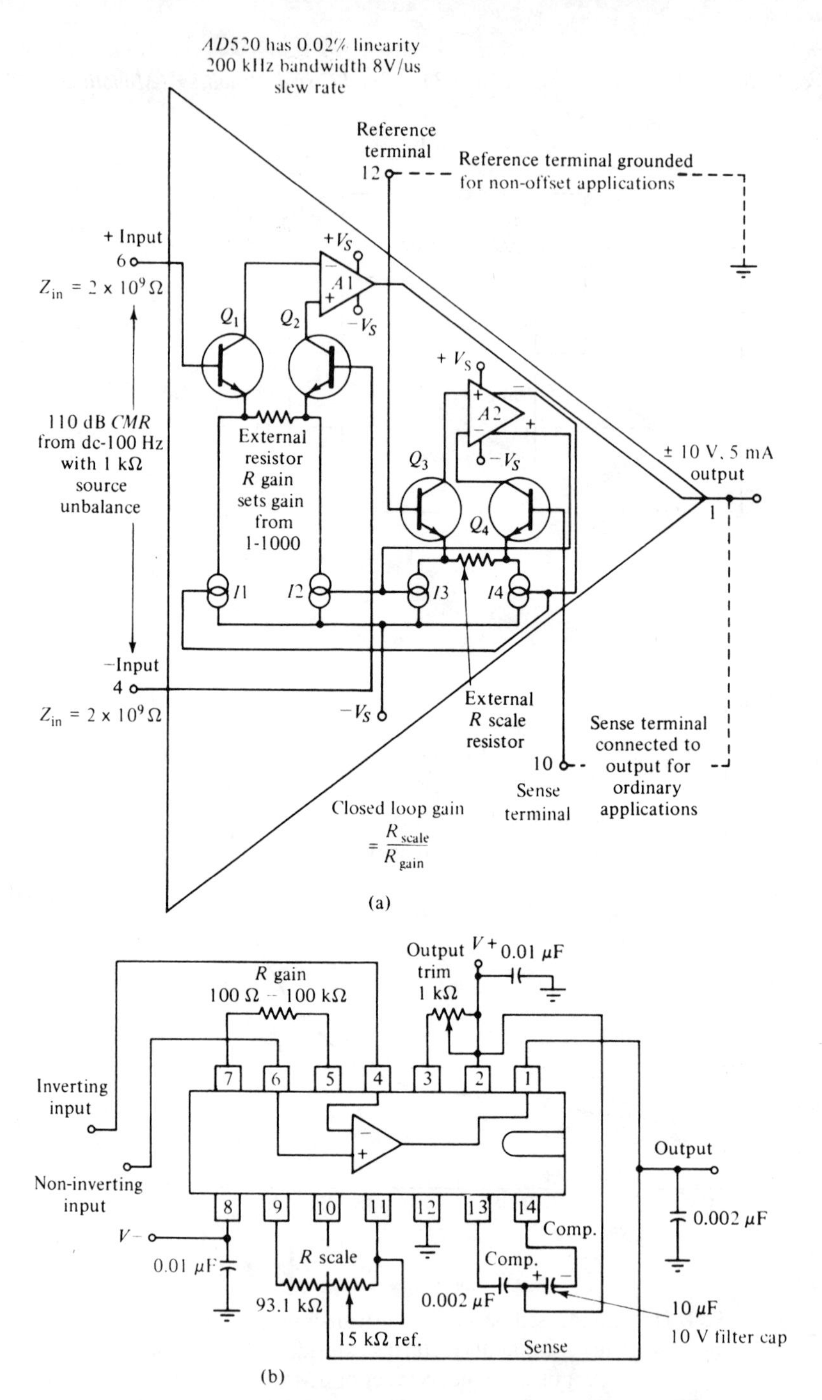

FIG. 11.10 Representative instrumentation amplifier AD520: (a) circuit configuration shows the two internal amplifiers with provision for external resistors R_{gain} and R_{scale}; (b) pin configuration of the 14-pin DIP, showing additional terminals for SENSE and REF connections used in the applications of Fig. 11.11. *(Analog Devices)*

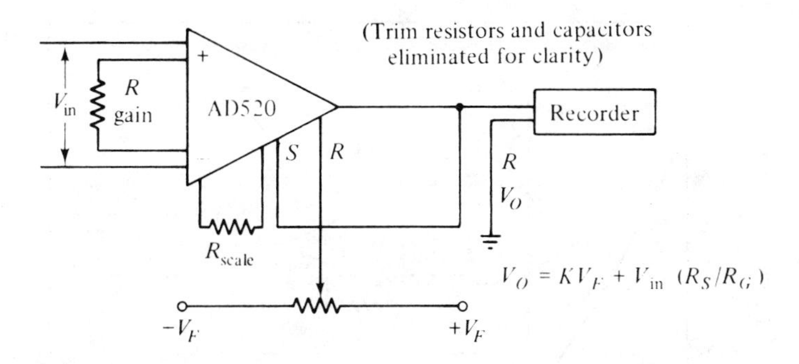

(a) Using reference (*R*) terminal

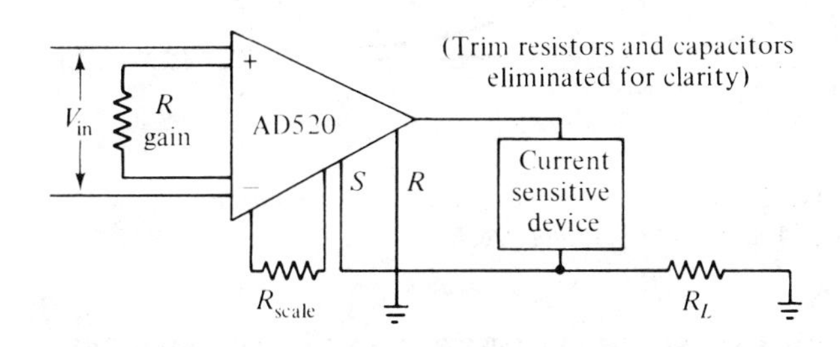

(b) Using sense (*S*) terminal

FIG. 11.11 Applications using AD520 optional terminals: (a) use of REF (R) terminal to position recorder pen; (b) use of SENSE (S) terminal to stabilize current through a floating component, such as a solenoid. *(Analog Devices)*

sense a voltage across a given load that is driven through a current-sensitive device. The current in this floating device, such as a sensitive relay or solenoid, is kept equal to the V_{out}/R_L, where V_{out}, as before, is $V_{in}(R_{scale}/R_{gain})$.

Noise

Low noise is obviously an important property in examining specifications for instrument amplifiers; it is generally given in the data sheet in the form of a graph as noise voltage (referred to input), as $mV/\sqrt{Hz}$, over a range of frequencies. This noise voltage is shown for the *AD520* in Fig. 11.12, reading in millivolts for unity gain and in nanovolts for a gain of 1,000.

Circuit applications for minimizing noise interference are discussed in Chapter 12.

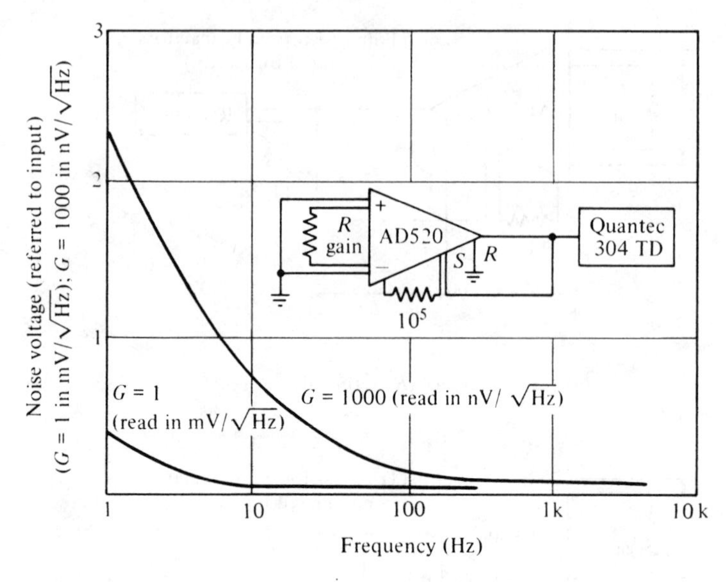

FIG. 11.12 Noise voltage graph, for instrumentation amplifier gains of 1 and 1,000. *(Analog Devices)*

11-11. MODULAR FORM OF INSTRUMENTATION AMPLIFIER

Most instrumentation amplifiers are packaged in modules of varying complexity. A representative modular example, *Teledyne/Philbrick model 4253*, is shown in block form in Fig. 11.13 for a low-drift FET instrumentation amplifier. The functional block diagram shows the use of five internal amplifiers (including an FET preamp), with the use of external gain resistor R_G providing an adjustable gain of 1 to 5,000, according to the formula

$$\text{gain} = 1 + \frac{100\ \text{k}\Omega}{R_G}.$$

Trim circuits for the reference and sense terminals (discussed in the previous section) are also shown.

Specifications for offsets are given in the data sheet[1] for three temperature ranges, because of the FET leakage-current dependency on temperature (2 × /10°C); also, noise voltage is given over three frequency ranges.

[1] Teledyne/Philbrick data sheet (see Appendix IV for address).

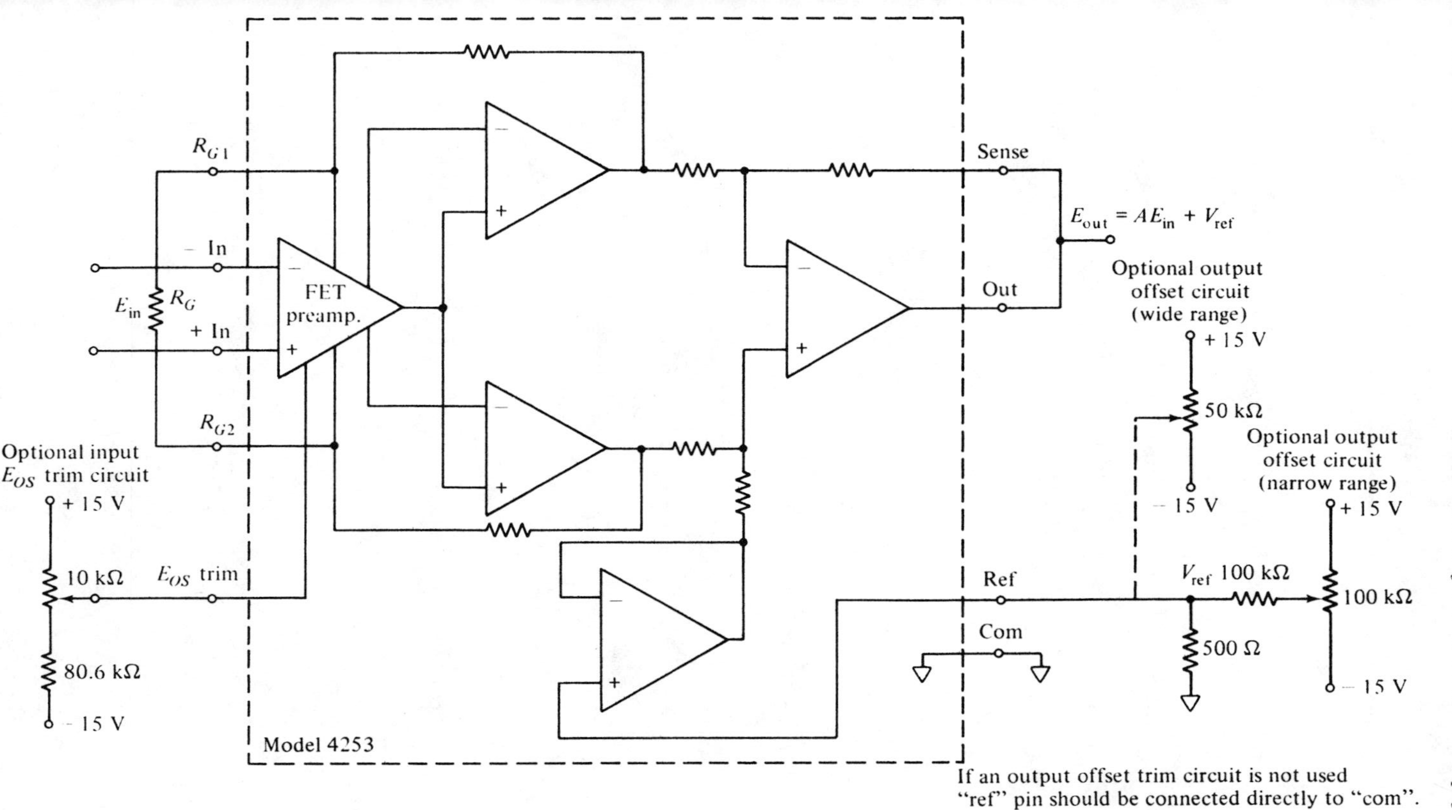

FIG. 11.13 Modular instrumentation amplifier, type 4253: in a functional block diagram form, showing FET-input preamplifier and four other OP AMPS, plus offset, reference and sense options. *(Teledyne/Philbrick)*

12

SPECIALIZED LINEAR-INTEGRATED-CIRCUIT APPLICATIONS

12-1. EXPANSION OF LINEAR-INTEGRATED-CIRCUIT FUNCTIONS

In the developments that continue to take place in both the design and the fabrication of linear ICs, we find newer capabilities appearing beyond the more or less "standard" functions. The previous chapters have indicated improvements that have been made in recent years in providing higher input impedances (as in the FET-input types), greater output-current ratings, and greatly refined specifications for improving the accuracy performance of instrumentation amplifiers.

A host of advanced application designs have been keeping pace with the device improvements, producing a large variety of new ways of using LICs. This chapter presents a sampling of selected specialized applications. In some cases the availability of these OP AMPS produces a simpler and more effective circuit, as in *active filters*, *electrometers*, and techniques for *low noise* in amplifiers of low-level signals. In other instances we find that the "linear" IC can be very profitably employed in nonlinear applications, such as *logarithmic amplifiers* and *function generators*. In addition to such selected circuits, a system application is presented in this chapter for a *digital voltmeter*

(DVM), which emphasizes the effective use of linear ICs as building blocks, thus serving to greatly simplify the design of an entire system.

One advanced type, an *op amp that can be programmed to fit a specific application*, offers an important concept in extending the flexibility of the OP AMP as a building block. Examples of the use of such programmable OP AMPS are presented in the next section.

see p 198

12-2. PROGRAMMABLE AMPLIFIERS

This group of OP AMPS offers amplifiers whose characteristics can be altered over a wide range by the selection of an external resistor to set the bias current. This feature was previously introduced in Section 11-8 and in the circuit of Fig. 11.4, where the *CA3078* and *CA3080* were given as examples of an *operational transconductance amplifier (OTA)*. This is a term used by RCA to indicate the capability of varying the transconductance (g_m) from 2 to 20,000 µmho by changing the amplifier bias current (I_{ABC}) through the selection of the desired external resistance.

In the Fairchild model *µA776*, this external resistor is called R_{set} for producing a low bias current (I_{set}), thus allowing *micropower operation* from two battery cells (± 1.2 V), for specialized battery applications, as shown in Fig. 12.1 (a).

This flexibility of operation is also extended to *high-output-current applications* by the *programmable amplifier/power switch*, model *CA3094/A* (this type was previously mentioned in Section 9-8 as a power-control-switch device). Its use here as a *linear power amplifier* is shown in Fig. 12.1 (b). While itself capable of delivering 3-W average power, it is shown in this circuit as a driver stage for a 12-W amplifier circuit feeding a complementary output stage of power transistors. The programming pin (terminal 5) permits varied special applications such as strobing, squelching, controlling gain (AGC) and mixing for amplitude modulation.

12-3. THE OPERATIONAL AMPLIFIER AS A VOLTAGE SOURCE

A simple and effective voltage source for producing a variable regulated-voltage output, using a simple Zener diode, is shown in Fig. 12.2. This circuit makes use of a basic OP-AMP property, where the negative feedback acts to keep a constant voltage difference between the two input terminals. The current from the V+ supply through resistor R_z establishes the Zener-regulated voltage as a reference voltage (V_{ref}), and this is applied to the non-inverting (+) terminal of the OP AMP. The application of negative feedback from the output-voltage divider determines the ratio R_2/R_1, and the output

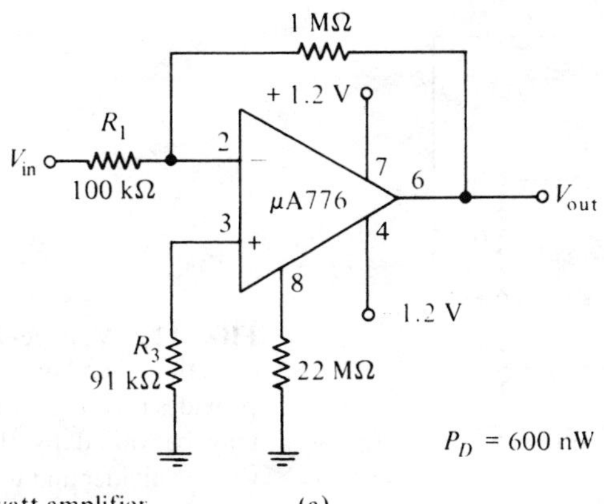

(a)

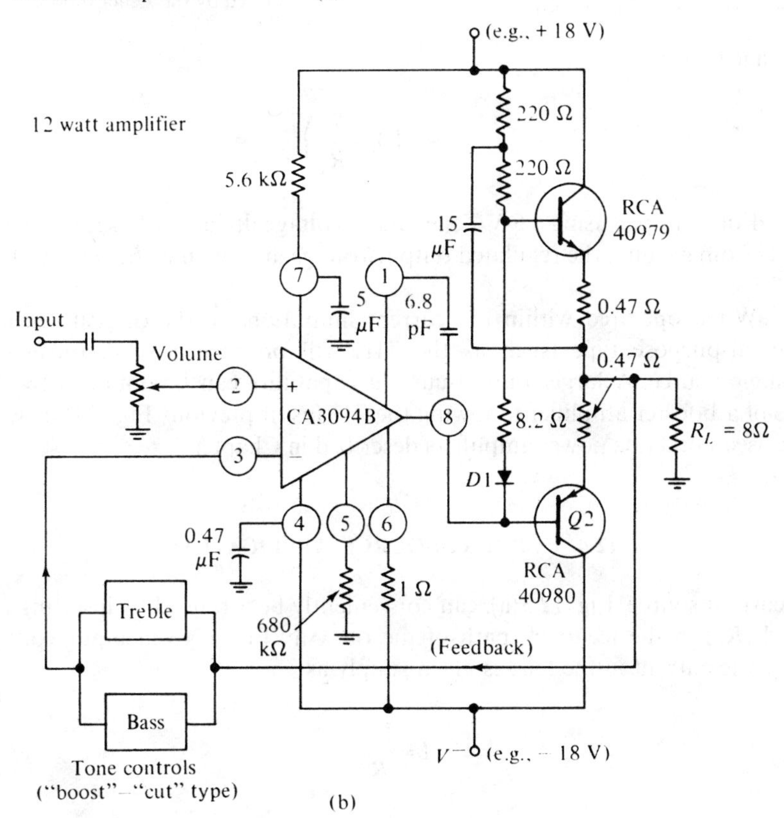

(b)

NOTE: Pin nos. apply to 8-pin package (either 8-pin metal can or 8-pin mini-DIP types).

FIG. 12.1 Programmable IC amplifiers: (a) nano-watt amplifier using the μA776 *(Fairchild Semiconductor)*, with setting resistor (R_{set})=22 MΩ; (b) programmable amplifier/power switch CA 3094 *(RCA)* in a 12-watt configuration with external transistors, and amplifier-bias-current resistor (R_{ABC})=680 KΩ.

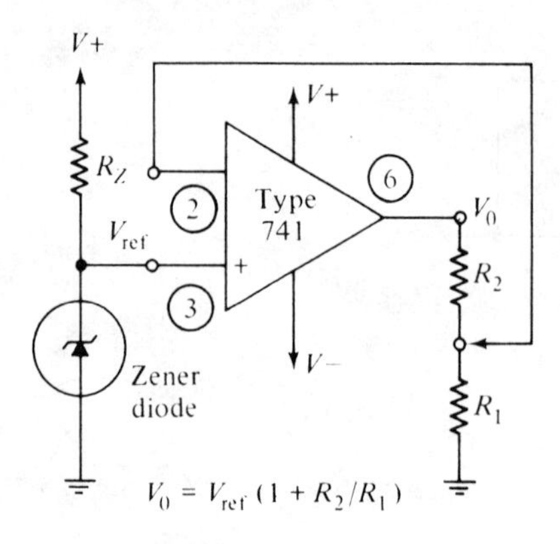

FIG. 12.2 Voltage-source circuit: the general purpose OP AMP provides a voltage source which may be varied by the output-voltage divider and which is regulated by the Zener diode.

voltage (V_o) is

$$V_o = V_{ref}\left(1 + \frac{R_2}{R_1}\right).$$

For example, using a 5-V Zener and a voltage divider of 15 kΩ tapped at 5 kΩ from ground, the regulated output voltage at V_o would equal 5 V (1 + 2), or 15 V.

When operated within the current limitations of the OP AMP, even a general-purpose type (such as the 741) will produce a very convenient voltage source. A larger output-current capability can be obtained by the use of a booster amplifier following the 741 (as in previous Fig. 7.1) or with the use of other dc power amplifiers described in Chapter 7.

12-4. CURRENT-SOURCE APPLICATION

A current source, Fig. 12.3 (a), can conveniently be arranged by inserting the load (R_L) in the feedback path of the OP AMP. For a given input voltage (V_{in}), the current in the load is given simply as

$$I = \frac{V_{in}}{R}.$$

A precision current-source application[1] is illustrated in Fig. 12.3 (b). Here the output of the OP AMP feeds an FET (as a source follower) to drive a bipolar output transistor, produces the output current I_o. A Darlington connection can be used in place of the FET bipolar combination in cases

[1] *Linear Applications*, a manual by National Semiconductor (see Appendix IV for addresses).

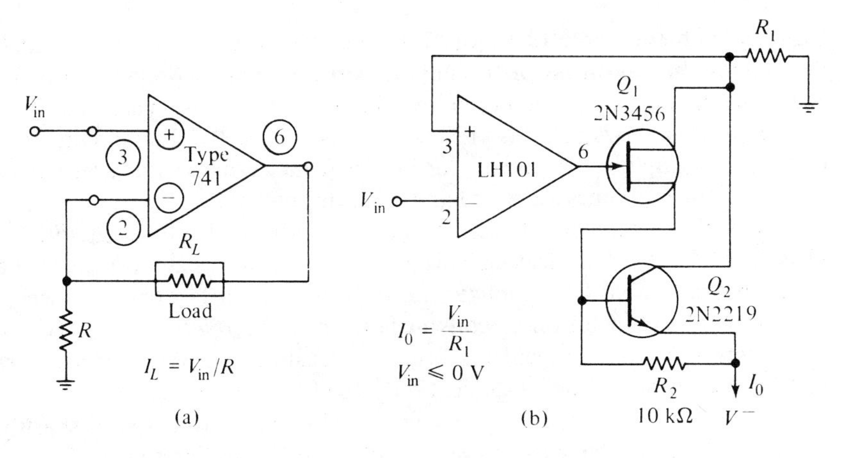

NOTE: Pin nos. apply to 8-pin package (either 8-pin metal can or 8-pin mini-DIP types).

FIG. 12.3 Current-source circuits: (a) simple arrangement for constant-current (I_L) through load; (b) precision current source, using LH101 *(National Semiconductor)* and external transistors (see text).

where the output current is high and the small base current of the Darlington input would not cause a significant error. For negative V_{in}, the output current (I_o) is again

$$I_o = \frac{V_{in}}{R_1}.$$

The OP AMP used here, the *LH101* is internally compensated for unity gain (as is the 741 type). If extended-bandwidth types are used (such as the LM**101**/A or μA**748**), they would employ an external compensation capacitor for this purpose (usually 30 pF for the types mentioned).

The impedance of this precision current source is essentially infinite for small currents, and is accurate for the usual cases where V_{in} is much greater than the offset voltage, and I_o is much greater than the bias current (I_B).

12-5. ACTIVE FILTER APPLICATIONS

In classical filter theory, the *passive components* of resistance, inductance, and capacitance (*RLC*) are traditionally used in network analysis for theoretical explanation and laboratory verification. This practice is educationally sound in forming clear concepts of frequency-selective circuits; it gives a good picture of how the increasing impedance with frequency of the coil *L* is effective in low-pass filters, along with the opposite action (decreasing impedance of capacitors *C*, as employed in high-pass filters). These factors, together with the value of Q (the ratio of reactance to resistance, which determines the sharpness of tuning), all form a solid theoretical basis for analyzing filter circuits, whether of low-, high-, or band-pass types.

In practice, however, it is preferable to use active RC filter circuits that make use of amplifiers, especially for low frequencies, where the large size of the inductance becomes awkward and often impractical.

With the convenient IC as the amplifier in an active filter circuit, the physical difficulty of fabricating coils on a chip remains as an obstacle to the use of the traditional RLC configurations. It thus becomes a practical necessity to avoid the inclusion of coils in favor of *RC circuits that use ICs in active filters.* Such arrangements are both feasible and very effective, by making use of the flexible amplification provided by OP AMPS.

The objective of producing an effective filter circuit (one that is sharply tuned and does not have excessive losses) and still avoiding the use of coils can be accomplished in two basic ways:

1. By the use of a *frequency-selective RC network in the feedback path of an amplifier;* this approach is discussed next.
2. By the use of a *gyrator circuit,* which replaces the L of the coil in the classical RLC circuits; this approach will be discussed in Section 12-6.

Passive Low-Pass RC Filter (Lag Network)

The passive low-pass RC filter of Fig. 12.4 (a) shows one section to consist of simply a resistor (R) in series with the load, and a capacitor (C) across a hypothetical load. The cutoff frequency ($f_c = 16$ kHz) at which the output of the filter is 70.7 percent of its dc value (or -3 dB down) is obtained from its transfer function as

$$f_c = \frac{1}{2\pi RC}, \qquad \text{in hertz},$$

where R is in ohms and C is in farads.

For a simple first-order arrangement as the one shown, with a single RC product, this -3 dB point occurs at the frequency where the reactance of the capacitor (X_c) is equal to resistance (R) and results in the widely used straight-line approximation of the Bode plot (with slope equal to -6 dB/octave).

The frequency depends (inversely) on the values chosen for R and C, and the resulting RC product, as shown in **Table 5.1** (Sec. 5-11); thus the values of 2 kΩ and 0.005 μF (shown in the circuit of Fig. 12.4) gives an RC product of $10 \times (10^{-6})$ or 10 μs, and a cutoff frequency f_c or $f_o = 16$ kHz.

When this passive low-pass type of RC filter is applied as frequency compensation for an OP AMP, it is known as a *lag network.* The shunt capacitor

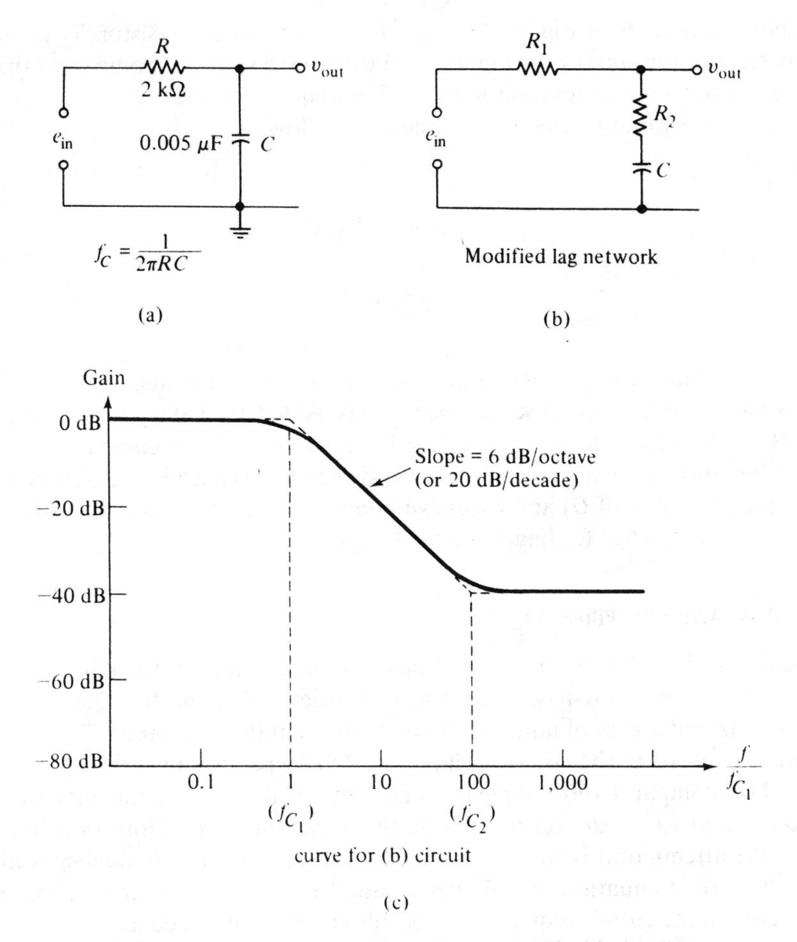

FIG. 12.4 Passive low-pass filters (lag network): (a) simple arrangement, with RC values for corner (or cutoff) frequency (f_C) of 16 kHz, with "roll-of" of 6 dB/octave; (b) modified arrangement with two corner frequencies f_{C_1} and f_{C_2}; (c) normalized frequency response curve for modified lag network of (b), with $f_{C_2} = 100\, f_{C_1}$.

called for in the lag network is sometimes self-contained in the OP AMP, as is the case of the *internally compensated 741* (and equivalent) types.

Modified Lag Network

In many other cases the lag network is made from external components. This is particularly the case where a modified lag network is used to shape the response curve, as in Fig. 12.4 (b) and (c). By adding the resistor R_2 to the capacitor in the shunt position, the -6 dB/octave slope is modified so that the response tends to level off at the higher frequencies. With the addition of R_2, two corner frequencies are obtained,[1] as follows:

$$f_{c1} = \frac{1}{2\pi(R_1 + R_2)\,C},$$

$$f_{c2} = \frac{1}{2\pi R_2 C},$$

To achieve more effective filtering, say, to reach as much as -15 dB down at 32 kHz, a second section needs to be added, providing an additional 6 dB for the octave, to produce a 12-dB/octave slope. When compared to an *LC* filter, this *RC* arrangement reveals its defects as having a lack of sharpness (a very low value of Q) and excessive attenuation (by the resistive voltage-divider action) when feeding a practical load.

Low-Pass Active RC Filter

The active circuit shown in Fig. 12.5 (a) fulfills a requirement for a maximally flat (Butterworth) low-pass filter of second order with a cutoff frequency (f_c) of 10 kHz and a gain of unity (zero attenuation) in the passband.[2] The unity gain is achieved in the voltage-follower *LM100* type OP AMP.

This comparatively simple arrangement has the filter characteristics of two isolated *RC* filter sections. Using the approximate relations of a Bode plot, the attenuation is roughly -12 dB at twice the cutoff frequency, with an ultimate attenuation of -40 dB/decade.[3] From the transfer function of this circuit, the cutoff frequency (f_c) of this circuit is obtained as[4]

$$f_c = 1/2\pi\sqrt{R_1 R_2 C_1 C_2},$$

and from this characteristic, the component values are determined as

[1] A. Barna, *Operational Amplifiers*, John Wiley & Sons, Inc., (Interscience Division), New York, 1971.

[2] National Semiconductor, LM110 data sheet.

[3] J. Eimbinder, ed., *Designing with Linear Integrated Circuits*, John Wiley & Sons, Inc., New York, 1969.

[4] Although the derivation of this (and more complicated) transfer functions is properly the task of texts in circuit analysis, a very straightforward derivation of this transfer function is given as an example in G. J. Deboo and C. N. Burrous, *Integrated Circuits and Semiconductor Devices*, McGraw-Hill Book Company, New York, 1971.

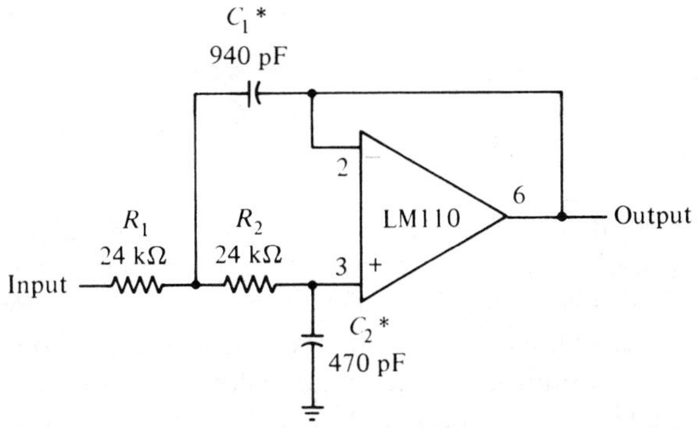

(a)

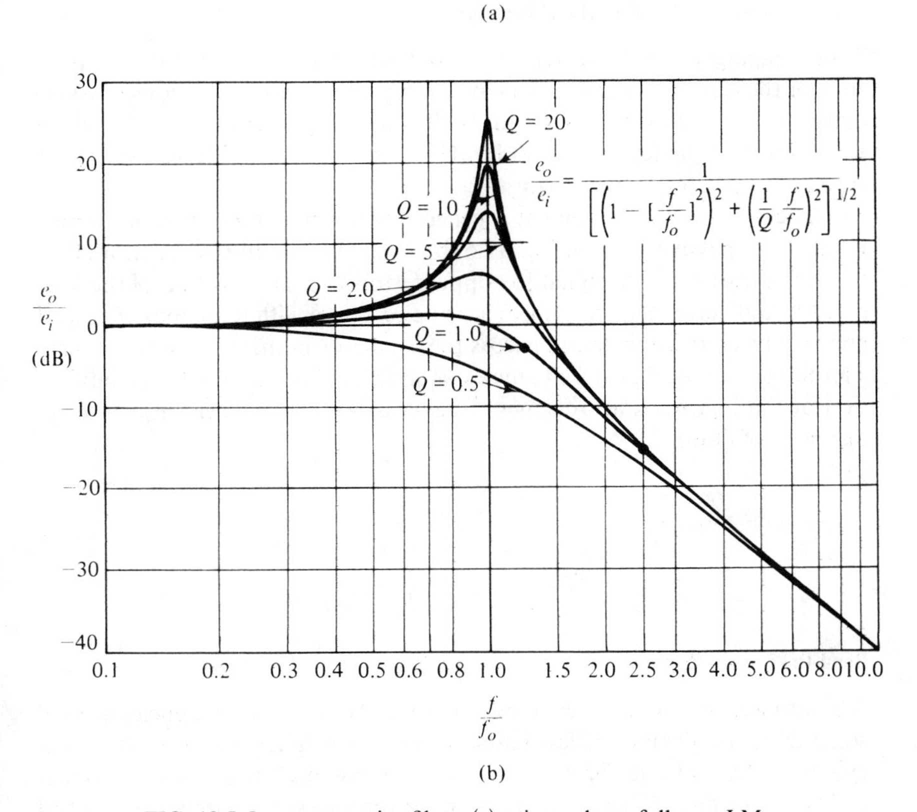

(b)

FIG. 12.5 Low-pass active filter: (a) using voltage-follower LM 110 type of OP AMP *(National Semiconductor)*, with values for high-frequency cutoff (f_C or f_o) of 10 kHz; (b) normalized frequency response plot for second-order low-pass active filter, showing effect of Q; where $Q = 0.7$ provides maximally flat response of Butterworth filter type. *(Beckman-Helipot Universal Active Filter, model 881)*

follows:

$$C_1 = \frac{R_1 + R_2}{\sqrt{2}R_1R_2(2\pi f_c)} \quad \text{and} \quad C_2 = \frac{\sqrt{2}}{(R_1 + R_2)\, 2\pi f_c}.$$

More complicated active filter circuits use more than a single OP AMP, along with multiloop feedback, to obtain desired characteristics, such as a higher Q. The *effect of the value of Q* on the low-pass function of a multiloop "universal" active filter (discussed in the next section) is shown in Fig. 12.5 (b). This is also a second-order filter action (slope of -40 dB/decade), and also indicates that a Q value of about 0.7 achieves the maximally flat characteristic.

Active High-Pass RC Filter (Lead Network)

A similar single-loop high-pass filter can be arranged by substituting capacitors for the resistors (and resistors for the capacitors) in the previous low-pass active filter. This high-pass circuit is shown in Fig. 12.6 (a) with the values for a cutoff frequency (f_c) of 100 Hz. Here again, the circuit uses the *LM110* type of voltage-follower OP AMP.

The high-pass arrangement is generally known as a *lead network*, which is shown in passive form in Fig. 12.6 (b). This puts the filter capacitor effectively in series with the signal [as opposed to the shunt position of the low-pass (or lag) network], and causes a rising response with frequency. The lead network has a resistor placed across the capacitor in this case, which causes a leveling off at the higher frequencies, as shown in Fig. 12.6 (c). This modification of the high-pass arrangement[1] again causes *two corner frequencies* to appear, as follows:

$$f_1 = \frac{1}{2\pi R_1 C},$$

$$f_2 = \frac{1}{2\pi C(R_1R_2/R_1 + R_2)}.$$

Band-Pass Active Filters

A band-pass filter is designed to pass a group (or band) of frequencies around some center frequency, while attenuating both the higher- and lower-frequency signals. One straightforward method for doing this is to *cascade a low-pass and high-pass filter*. This method has the advantage of attaining a very broad passband, and of separately tuning the high and low cutoff frequencies of the filter.

[1] Eimbinder, *op. cit.*

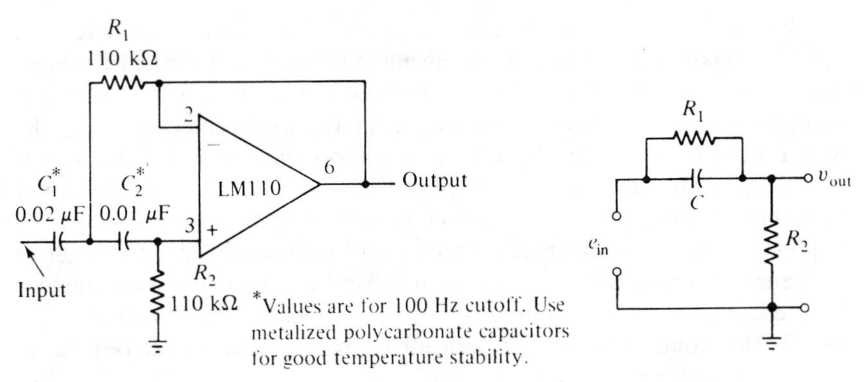

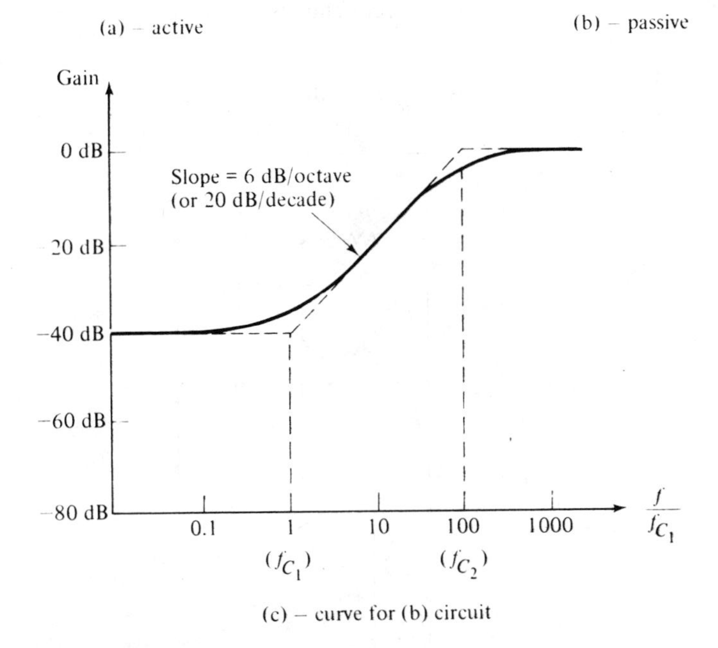

FIG. 12.6 High-pass filters (lead network): (a) active high-pass filter, where the position of resistors and capacitors of low-pass (Fig. 12.5) are interchanged, and values are for low-frequency corner of 100 Hz *(National Semiconductor)*; (b) passive high-pass (lead) network; (c) normalized frequency response curve for (b), with corner frequency f_{C_2} equal to 100 f_{C_1}.

For use with a single OP AMP in a single-loop feedback, the circuit of Fig. 12.7 makes use of a twin-T arrangement.[1] Analysis of the transfer function of the twin-T arrangement shows that currents in the upper T and lower T are equal and of opposite phase. Accordingly, at the center frequency the twin-T network in the feedback loop passes no current, and so the center frequency is not attenuated. For frequencies on either side of the center frequency, however, feedback current progressively increases, causing the response to fall off on either side of the center frequency. Even with the use of the general-purpose 741 OP AMP, the passband can be made quite narrow. (This effect is the opposite of the *notch filters*, which use the twin-T arrangement in the input, rather than in the feedback path, and hence can be made to have a very sharp rejection response.)

For either the pass or rejection function of the twin-T arrangement, the design formulation for the center frequency is the same:

$$f = \frac{1}{2\pi RC}, \quad \text{in hertz,}$$

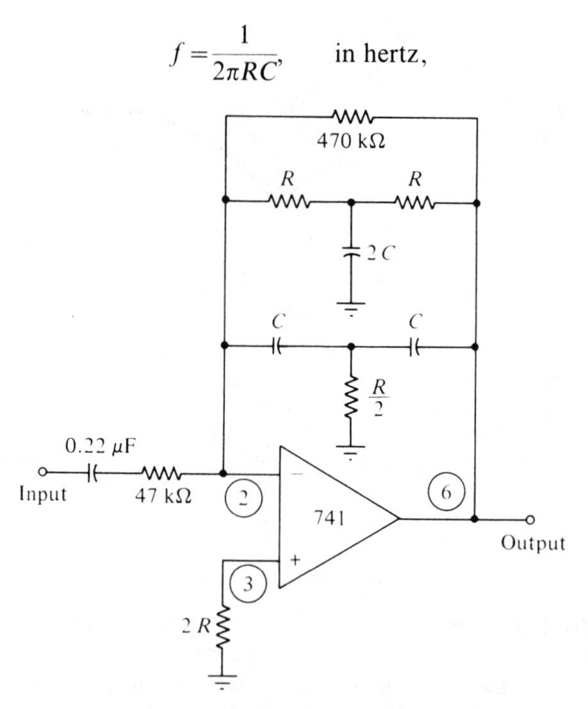

FIG. 12.7 Band-pass active filter: a single-loop feedback arrangement, using a twin-T as the feedback element.

[1] R. Melen and H. Garland, *Understanding IC Op Amps*, Howard W. Sams Company, Inc., Indianapolis, 1971.

The aid for determining component values in this formula by use of their *RC* product is given in Table 5-1 and also applies here.

12-6. GYRATORS AND Q-FACTOR IN ACTIVE FILTERS

In discussing the *RC* type of active filter, we should not neglect the massive storehouse of circuit-analysis theory that is based on the use of coils in the classical *RLC* filters. It is possible to *simulate the action of an inductor by a gyrator circuit*, which uses an active *RC* circuit to give the effects of coil action. The gyrator effect can be shown most clearly in the circuit of Fig. 12.8, where two op amps are used to provide a simulated inductor, given by the formula[1]

$$\text{simulated inductance } L = \frac{R_1 R_2 R_4 C_1}{R_3},$$

where L is in henries, R is in ohms, and C is in farads.

As a simple illustration of this action[2], if we assume that $R_1 = R_2 = R_3 = R_4 = 10\ \text{k}\Omega$ and $C_1 = 0.01\ \mu\text{F}$, we find

$$\begin{aligned} L &= CR^2 = 0.01(10^{-6})\ 100(10^6) \\ &= 1\ \text{H}. \end{aligned}$$

It should be noted that the inductance is simulated with one end of the input at ground, and therefore this kind of circuit cannot be used to simulate a "floating" inductance, where neither end is grounded. Another, more complicated method, is used, employing the concept of negative resistance to get around this difficulty, especially for low-pass active circuits, where the equivalent coil must be in a series position, and is therefore ungrounded.

Effect of Quality Factor (Q)

The sharpness of tuning in a filter circuit, especially for band-pass filters, is generally given by the relation for the -3-dB bandwidth, [BW(3 dB) = width between the -3-dB points on the low and high sides of the center frequency (f_c)] as follows:

$$\text{BW}(3\ \text{dB}) = \frac{f_c}{Q}.$$

[1] *Ibid.*

[2] The derivation of this action, beyond the scope of this book, may be found in Deboo and Burrous, *op. cit.*

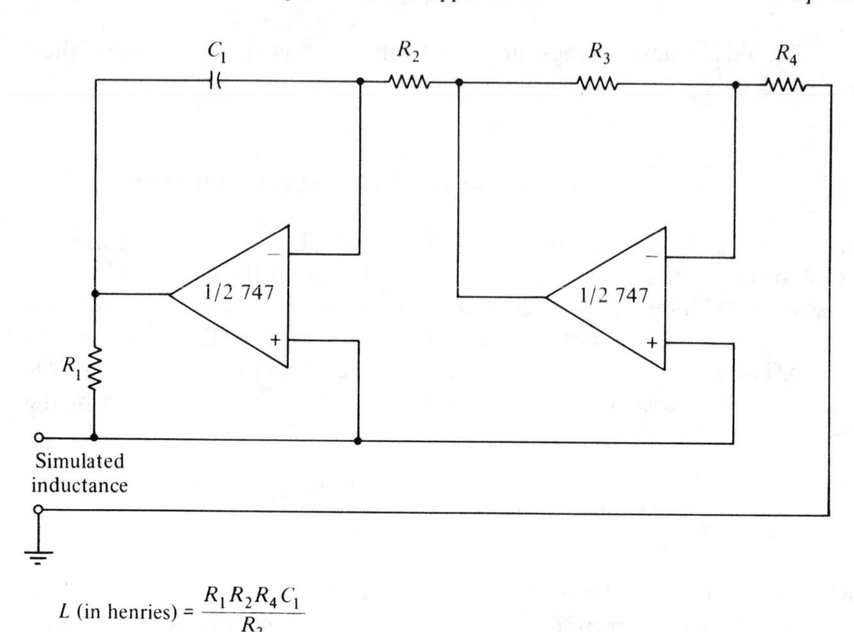

FIG. 12.8 Gyrator circuit: with four equal resistors (R), the simulated inductance (L) equals CR^2, in henries.

The effect of Q on the bandwidth of an active filter can be illustrated by an example of a universal active filter (*Beckman Helipot, model 881*). This hybrid circuit Fig. 12.9(a) uses three OP AMPS (K_1, K_2, and K_3) in a multiloop feedback circuit, and K_4 as an uncommitted 741 amplifier for gains greater than unity. Functionally, K_1 is a summing amplifier, and K_2 and K_3 are integrators, providing simultaneous low-pass, high-pass, and band-pass outputs. Two external resistors set the desired cutoff frequencies (f_o), while three other external resistors determine gain (A_{VCL}) and the Q of the overall filter. The effect of the selected Q on the bandwidth of the band pass filter is shown by the graph of Fig. 12.9(b) which is in normalized form for unity gain and center frequency (f_o). For sharp tuning, values of Q greater than 10 are generally used.

12-7. LOGARITHMIC AMPLIFIER

The "linear" OP AMP can be combined with a non-linear element (such as the base–emitter diode of a transistor) to achieve the very useful nonlinear *logarithmic function* (Fig. 12.10). In this circuit the *output voltage* (v_o) *is proportional to the logarithm of the input voltage* (e_{in}):

$$v_o \text{ varies as } \log(e_{in}).$$

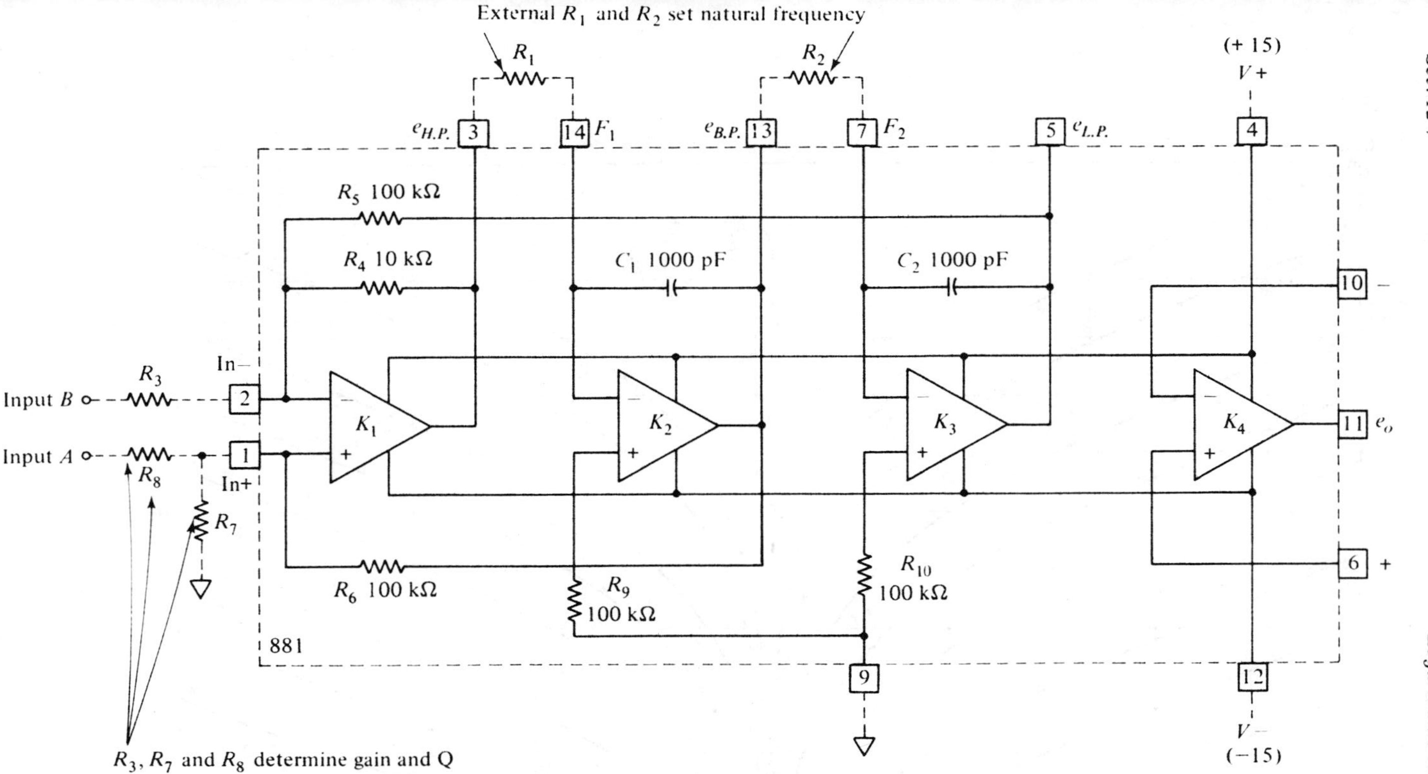

FIG. 12.9 Universal active filter, a hybrid, multiloop feedback circuit (second-order), uses three OP AMPS for simultaneous low, high, and band-pass outputs (K_4 is uncommitted gain amplifier); (b) bandwidth ($BW_{(3\,dB)}$) of band-pass, as determined by selection of the value for Q. *(Beckman-Helipot, model 881)*

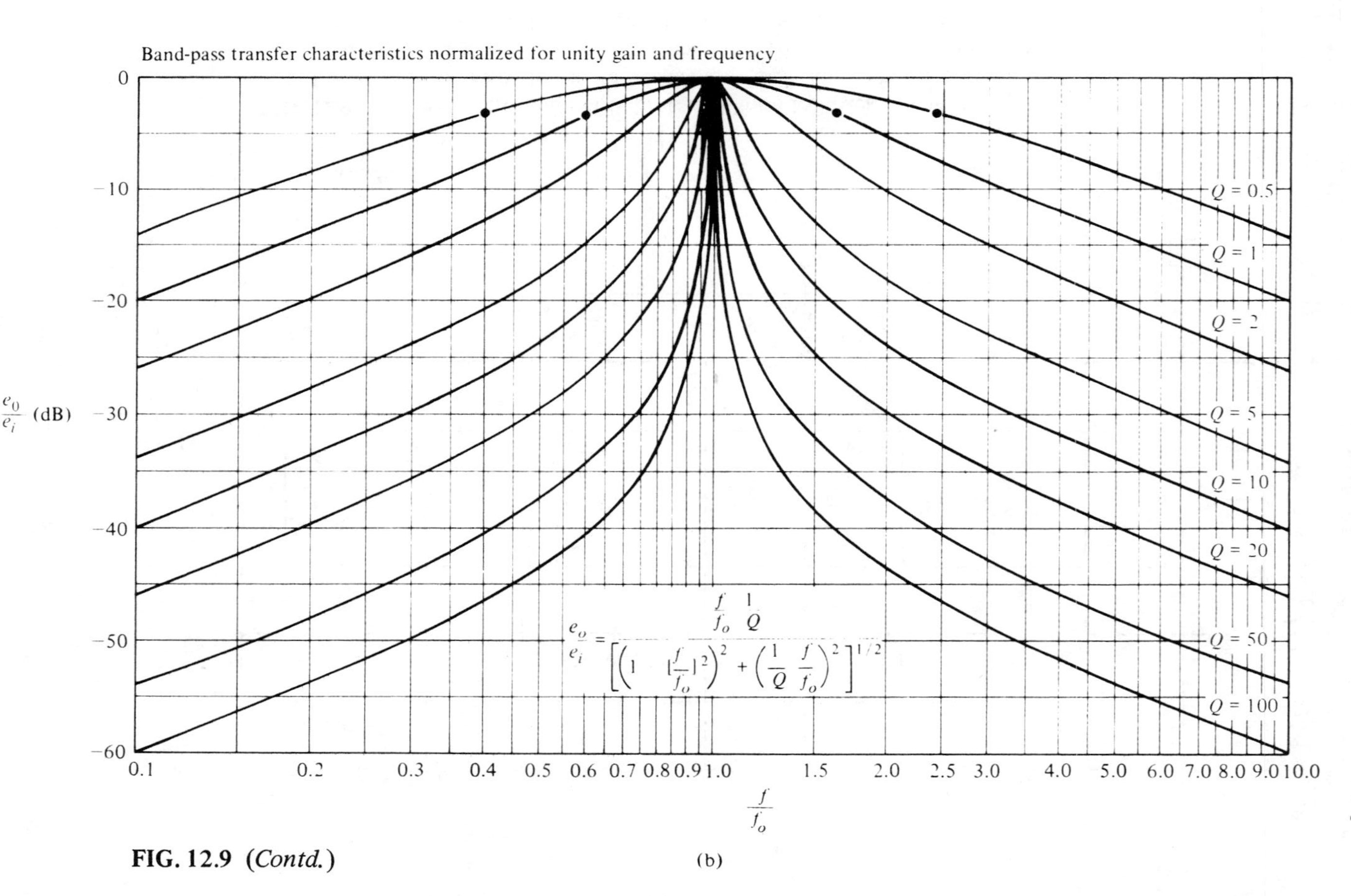

$$\frac{e_o}{e_i} = \frac{\frac{f}{f_o}\frac{1}{Q}}{\left[\left(1 - \left(\frac{f}{f_o}\right)^2\right)^2 + \left(\frac{1}{Q}\frac{f}{f_o}\right)^2\right]^{1/2}}$$

(b)

FIG. 12.9 ***(Contd.)***

As a result of this logarithmic relation, the output is greatly compressed, so the response of a meter across the output will act like a *decibel meter*; that is, the meter scale will cover a very wide dynamic range (such a scale might be −40 to +20 dB, a range of three orders of magnitude or 1,000:1).

The circuit action is based on the exponential relation of the base-to-emitter junction of the transistor (I_B versus V_{BE}) and the resultant feedback current that flows through the collector. Note that the transistor is employed in a common-base configuration, with the collector at a practically zero voltage, since it is tied to the virtual ground at the inverting terminal. Keeping in mind the fact that the OP AMP keeps the feedback current equal to the input current (e_{in}/R), we can express the output voltage ($v_o = V_{BE}$) as

$$v_o = K(\log e_{in}),$$

where the constant K includes the familiar exponential relation of diode leakage current (kT/q) and also the effect of the fixed resistance R.

The OP AMP used (*Intersil ICL8007*) is of the FET-input type, which is internally compensated (pin compatible with the 741 type) and which, in

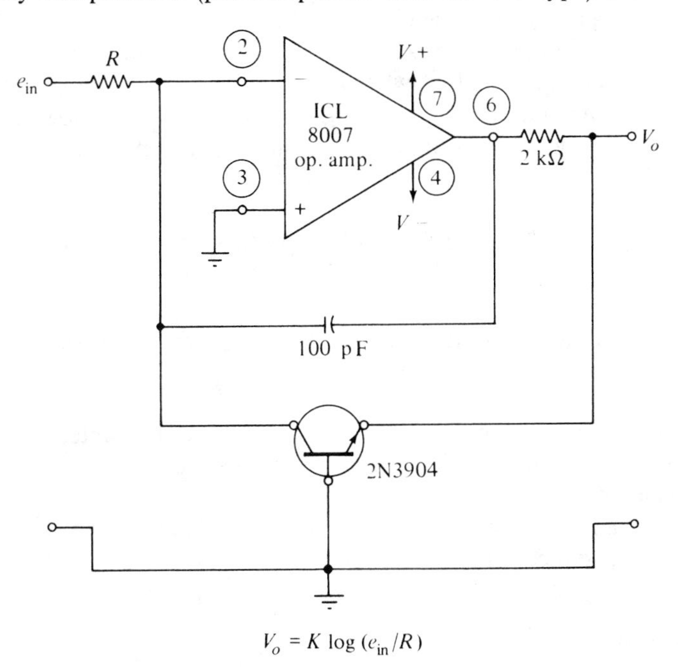

NOTE: Pin nos. apply to 8-pin package (either 8-pin metal can or 8-pin mini-DIP types).

FIG. 12.10 Logarithmic amplifier circuit: the ICL8007 OP AMP is of the FET-input type for high input impedance and has a high slew rate (6 V/μs) to handle the large dynamic range of input signals.

addition, has the desirable properties of a very high input impedance ($10^{12}\ \Omega$, or 1,000,000 MΩ) and an improved slew rate of 6 V/μs, thus allowing it to accommodate the wide range of input voltages that is necessary here.

For use as a decibel meter, the position of 0 dB on the scale can be arranged (by means of the offset-null provision) by applying an input voltage that corresponds to the 0-dB reference voltage (a much used reference level, the 0 dBm for 1 mW in 600 Ω, calls for an input of positive 0.775 V).

The logarithmic circuit is also often adapted to use as a *speech compressor*. Employed in this way, it overcomes a basic limitation of microphone amplifiers that are required to respond to both low- and high-level. In the presence of loud sounds, such amplifiers are prone to distort, or in the case of modulators, to overmodulate a transmitter. When the speech compressor is inserted between the microphone and its amplifier, the latter can be set to amplify a fairly low sound, and then, when a loud sound is applied, the output level is compressed so that it is not that much greater, thus preventing overdriving of the amplifier.

Another interesting application, functioning as a special kind of *analog multiplier*, involves the use of two log amplifiers into a summer, which then feeds an antilog amplifier. (The circuit for the antilog amplifier is made very simply by transposing the transistor into the input circuit and the resistor into the feedback circuit.) Then, when e_1 and e_2 are respective inputs to each log amplifier, the summer presents $\log e_1 + \log e_2$ to the antilog amplifier, which then produces the antilog of this sum, as the product of the two inputs[1]:

$$\text{antilog}(\log e_1 + \log e_2) = e_1 \times e_2.$$

12-8. LOW-NOISE AND HIGH-IMPEDANCE (ELECTROMETER) APPLICATIONS

The problem of extracting a weak signal from various unwanted interfering effects (generally lumped together as noise signals), requires the selection of special types of linear ICs. These include *low-noise types* in those instances where the signal is very weak, or *electrometer types* in instances where a very high input impedance is necessary.

Low-Noise Types

Of the many types of unwanted signals that interfere with the amplification

[1] This and other advanced multiplying functions are described in G. E. Tobey "Analog Modules," *Electronic Products*, Feb. 19, 1973.

of the desired signal, the rejection of *common-mode signals* has already been dealt with in Chapter 11. While an intensive discussion of many other types of noise are beyond the scope of this book, the practical aspect of selecting an OP AMP that minimizes *burst (or "popcorn") noise*, along with the "1/f noise," can be presented as a straightforward method[1].

For such noise reduction in the general-purpose 741 type, RCA offers premium type CA6**741**, which is tested for rigid low-noise standards for the low-burst-noise property and good performance for the $1/f$ noise, which is especially pronounced at the low frequencies.

In a similar manner, for the more refined programmable (or micropower) types of op amp that might be used for *low-level signals* (in medical electronics, for example), the *CA6078A* offers low-noise specifications that are improved over the general micropower *CA3078A* type.[2]

Electrometer Types

Currents as low as a fraction of a picoampere (10^{-12} A) are typical of signals encountered in chemical (pH) and medical (EEG) electronics, and, for such weak-signal measurement, require the special *very high impedance type of amplifier* known as an electrometer. This necessity arises from the fact that a very high input impedance is necessary in the device in order to develop a significant voltage drop from such a minute current, along with the fact that most of the sources being measured are themselves of the very high internal impedance type (as, for another example, the piezo-electric or capacity type of transducer).

These conditions naturally suggest the use of field-effect transistors (FETs) to obtain the necessary high input impedance. The FET-input types of OP AMPS, having junction FETs (JFETs) integrated on the IC chip, satisfy this requirement, and examples of such monolithic forms will be presented, offering at least 100 GΩ (1×10^{11} Ω) and more for the input resistance and corresponding low leakage currents. [Where still better values are required, *dual MOSFETs* may be used externally; these can

[1] The following references are available in the commercial literature for more complete discussion of noise-reduction techniques:

(a) For a general discussion of noise sources and shielding techniques, see Letzter and Webster, "Noise in Amplifiers," *IEEE Spectrum*, Aug. 1970.

(b) For signal-averaging and correlation techniques, see the *Hewlett-Packard Journal* of April 1968 (on the 5480A signal analyzer) and of Nov. 1969 (on the 3721-A correlator).

(c) For phase-locking methods, see the Keithley 822 phase-sensitive detector and the *Princeton Applied Research Technical Bulletin 109* (for the JB-5 lock-in amplifier).

[2] Op Amp Data Bulletins (File Nos. 531 and 535) give additional application and test data (RCA).

provide such values of input resistance as $R_{in} = 10^{15}\ \Omega$, and leakage currents as low as 1 fA (10^{-15} A)].[1]

Measuring Picoamperes

An electrometer circuit that can measure current in the picoampere (10^{-12} A) range is shown in Fig. 12.11. The OP AMP used is the National type *LH0052* having an input resistance (R_{in}) of 1,000,000 MΩ (1×10^{12} or 1 TΩ), paralleled by 4 pF; the input-offset current is in the range of femto-amperes (10^{-15} A). Since it is internally compensated, it is pin-compatible with the 741 type. The diagram shows the need for using shielded leads for the input probe and the very high feedback resistor of 100,000 MΩ (100 GΩ or 10^{11} Ω). This circuit is suggested for use in pH meters and radiation detectors, as examples of sensors that produce very small signal currents.

Guarding Circuits

In mounting electrometer-type OP AMPS, it is very possible for the leakage current across the sockets (or printed-circuit boards) to exceed the pico-ampere bias current of the device. Such leakage paths, which are essentially unbalanced, must be avoided because of the relatively large offset voltages that can be developed in the high impedances of the source and rest of the circuit. In using the *Burr–Brown model 3521* as an electrometer, for example, it is suggested that the inputs be connected to Teflon standoffs, or, if mounted directly, to use a *"guard" pattern,* as shown in printed-circuit form in Fig. 12.12 for the metal-can package of the *3521.* By means of this guarding technique, when the guard pattern is connected as shown, leakage currents

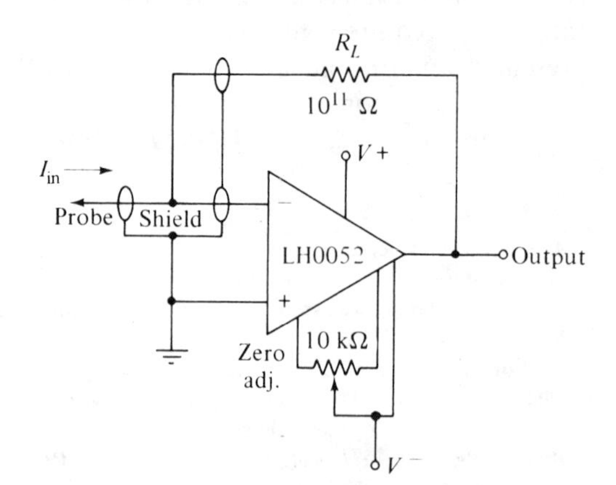

FIG. 12.11 Picoampere (10^{-12} A) amplifier, used as an electrometer, suitable for pH and radiation detector applications. *(National Semiconductor)*

[1] Discussed in greater detail for advanced electrometer use in Deboo and Burrous, *op. cit.*

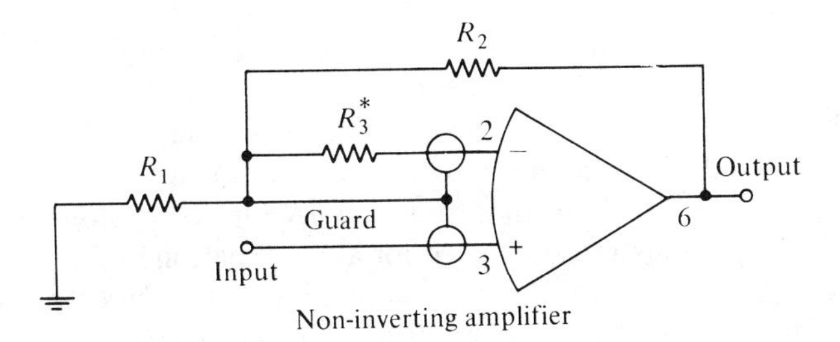

* R_3 may be used to compensate for very large source resistances ($R_S > 5$ MΩ) $R_3 = R_S$)

Note: $\frac{R_1 R_2}{R_1 + R_2}$ must be LOW impedance

(a)

V+
Output
7 8 1
3
2
5
4
V−
Guard
Inputs

(Bottom view)
Board layout for input guarding with TO-99 package.

(b)

FIG. 12.12 Guard connections for very high input impedance amplifiers; minimizes effect of leakage currents at mounting terminals (in addition to conventional shielding of input and feedback leads, when amplifying weak signals). *(Burr–Brown, BB3521 series)*

are rendered relatively harmless, since leakage paths from the board terminals are placed at practically the same potential as the input connection.

Drift Considerations

In addition to the precautions mentioned as to shielding and guarding, it is important that the electrometer type have low-drift characteristics, even after the initial voltage offset has been nulled. These characteristics are carefully spelled out in the data sheets for the *LH0052* and *BB3521* examples cited. In the same category, the *Analog Devices AD523*, which has an initial bias current of less than 1 pA, is also carefully specified to account for any drift with temperature, and/or with time.

12-9. SAMPLE-AND-HOLD APPLICATIONS

Among the digital-interface circuits mentioned in Chapter 10, the special type of sample-and-hold circuit serves as an example of a *storage application*, generally associated with analog-to-digital (A/D) conversion and peak-detector circuits. It illustrates the need for a combination of many desirable properties in one device, that is, especially good *input characteristics* with regard to high Z_{in}, low offset (V_{io}) and drift, all of which are difficult to obtain when *combined with high speed* (*or slew rate*) along with reasonable settling times.

A fairly simple sample-and-hold circuit is shown in Fig. 12.13(a), using the *LM 102* voltage follower. (Although the characteristics of $Z_{in} = 10^{10}\ \Omega$ and a slew rate of 10 V/μs are good for a bipolar device, improvement in both figures can be obtained from FET-input types, such as the *AD516* or the *AD505*, with slew rates of 20 and 120 V/μs, respectively). In the circuit operation of part (a), the sampling signal closes the MOSFET switch (Q_1) and charges the storage capacitor C_1 to the input value. When the switch is opened, the voltage-follower action makes and holds the output equal to the sampled input. As a necessary consequence, the dielectric of the holding capacitor must be of a high-quality type, such as Teflon, polycarbonate, or the like.

For the higher-performance type of device, a complete sample-and-hold circuit that includes all the required buffering and FET sampling-gate elements in a single 12-lead, TO-8 metal-can package is contained in the *LH0023* OP AMP (or, for inverted logic, in the *LH0043* OP AMP). The functional circuit of the elements enclosed in the *LH0043* package (except for storage capacitor) is shown in part (b) of the figure. This represents a significant size reduction from the more conventional (and costly) discrete or module designs.

12-10. MULTIPLE (TRIPLE AND QUAD) OPERATIONAL AMPLIFIERS

The integration of three (and four) operational amplifiers on one chip affords great flexibility in a variety of special applications.

Triple Op Amp

The *CA3060*, for example, consists of three operational amplifiers of the operational-transconductance-amplifier (OTA) type, previously described in discussing the single *CA3078*/A of that type. In such OP AMPS, the transfer function is best described in terms of the ratio of its current output against the input voltage, that is, in terms of its *transconductance* (g_m); this arises from the fact that, unconventionally, the output impedance of this amplifier

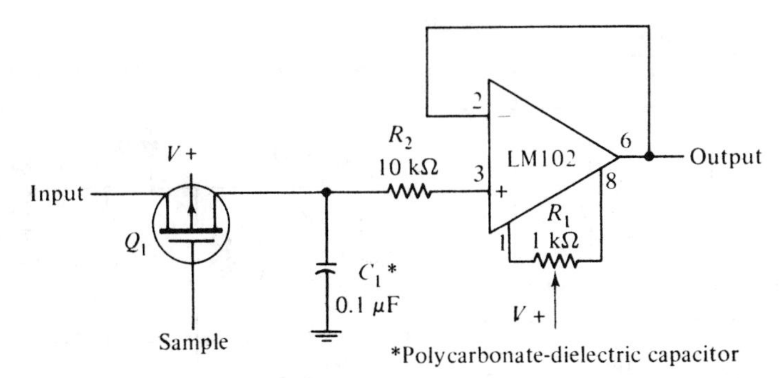

(a) Simple circuit

NOTE: Pin nos. apply to 8-pin package (either 8-pin metal can or 8-pin mini-DIP types).

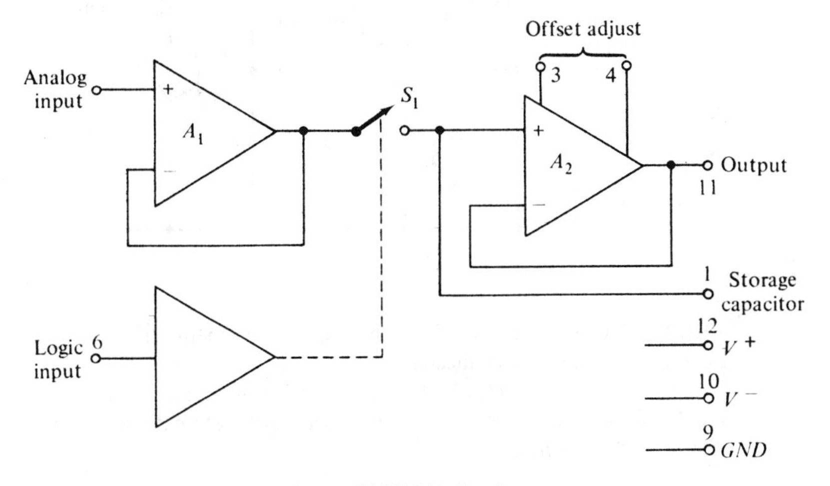

(b) Advanced LH0043 circuit

FIG. 12.13 Sample-and-hold circuits: (a) simple circuit with MOSFET sampling switch and LM102 voltage-follower; (b) advanced circuit of LH0043 OP AMP, self-contained in 12-lead TO-88 can, requiring only external storage capacitor. *(National Semiconductor)*

type is of a high value, and also because its resulting transconductance can be varied over a wide limit by the choice of an external resistor that changes the amplifier-bias-current (I_{ABC}) value.

Thus, by varying the external resistor (R_{ABC}), the bias current can be made to vary from 1 to 100 μA, with *corresponding g_m values of 300 to 30,000* μ℧. This biasing is independent for each of the three amplifiers, and the functional schematic diagram in Fig. 12.14 shows the circuit of one of them. (Also shown in the diagram is the self-contained Zener-diode system for regulating the bias-current supply.) All the circuitry for the three amplifiers

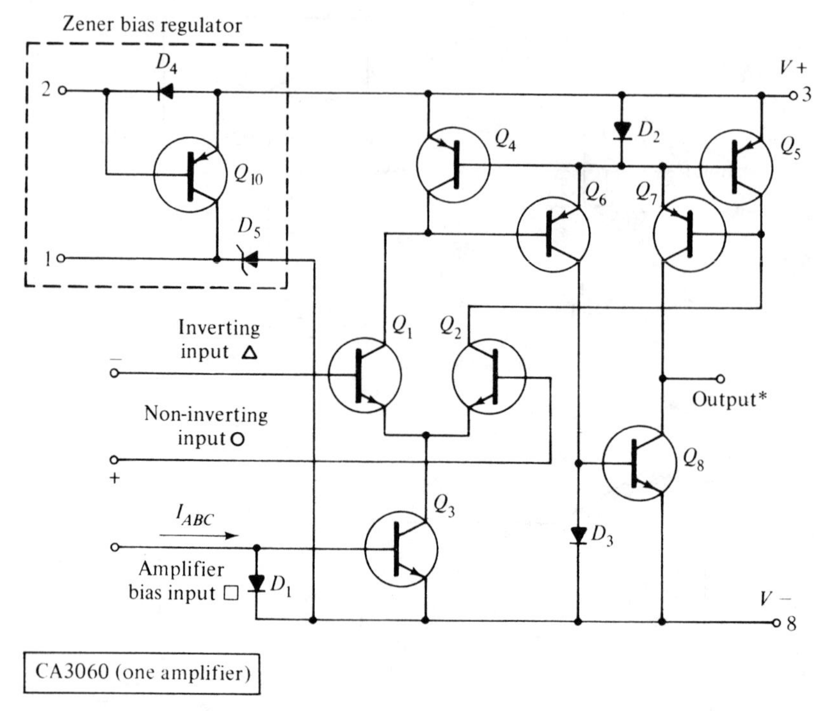

FIG. 12.14 Circuit of one of the three amplifiers in the triple operational-transconductance-amplifier (OTA), CA3060; the amplifier-bias-current (I_{ABC}) of each independent amplifier in the 16-pin DIP package can be varied by a separate external resistor. *(RCA Solid-State Division)*

and the Zener bias regulator is contained in a 16-lead DIP package.

Aside from the variable-bias system and the high output impedance, each of the three amplifiers has characteristics similar to the conventional OP AMP when feeding a relatively high resistance load. Some of the versatile applications of the *CA3060* include multiple-amplifier arrangements for active filters, micropower, and various mixing or modulation functions.

Quad Op Amps

The inclusion of four OP AMPS in one IC device is featured by the *LM3900*, or the *MC3301/3401* quad OP AMPS.

The *LM3900*, for example, consists of four independent internally compensated amplifiers, each similar to a typical dual-input OP AMP. It is packaged in a 14-lead DIP (8 input, 4 output, and 2 power-supply leads), and is designed to operate from a single power supply of 4 to 36 Vdc. Among the many possible configurations, Figs. 12.15(a) and (b) illustrate cases utilizing all four amplifiers of the device. Part (a) shows the circuit of an audio mixer (or channel selector) for three separate amplifier channels feeding the fourth amplifier as a mixer to furnish the desired output. In part (b) each quarter of the IC is used to generate a different output signal, as

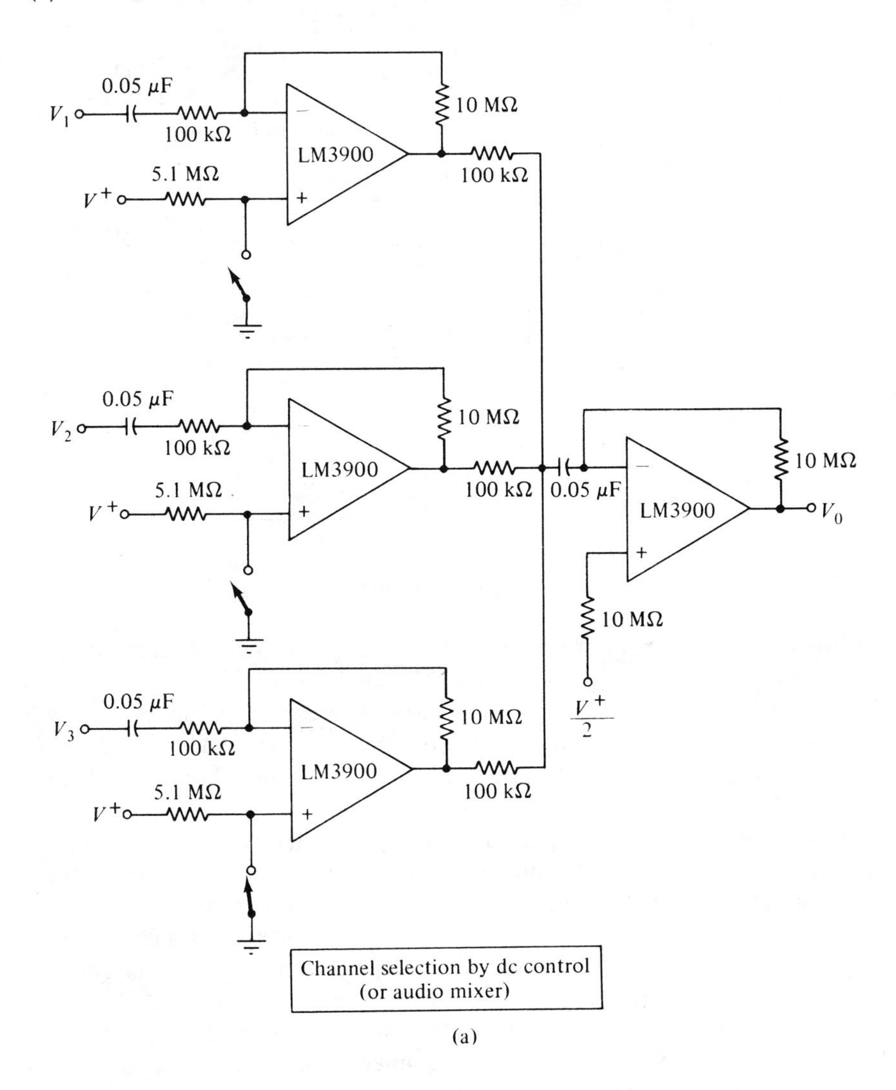

FIG. 12.15 Quad amp (LM3900) applications: (a) the four independent OP AMPS of the 14-lead DIP package used as a 3-channel mixer (or channel selector); (b) samples of wave-forms obtainable from each $\frac{1}{4}$ of the LM3900. *(National Semiconductor)*

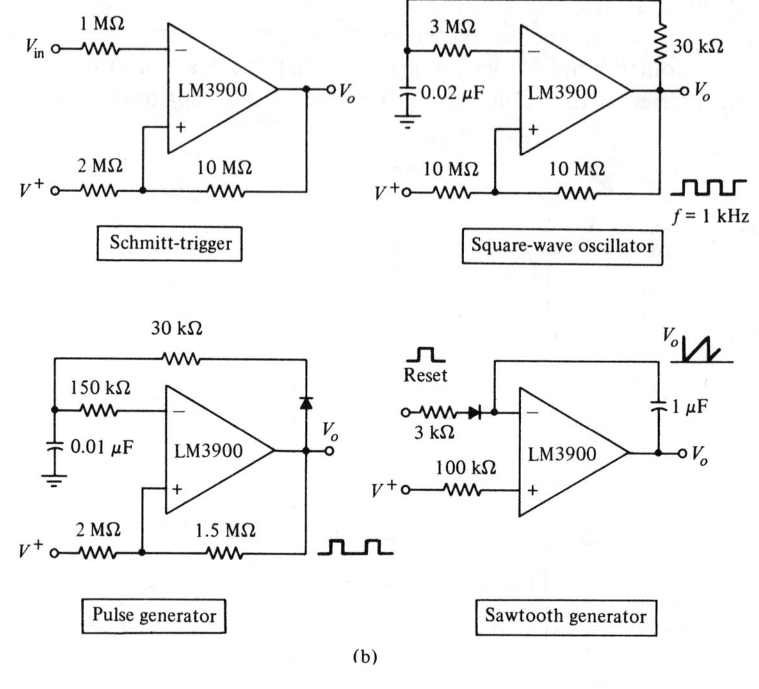

(b)

FIG. 12.15 (*Contd.*)

shown, producing, respectively, a Schmitt trigger, square-wave oscillator, pulse, and sawtooth generator, all from the single device.

Other applications are suggested by the circuit diagrams of *over 40 typical applications* included in the data sheet[1] of the *LM3900*. Among other suggested applications are *RC* active filters and low-speed high-voltage digital logic gates.

Another version of the quad OP-AMP type, the *MC3301P*, is specifically designed for single positive supply operation, such as in +12-V automotive applications. It is offered in the 14-pin DIP package (or as *MC3401P* in a 16-pin DIP package) with four internally compensated independent amplifiers. An application using all four amplifiers in a tachometer system is shown in Fig. 12.16, illustrating the ease with which each amplifier can be used functionally as part of an overall system.

Another example[2] illustrating the integration of multiple OP AMPS

[1] *Linear Integrated Circuits*, a manual commercially available from National Semiconductor (see address in Appendix IV).

[2] "Analog Switching without FETs," *Electronic Products*, Oct. 18, 1971.

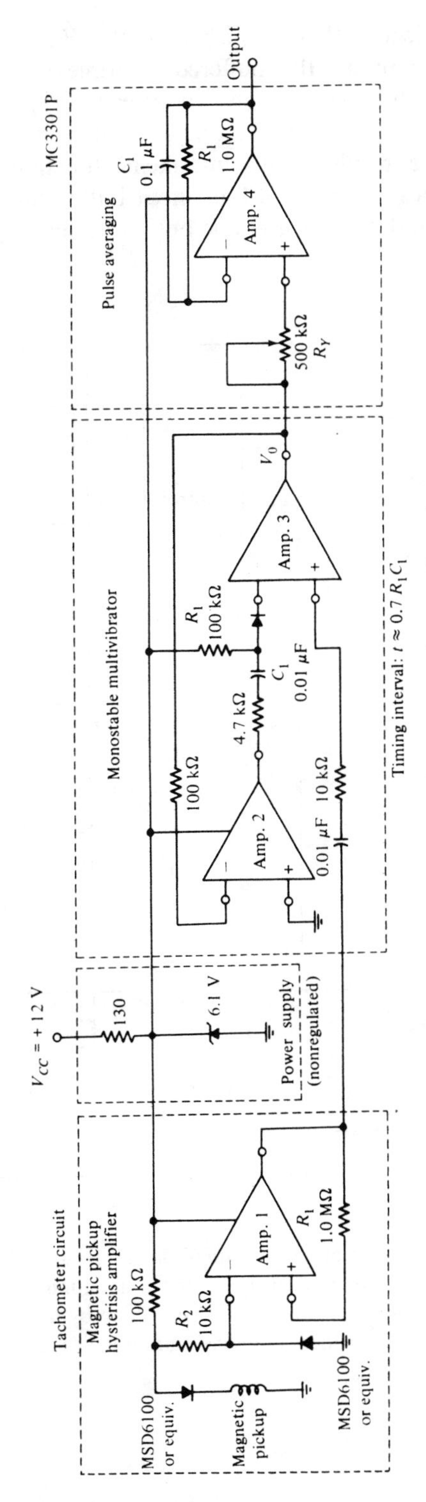

FIG. 12.16 Quad OP AMP (MC3301 P), used in an automatic tachometer system; each of the four internally compensated independent amplifiers is employed for its particular function. (*Motorola Semiconductor*)

in a single monolithic 16-lead DIP is furnished by the *Harris HA2400*, a gate-controlled quad OP AMP. In the buffered multiplexing application shown in Fig. 12.17, the feedback signal places the selected amplifier channel in a voltage-follower configuration. The decode/control terminals are used for channel selection. As a result of this interfacing function, the single quad op amp package takes the place of four input buffer amplifiers, four analog switches with digital decoding, and one output buffer amplifier.

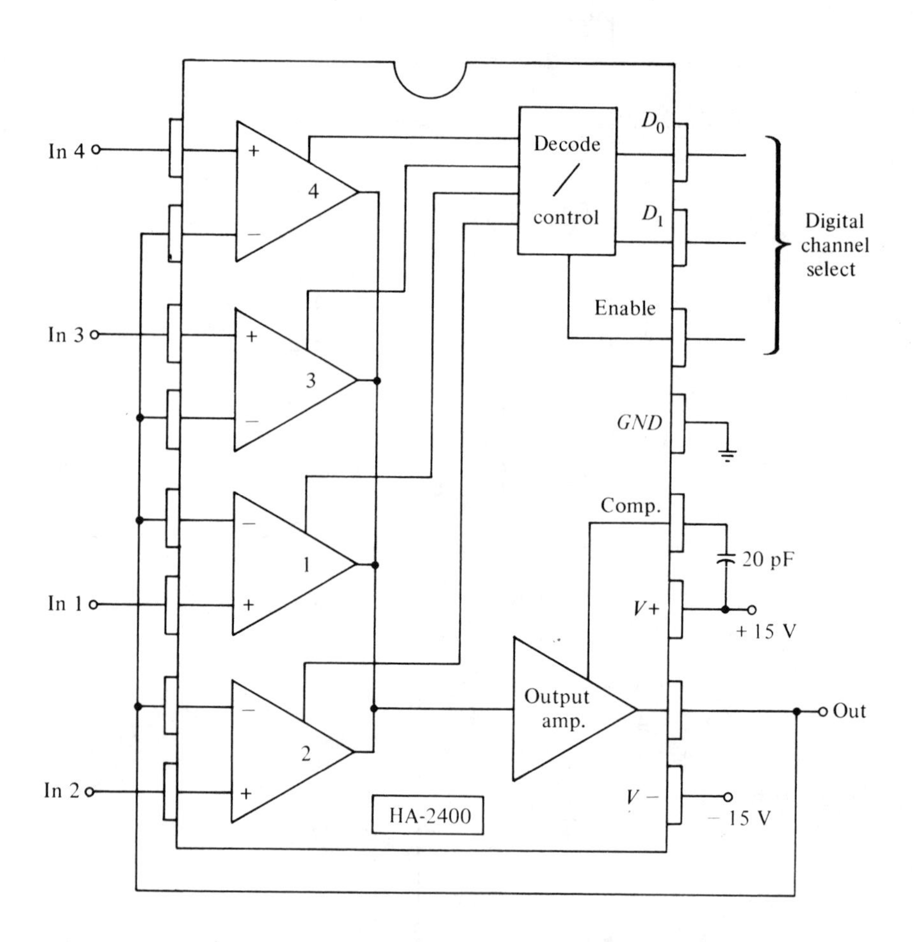

FIG. 12.17 Gate-controlled monolithic quad OP AMP (HA2400): in this buffered multiplexing application, the four amplifiers, used as buffers, are combined with analog switching by digital selection and an output amplifier, all in one monolithic package. *(Harris Semiconductor)*

12-11. THE VOLTAGE-CONTROLLED OSCILLATOR (VCO) AND WAVE-FORM GENERATORS

While the main function of most linear ICs is generally regarded as some form of amplifier action, it is well to keep in mind that OP AMPS (and other LIC types) are equally useful as *oscillators.*

RC Oscillators

As was illustrated in the quad OP AMP of Fig. 12.15, any single OP AMP can easily be configured to generate *square waves, sawtooth waves,* or *pulses* with a minimum amount of external components. Additionally, the production of *sine waves* is also accomplished easily in the popular *Wien-Bridge arrangement* shown in Fig. 12.18(a), which uses the internally compensated *LM107* (or its equivalent 741 type of OP AMP). In part (b) two OP AMPS are used to provide a dual-wave function generator, which produces a *triangle-wave* output by means of using the second OP AMP (*LM107*) to integrate the square-wave output from the first OP AMP (*LM101A*).

Voltage-Controlled Oscillator

All the preceding generators, it will be noted, employ *RC (or multivibrator) circuits,* where the frequency is determined by the *RC* time constant. [This can be seen particularly in the function generator of Fig. 12.18(b), where resistor R_3 is labeled as the frequency control.] However, there are many applications where the need is to control the frequency by means of an applied (or control) voltage. This function is achieved in the *voltage-controlled oscillator (VCO).* A typical VCO (*SE/NE566*) is illustrated in block-diagram form in Fig. 12.19(a), and in schematic form in part (b).

In this arrangement, the external capacitor (C_1) is alternately charged in a linear fashion by one current source and then linearly discharged by another to provide the buffered triangular output. The Schmitt-trigger action determines the charge–discharge levels and provides the buffered square-wave output.

The connection circuit for the VCO in Fig. 12.19(b) shows the R_1C_1 time constant that determines the free-running frequency. Additionally, the control voltage (V_c) at terminal 5 is set by the voltage divider consisting of R_2 and R_3. The initial bias for the control voltage (V_c) must be set in the range between $\frac{3}{4}$ V+ and V+; for example, for a 12-V supply, V_c may vary between 9 and 12 V and thus exerts its influence in determining the output frequency. In all, then, there are three frequency-determining elements (R_1, C_1, and V_c); when keeping C_1 constant and at a fixed V_c, the frequency is *adjustable over a 10:1 frequency range,* by choice of R_1, up to a maximum

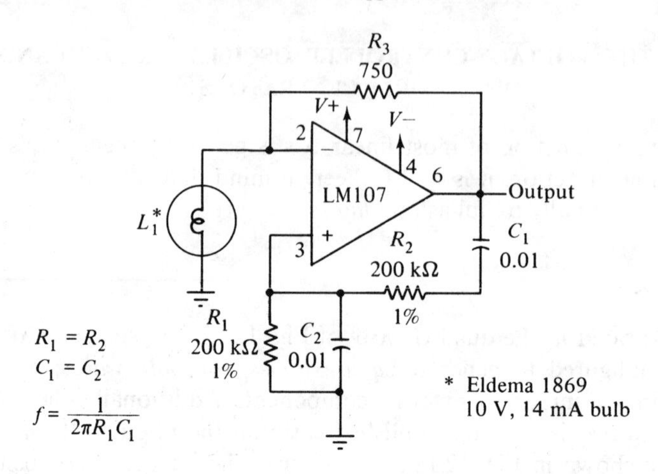

(a) Wien-bridge sine-wave oscillator

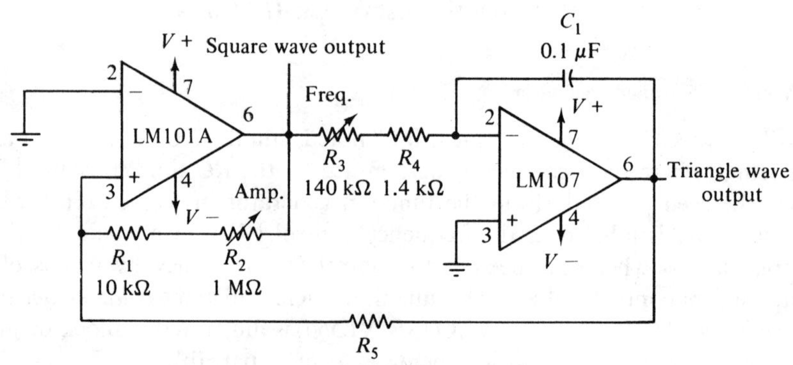

(b) Function generator

FIG. 12.18 Additional RC oscillators [see also Fig. 12.15(b)]: (a) sine-wave oscillator, using the frequency selective R_1C_1 elements of the Wien-Bridge arrangement; (b) two OP AMPS provide square- and triangular-wave outputs. *(National Semiconductor, Linear Applications, AN-31)*

frequency of 1 MHz. In a similar manner the frequency can be modulated over a 10:1 range by the control voltage V_c.

VCO Design Values

As an example, we may use a typical single-supply voltage of +12 V; using

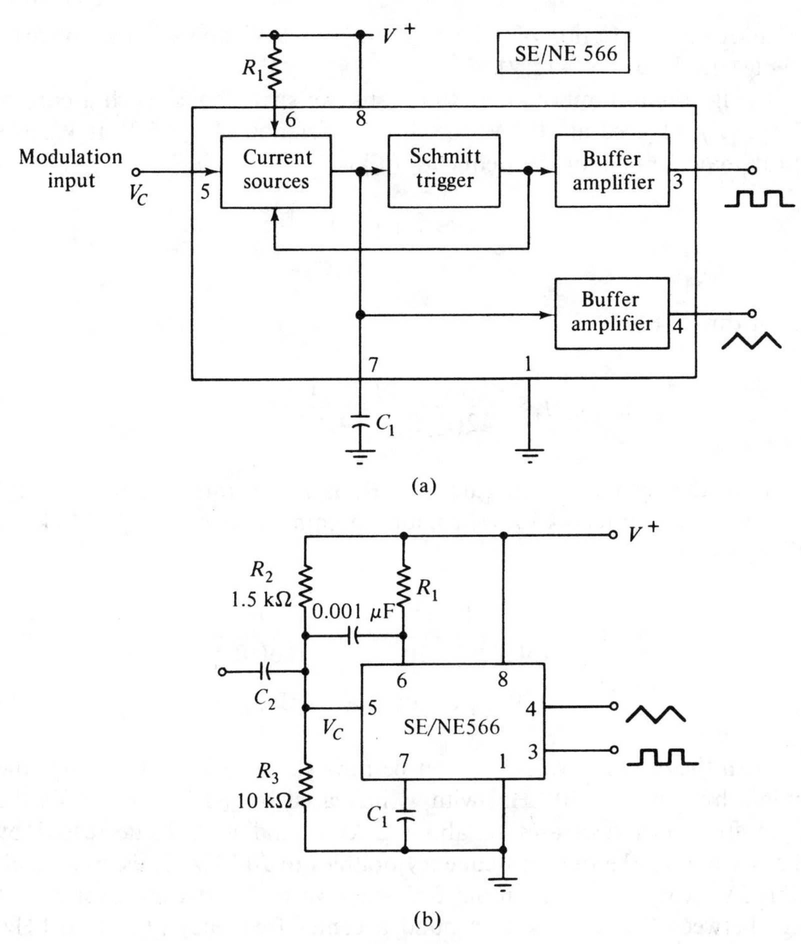

FIG. 12.19 Voltage-controlled oscillator (VCO): (a) block diagram of function generator SE/NE566, whose frequency can be controlled by control voltage V_c; (b) connection diagram for initial bias at V_c (terminal ⑤) and R_1C_1 elements for simultaneous production of square and triangular waves. *(Signetics LIC Manual, Vol. I)*

the values given in Fig. 12.19 we have

$$V_c = V\frac{R_3}{R_2+R_3}$$

$$= 12\frac{10\text{ k}\Omega}{11.5\text{ k}\Omega} = 12(0.87)$$

$$= 10.4\text{ V}$$

This allows a V_c variation of -1.4 V (down to $+9$ V), and an upward variation of $+1.6$ V (up to $+12$ V).

Using round-number approximations, we shall choose such a control voltage (V_c) centered at 10.5 V (to allow a variation of ± 1.5 V at V_c). The approximate fomula for the frequency (f_o) is

$$f_o \approx \frac{2(V^+ - V_c)}{V^+} \cdot \frac{1}{R_1 C_1},$$

and for this case

$$f_o \approx \frac{2(12-10.5)}{12(R_1 C_1)} = \frac{1}{4R_1 C_1}$$

Since the recommended value for R_1 is in the range between 2 and 20 kΩ, we shall use $R_1 = 4$ kΩ. Then, for a nominal frequency (f_o) of 10 kHz, with $R_1 = 4$ kΩ,

$$C_1 = \frac{1}{16(10^3)\,10(10^3)} \quad \text{or} \quad \frac{1}{160(10^6)}$$
$$= 0.0062\ \mu\text{F} \quad \text{or} \quad 6{,}200\ \text{pF}$$

With these design values, V_c can be now varied ± 1.5 V to change the nominal frequency of 10 kHz; with V_c increased by ± 1.5 V (to 12 V), the output frequency decreases to about 2 kHz; and with V_c decreased by -1.5 V (to 9 V), the output frequency doubles to 20 kHz. Thus, as a result of this 3V swing in V_c, the output frequency varies (or sweeps) over a 10:1 range between 2 and 20 kHz around a center frequency (f_o) of 10 kHz.

In a similar manner, the center frequency can be shifted from 10 kHz by changing R_1 from its initial value of 4 kΩ down to 2 kΩ (for a new higher $f_o = 20$ kHz), or we can increase R_1 by 5 times to 20 kΩ (for a new lower $f_o = 2$ kHz).

Consequently, for the values given (and within the 1 MHz maximum for this VCO), the available frequencies can be shifted by these particular choices from one fifth of 2 kHz (or 400 Hz) up to 40 kHz (in either a square or triangular wave form).

The ability to vary output frequency by means of an input voltage naturally suggests the use of the VCO for a *sweep frequency*, as might be used for testing (and displaying) the frequency response of an amplifier under test. Similarly, the VCO output provides a convenient method for *frequency-modulation (FM)* applications. Still another important development is the use of the VCO as an important element in the widely used *phase-locked-loop (PLL)* circuits (which were discussed in Section 8-8).

Wave-Form Generators

The integration of a *complete signal-generator capability* has been successfully accomplished in a single DIP package, as exemplified in the 14-lead *Intersil ICL8038*, and in the *16-lead XR205 model from Exar Integrated Systems.*

The functional block diagram of the *ICL8038* [Fig. 12.20(a)] shows the internal elements, including two current sources, with two associated comparators and a flip-flop. The current sources (*I* and 2*I*) provide equal charge and discharge times to external capacitor *C*, producing the *triangular wave*; simultaneously, the flip-flop produces the *square wave.* Each of these wave forms has its own buffer amplifier to provide isolated outputs. The sine-converter section (itself composed of 16 transistors) functions in a nonlinear manner to change the triangular input into sine waves by providing a decreasing shunt impedance as the potential of the triangle moves toward the two extremes.

In addition to the three basic wave forms shown, the production of a wide-range (1,000: 1) *sweep output* is shown by the connections in Fig. 12.20(b). The frequency (or repetition rate) for the various wave forms is selected with a minimum of external components. The stable wave forms cover a range of operation from below 0.001 Hz up to a bit more than 1 MHz.

Another type of monolithic wave-form generator (model XR205) is shown in system block form in Fig. 12.21(a). The internal elements in this model are seen to include an analog multiplier that connects externally to an uncommitted VCO and a separate buffer amplifier. While suitable for split-supply operation, the external connections for single-supply operation are shown in part (b).

The outputs include the three basic *sine, square, and triangular* wave forms, and also provision for *ramp or sweep voltages.* In addition, a variety of *modulation outputs* can be obtained, including AM, FM, phase-shift-keyed (PSK), frequency-shift-keyed (FSK), and tone-burst outputs. (The XR205 data sheet shows a comprehensive series of 19 oscilloscope photos illustrating the large variety of modulated outputs available.)

The upper frequency limit (for sine waves) extends to 4 MHz, with 20-ns rise and fall times for the square waves. The versatile modulation characteristics (which also include double side-band with suppressed carrier) provide the basis for very flexible signal generation. (A kit for a self-contained AM/FM signal generator, composed of two *XR205s*—one for modulation and one for carrier—is provided with a prepared printed-circuit board and all controls in the form of a Waveform Generator Kit, *XR205K*. Also offered independently, as *XR2207*, is a highly stable and versatile VCO.)

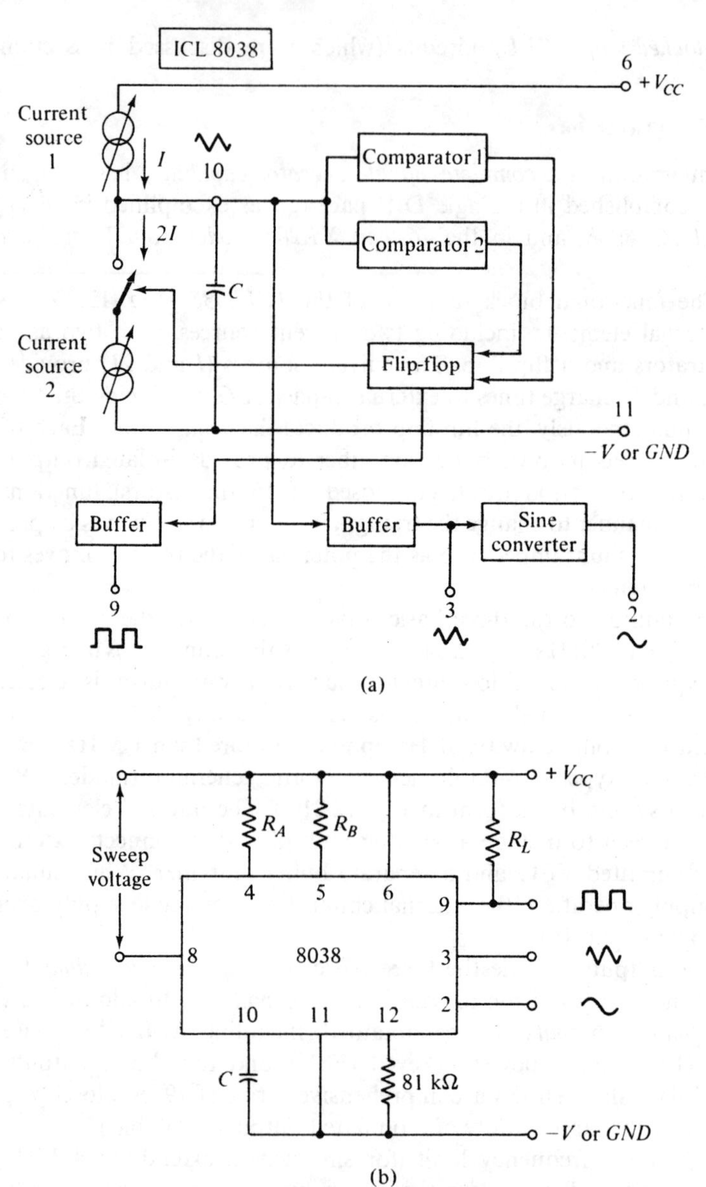

FIG. 12.20 Wave-form generator (ICL8038) in 14-pin DIP package: (a) functional block diagram, showing internal elements; (b) external connections for producing basic square, triangular, and sine waves. *(Intersil, ICL8038, Application Bulletin, AO12)*

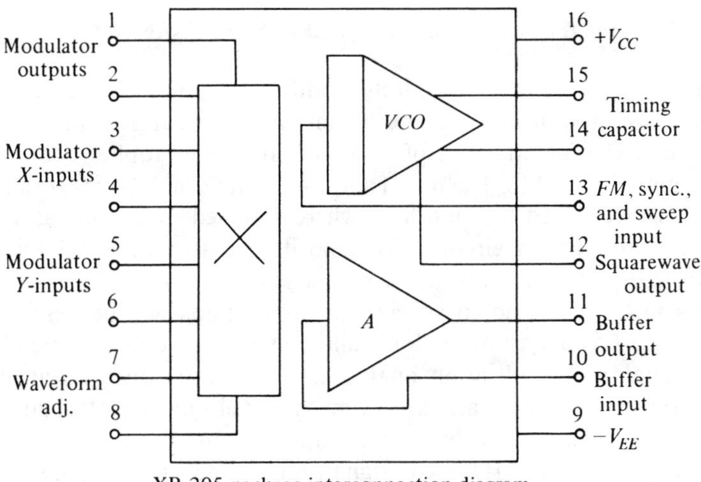

XR-205 package interconnection diagram

(a)

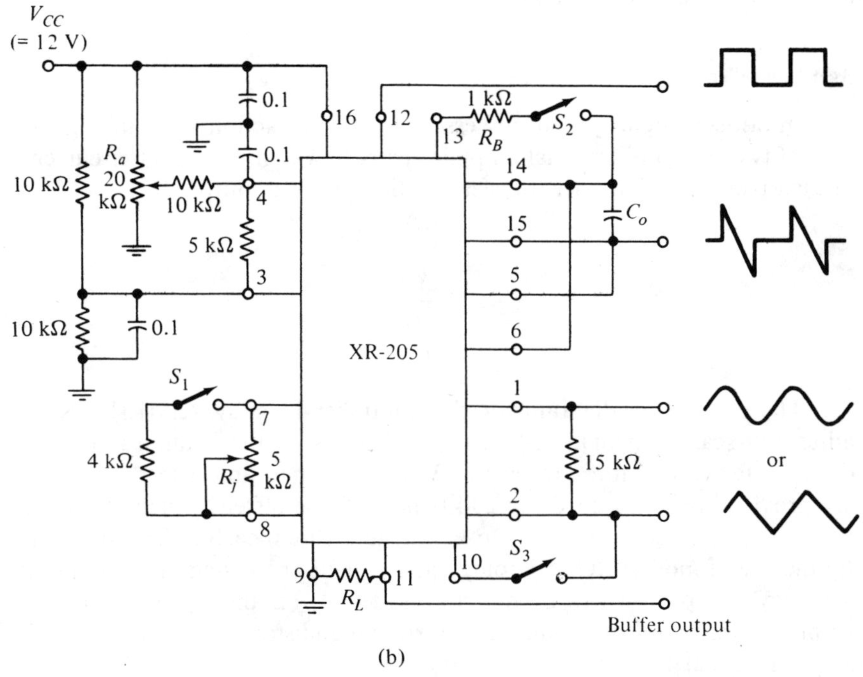

(b)

FIG. 12.21 Wave-form generator (XR205), in 16-pin DIP package: (a) system block diagram, showing analog multiplier, VCO, and buffer amplifier elements; (b) external connections for single-supply operation and basic wave-forms (not including modulated wave outputs). *(Exar Integrated Systems, XR205 Data Bulletin)*

12-12. MULTIPLIERS, DIVIDERS, SQUARERS, SQUARE ROOTERS

The multiplying function (along with its modifications) originally developed for analog computation has become an important linear IC application, especially since the introduction of an entire analog multiplier contained on a single monolithic chip. The use of a four-quadrant multiplier (*MC1596*) was previously described (Section 8-7), where it served as a modulator (or mixer) in communication circuits. Its opposite use as a demodulator (or product detector) was also introduced in that section.

In this section, additional specialized applications will be described, using the *AD530* as a typical analog multiplier of the versatile type (Fig. 12.22). The analog multiplication in this device is based on the principle of *variable transconductance*, and serves many useful purposes. In addition to the multiplying function, it has the capability of dividing, squaring, and square rooting (hence its MDSSR designation); moreover, it also finds use under other functional names as previously mentioned in connection with modulation and demodulation, all of which makes it almost as versatile an LIC device as the IC OP AMP.

As a Multiplier

The multiplier circuitry depends essentially on its action in changing the gain of two sets of differential amplifier pairs by varying the emitter current of each pair. The multiplier transfer function yields an output (E_o):

$$E_o = \frac{XY}{10}$$

The connection diagram for the multiplier [Fig. 12-22(a)] shows an adjustable scale control (5 kΩ) for the Y input, while the X input is applied directly. Provisions for zeroing the X and Y input, and also the output, are provided (when needed) by 20-kΩ pots. These offsets, which affect the dc accuracy, are dependent on close matching within each differential pair. By the use of modern IC technology and careful processing, the overall dc accuracy is kept down to production grades of ± 2 and ± 1 percent. The amplitude response with frequency (maximum undistorted output of ± 10 V) is down 3 dB at 1 MHz.

It is interesting to compare this method of multiplying X and Y with the older *quarter-square method*, which can be implemented with a number of amplifiers acting as summing, difference, and squaring amplifiers, with the squarer using the principle of square-law response. This method was

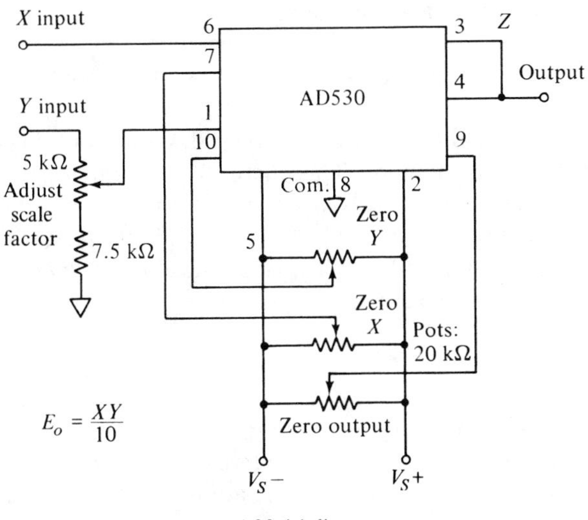

(a) Multiplier

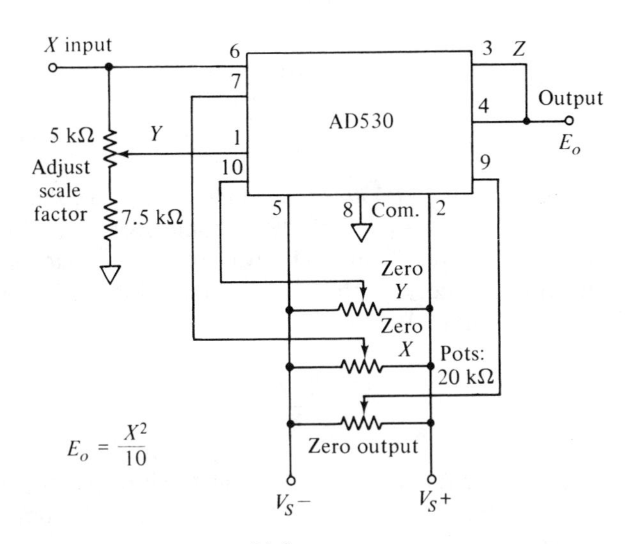

(b) Squarer

FIG. 12.22 Analog multiplier (AD530): (a) connected as a multiplier, to yield a product $XY/10$; (b) with X and Y inputs tied together, it is connected as a squarer, yielding $X^2/10$ as an output. *(Analog Devices, AD530 Data Bulletin)*

widely used in the past, based on the equation

$$(X+Y)^2-(X-Y)^2=X^2+2XY+Y^2-X^2+2XY-Y^2=4XY.$$

The remaining $4XY$ is then divided by 4 to yield XY (accounting for its name of the quarter-square method). This method, although accomplishing its purpose, seems tortuously involved compared to the more direct method using the LIC multiplier to yield XY/K, which provides much simpler external circuitry for not only multiplying, but also for dividing, squaring, and extracting the square root, as shown next.

As a Squarer

The connection diagram for a squarer [Fig. 12.22(b)] shows that the squaring function is accomplished very simply by tying the X and Y inputs together. It is done here by connecting the scale-adjust control (for the Y input) to the X input, and yields an output (E_o):

$$E_o=\frac{X^2}{10}$$

As a Divider

In the divider mode of the *AD530*, shown in Fig. 12.23(a), the numerator input signal is fed to the Z input and the denominator to X. The output terminal is connected to the Y input by way of the scale-adjust control. The transfer function for normal operation (where positive X signal is first inverted) yields an output (E_o):

$$E_o=\frac{10Z}{X}$$

It is necessary to invert a positive X signal, since positive X as the denominator in this mode would cause the feedback to become positive, and cause an overrange (or "pegging") output. Although not harmful to the device, the normal division operation is done with an inverted positive X signal.

As a Square-Rooter

The connections for the square-root function are shown in Fig. 12.23(b). Here the X and Y inputs are tied together, as in the squaring function, and the connections are arranged so that the Z input is divided by the squared

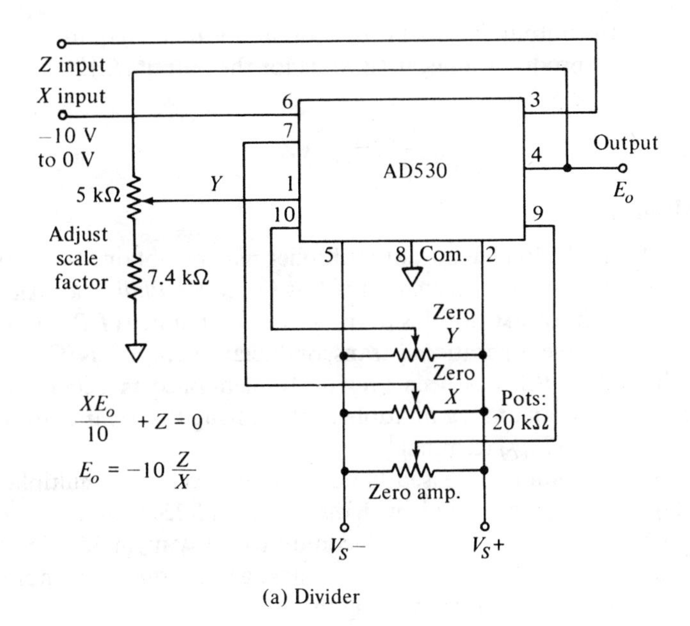

(a) Divider

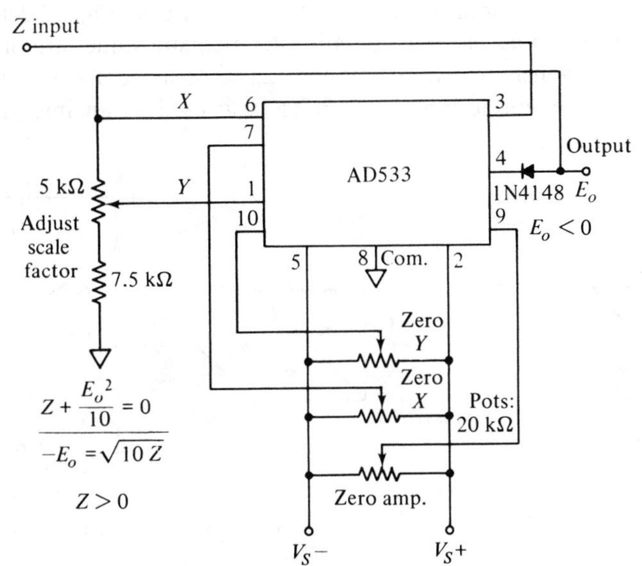

(b) Square-root function

FIG. 12.23 Additional analog multiplier functions: (a) connected as a divider, to yield $10Z/X$; (b) connected for a square-root function to yield $\sqrt{10Z}$. *(Analog Devices, AD530 Data Bulletin)*

magnitude in the output. Normal operation in this mode requires a positive input for Z, and produces a negative value for the output (E_o):

$$-E_o = \sqrt{10Z}$$

Analog Multiplier Models

In module form, the four-quadrant multiplier may be obtained as a module that requires an external OP AMP for dividing (*Teledyne/Philbrick model 4450*) or one in which the OP AMP for this purpose is self-contained (*T/P model 4452*).

The use of the operational-transconductance-amplifier (OTA) type, such as the *CA3080/A*, has been previously mentioned in Section 11-8. It can also be used as an analog multiplier by applying the Y input to the amplifier-bias-control (I_{ABC}) terminal[1].

A model providing increased bandwidth as a versatile multiplier in a 16-lead DIP package is offered by *Exar, model XR2208*. Since the device contains a buffer amplifier plus an uncommitted OP AMP, in addition to the four-quadrant multiplier, all in the one package, it reduces the number of external components required in many instances.

The general symbol for a multiplier is shown in Fig. 12.24(a), which is an example of a *mean-square* circuit. Mean-square values are often useful for evaluating nonsinusoidal signals and can be obtained using the multiplier in a squaring mode, followed by an OP AMP used as an integrator [Fig.

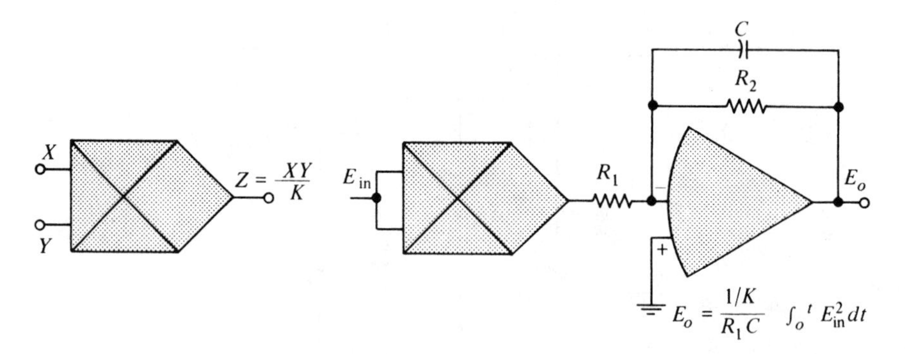

FIG. 12.24 Symbol representation of analog multiplier: (a) general symbol for multiplier; (b) symbol used in a mean-square function, yielding an average value of E_{in}^2.

[1] Details are given in IC Application Note (ICAN-6668) in the *RCA Solid State* Databook Series: SSD-202A.

12.24(b)].[1] The output E_o is given as

$$E_o = \frac{1}{KR_1C \int_o^t (E_{in})^2 \, dt}$$

Root-mean-square (true rms) measurement can be obtained by applying this mean-square value to the square-rooter connection shown previously.

The special applications shown here do not begin to exhaust the analog multiplier possibilities, which also include function generation, automatic gain control (AGC), and phase-comparator functions, among others. This versatility accounts for the presence of so many multiplier models, as is also the case for the op amp segment of the linear ICs.

For clearing up some of the possible confusion resulting from the multiplicity of models (as, in this instance, *MC1596, AD530, 4452, CA3080/A*, and *XR2208*), attention is called to the cross-referenced index in Appendix III, which identifies the manufacturers' designations and describes the LIC type. Appendix IV gives the addresses for obtaining the valuable technical data offered by the manufacturers for the specific models.

12-13. DIGITAL-VOLTMETER SYSTEM

As an example of a system application, the increasingly popular digital voltmeter (DVM) serves as a good illustration, showing how linear ICs are combined with digital-interface and digital ICs, all forming an operating voltmeter system.

The heart of the most widely used type of DVM is an interface linear IC, an A/D converter that uses a *dual-slope integration method.* This method operates essentially as a voltage-to-time conversion, but it accomplishes the conversion in a manner that is significantly better than a simple integrator. As opposed to the linear ramp output that is obtained from a single integrator, the dual-slope method produces a time measurement that encompasses both the charge and discharge times of an integrating capacitor, and consequently it produces an output that is more accurately proportional to the ramp slopes.

The operation of the dual-slope method of integration is shown in Fig. 12.25. At the start of each measuring cycle, the unknown input voltage (V_{in}) is applied to a buffer amplifier, before being integrated by the OP-AMP integrator. With C_1 in the feedback loop, the capacitor is charged by V_{in}, producing a rising ramp for a fixed time (T_1), which is set for a predetermined

[1] J. Pepper, "Analog Multiplier Applications," *Instruments and Control*, June 1972.

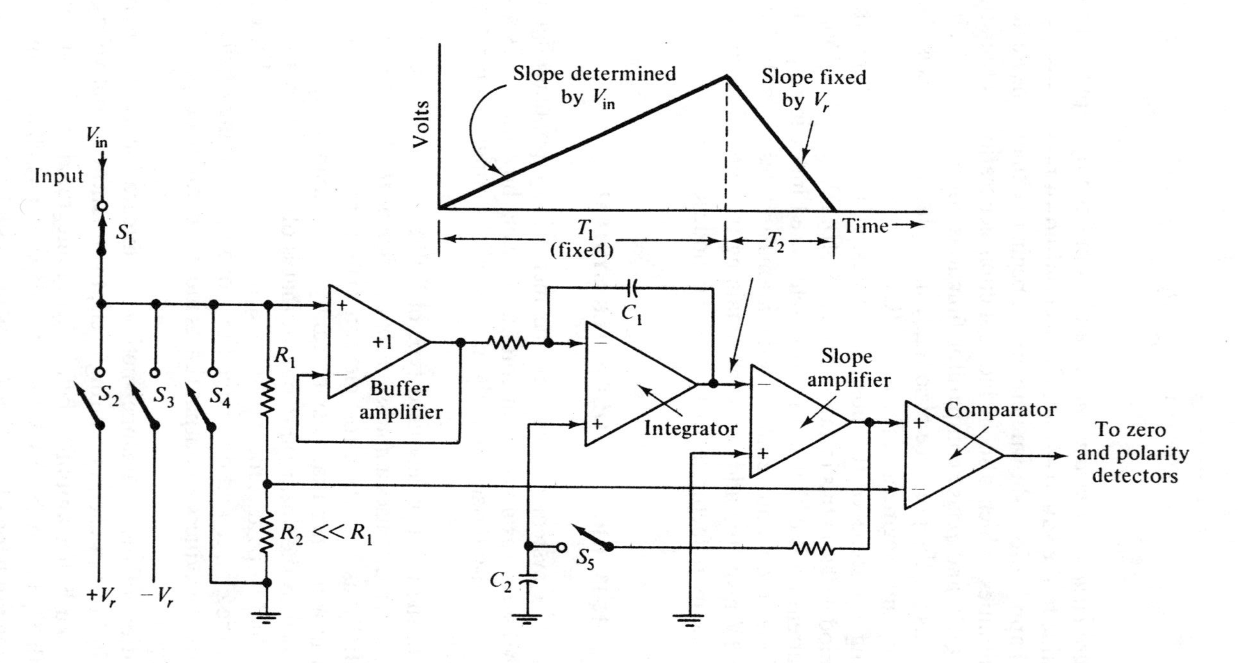

FIG. 12.25 Dual-slope integration system used in a digital voltmeter (DVM); the input voltage V_{in} determines the height of the ascending ramp in the fixed time (T_1), while the descending ramp, initiated by reference voltage V_r, establishes the discharge time (T_2), which is proportional to the unknown V_{in}.

amount of clock pulses. When this fixed number of pulses has been counted, the control circuits open switch S_1 and close either of the S_2 or S_3 switches to apply a reference voltage ($+V_r$ or $-V_r$) of the opposite polarity to the input V_{in}. This discharges capacitor C_1 to produce the descending ramp, during which time the amount of pulses is counted as T_2 until the integrator output falls to zero, and the control circuit (zero detector) stops the count. Since the charge time (T_1) is fixed, the discharge time (T_2) caused by the known reference voltage (V_r) is determined solely by the charge acquired from the unknown voltage (V_{in}). For example, if the charge time is made equal to 10,000 clock pulses and V_r equal to 1 V (+ or − as required), the count accumulated during discharge time (T_2) can be displayed digitally as the measured voltage.[1]

Advantages of Dual-Slope Method

This method, while retaining the basic advantage of an integrating-conversion method—in which *noise and other interference are averaged out*—also offers an important advantage in that other *potential error sources are self-canceling*. Where the accuracy of the single-integration method must suffer from any long-term changes in the clock rate or in the passive *RC* elements, such changes in the dual-slope method affect both up-and-down slopes equally, and therefore do not affect the count ratio.

In determining the overall accuracy of the digital voltmeter, it is necessary to consider other noncancelling errors, such as offsets in the ICs and nonlinear delays in the zero-detection process. Various methods are used to minimize such accuracy-degrading sources.[2] However, with the liberal use of the flexible and advanced ICs that are available, the net accuracy figures offered by even the relatively inexpensive DVMs are highly satisfactory, coupled with the very convenient small size that results from the high-count MSI and MOS/LSI forms of integration.

Commercial DVM (and DMM) Models

The *Weston model 4400* is a $3\frac{1}{2}$-digit multimeter (DMM) [Fig. 12.26(a)]. It is designed as a relatively inexpensive instrument for field use, since it includes a self-contained rechargeable battery pack. On its most sensitive range (200 mV), the display reads up to 199.9 mV and down to 0.1 mV. (The $\frac{1}{2}$ digit refers to the 100 percent overrange for the display of the one

[1] "HP3470 Digital Multimeter System," *Hewlett-Packard Journal*, Aug. 1972.

[2] Further details of DVM operation can be obtained from "HP3470 Digital Multimeter System," and also from S. Prensky, *Electronic Instrumentation*, 2nd ed., Prentice-Hall, Inc., Englewood Cliffs, N.J., 1971, Chap. 17.

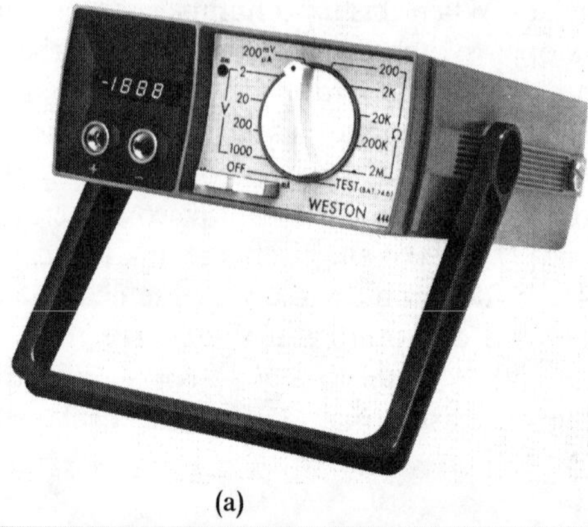

(a)

(b)

FIG. 12.26 Digital multimeters (DMMs): (a) A $3\frac{1}{2}$-digit portable multimeter with rechargeable battery pack *(Weston, model 4440)*; (b) A $4\frac{1}{2}$-digit multimeter, also with rechargeable batteries, showing small-size made possible by use of MOS/LSI chip, even though it offers 21 measurement ranges. *(Data Precision. model 245)*

digit at the extreme left, when the measured voltage is greater than 99.9 mV.)

Most of the available $3\frac{1}{2}$-digit DVMs offer accuracy figures close to 0.1 percent ± 1 digit, for dc voltage. (The digital multimeters in this class offer accuracies up to 0.5 percent for ac resistance and current readings.)

An example of the $4\frac{1}{2}$-digit multimeter (DMM) is shown in Fig. 12.26(b) for the *Data Precision model 245*. Here, again, the result of the use of large-scale integration (of the MOS/LSI chip type) can be seen in its compact size, even though it offers 21-range versatility with 0.005 percent resolution. While essentially of the dual-slope type of integration, it uses the (trademarked) designation of *Tri-Phasic* to indicate a third phase of operation (in addition to the two phases of dual-slope operation) that up-dates each auto-zero cycle by an error-integrator/memory circuit, so that noncancellable errors are reduced when the analog input is integrated together with the stored error.

As a final example of the growing tendency for integrating more and more discrete functions on a single chip, a *single array that combines all the logic functions of a $4\frac{1}{2}$-digit voltmeter* is the *3814 DVM array* illustrated in Fig. 12.27. This is an MOS type of digital IC (*Fairchild model A7R-3814-*

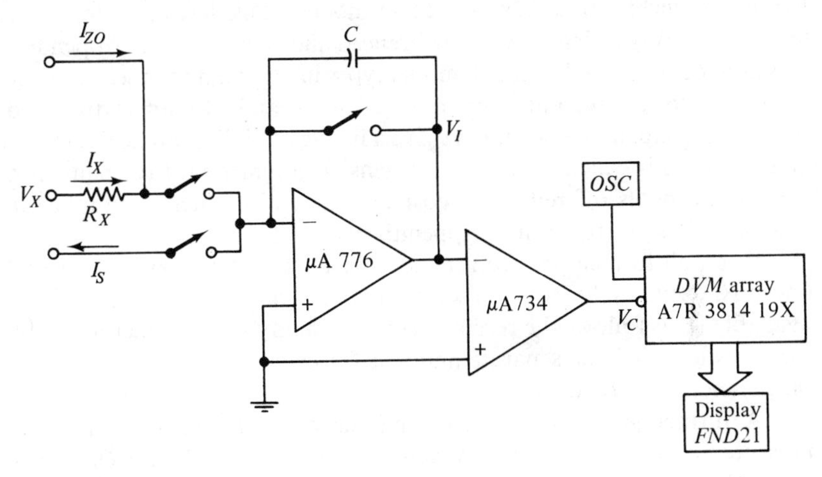

FIG. 12.27 Digital voltmeter (DVM) array on a single chip; when connected to three OP AMPS, as shown in functional form, it operates as a complete DVM, with an LED (FND21) display unit. *(Fairchild Semiconductor)*

19X). The complete DVM circuit is formed when the array is combined with the three linear ICs shown (**776** type OP AMP as integrator, **734** comparator type, and an *IC oscillator*). The output of the 3814 DVM array then feeds the multiplex display module (*FND21*) for the digital readout.[1]

In summary, in the *easy-to-read DVM*, we can also observe the great improvement in the *convenient compactness* and the *enhanced reliability and accuracy* that results from the strategic use of the combination of linear and digital ICs. While the pointer type (or analog type of meter), with its usual ±2 percent accuracy, still has its valid uses, the digital type of instrument—employing a number of ICs—is making steady progress in achieving a firm place for itself, spreading over the entire field of electronics.

[1] Fairchild Semiconductor 3814 data sheet and application note.

12-14. SOURCES FOR SPECIFIC TECHNICAL INFORMATION

When one considers the rapid pace at which developments in linear ICs have been developed and how many new models have been (and still are being) introduced as a result, it becomes clear that one of the most helpful aspects of a LIC manual will be to supply a convenient guide to reference material for further technical details of specific models. This has been done in this text in two ways: first, a cross-referenced index of LICs in Appendix III lists and identifies well over 300 model types in alphanumeric sequence, with a notation for the section where most of these can be found as discussed in the text. Second, in this section we give a listing of bibliographical references, and attention is called to the comprehensive commercial LIC manuals and application notes offered at no cost by the semiconductor manufacturers (whose addresses are given in AppendixIV).

It is well to reemphasize a major objective of this text in treating the linear ICs separately from the even greater welter of digital ICs that are available. This follows the recent practice of many semiconductor manufacturers, who now issue separate manuals for each category of discrete, digital IC and linear IC devices.

A further subdivision, separating the OP AMPS from other LICs, is presented in the form of a *Quick Reference Selection Guide for Op Amps*, in Appendix II. Admittedly a partial listing, it serves as a further *simplification for initial identification* of the many available types that are typical of this versatile LIC device. In this connection, the *caveat* bears repeating that many additional devices are offered by *other manufacturers, whose names and addresses are listed* in Appendix IV. Much valuable technical data and applications may be gained from their informative catalogs.

12-15. Bibliography List I

Altman, L., "Bridging the Analog and Digital Worlds with Linear IC's," *Electronics*, Vol. 45, No. 12, June 5, 1972.

Barna, A., *Operational Amplifiers*, John Wiley & Sons, Inc., (Interscience Division), New York, 1971.

Deboo, G. J., and C. N. Burrous, *Integrated Circuits and Semiconductor Devices*, McGraw-Hill Book Company, New York, 1971.

EDN Seminar, *Linear Integrated Circuits, Applications and Innovations*, Articles reprinted from an engineering seminar sponsored by EDN Magazine, Cahners Publishing Company, Denver Division, 1969.

Dobkin, R. C., "New Developments in Monolithic Op Amps," *Electronics World*, July 1970.

Eimbinder, J., ed., *Application Considerations for Linear Integrated Circuits*, John Wiley & Sons, Inc., New York, 1970.

Eimbinder, J., ed., *Designing with Linear Integrated Circuits*, John Wiley & Sons, Inc., New York, 1969.

Electronic Products, "Analog Switching Without FET's," Oct. 18, 1971.

Graeme, J. G., *Applications of Operational Amplifiers: Third Generation Techniques*, McGraw-Hill, New York, 1973.

Hewlett-Packard Journal, "5480A Signal Analyzer," April 1968.

Hewlett-Packard Journal, "3721-A Correlator," Nov. 1969.

Hibberd, R. G., *Integrated Circuits*, McGraw-Hill Book Company, New York, 1969.

Hnatek, E. R., *User's Handbook of Integrated Circuits*, Wiley Interscience, New York, 1973.

Kahn, M., *The Versatile Op Amp*, Holt, Rinehart and Winston, Inc., New York, 1970.

Lenk, J. D., *Handbook of Digital and Linear Integrated Circuits*, Reston Publishing Company, a subsidiary of Prentice-Hall, Inc., Englewood Cliffs, N.J., 1973.

Lenk, J. D., *Handbook of Simplified Solid-State Circuit Design*, Prentice-Hall, Inc., Englewood Cliffs, N.J., 1971.

Letzter and Webster, "Noise in Amplifiers," *IEEE Spectrum*, Aug. 1970.

Melen, R., and H. Garland, *Understanding IC Op Amps*, Howard W. Sams Company, Inc., Indianapolis, 1971.

Millman, J., and C. C. Halkias, *Electronic Devices and Circuits*, McGraw-Hill Book Company, New York, 1967.

Millman, J., and C. C. Halkias, *Integrated Electronics*, McGraw-Hill Book Company, New York, 1972.

Pepper, J., "Analog Multiplier Applications," *Instruments and Control*, June 1972.

Rogen, B. R., "Op Amps at Work, 10 Audio Circuits," *Radio Electronics*, Dec. 1972.

Sclater, N., "On Op Amps," *Electro-Procurement*, Dec. 1972.

Smith, J. I., *Modern Operational Circuit Design*, John Wiley & Sons, Inc., New York, 1971.

Sorkin, R. B., *Integrated Electronics*, McGraw-Hill Book Company, New York, 1970.

Stern, L., *Fundamentals of Integrated Circuits*, Hayden Book Company, Inc., New York, 1968.

Textronix, *Tekscope*, Nov. 6, 1972.

Tobey, G. E., "Analog Modules," *Electronic Products*, Feb. 19, 1973.

Tobey, G. E., L. P. Huelsman, and J. G. Graeme, *Operational Amplifiers, Design and Applications*, McGraw-Hill Book Company, New York, 1971.

Commercial LIC Manuals and Application Notes

(Available at no cost—manufacturers' addresses are given in Appendix IV.)[1]

Analog Devices: *Product Guide*
Burr–Brown Research: IC catalog
Fairchild Semiconductor: *Linear Integrated Circuits Data Manual*
Motorola: *Linear Integrated Circuits Data Book*
National Semiconductor: *Linear Integrated Circuits* and *Linear Applications*
RCA Solid-State Data Book Series: SSD-201A—*Linear Integrated Circuits, Selection Guide Data*, and SSD-202A—*Linear Integrated Circuits Application Notes*
Signetics: *Linear Integrated Circuits*, Vol. I, *Integrated Circuits, Digital, Linear MOS Manual*, and *Application Memos*
Teledyne/Philbrick: *Product Guide*
Texas Instruments: IC catalog

[1] A comprehensive listing of practically all available linear ICs is offered (on a two-issue-per-year subscription basis) as the *Linear IC D.A.T.A. Book*, from D.A.T.A. Inc., 32 Lincoln Ave., Orange, N.J. 07050.

APPENDIX I

DECIBEL CONVERSION CHART FOR APPROXIMATE VOLTAGE-GAIN RATIOS

Voltage-gain definition: $\text{No. of decibels (dB)} = 20 \log \frac{V_{out}}{V_{in}}$

No. dB		*Numerical Gain Ratio*	*No. dB*	*Numerical Gain Ratio*
Reference	0	=1.0		
	1	=1.12	13	=4.47
	2	=1.25	14	=5.0
	3	**=1.41**	15	=5.62
	4	=1.58	16	=6.31
	5	=1.78	17	=7.08
	6	**=2.0**	18	=8.0
	7	=2.24	19	=8.91
	8	=2.51	**20**	**=10.0**
	9	=2.82	**40**	**$=10^2=100$**
	10	=3.16	**60**	**$=10^3$**
	11	=3.55	**80**	**$=10^4$**
	12	=4.0	**100**	**$=10^5$**

DECIBEL CONVERSION TABLE

Note: The decibel values were historically employed to express *power ratios*; but since then they have come into common use to express equivalent *voltage ratios, where, it should be noted, the formula relationship is different from the power ratio relationship.*

Since the decibel values in this text are used practically exclusively to express *voltage* amplification ratios, only these equivalent ratios[1] are given. The distinction is shown by the following definitions:

For voltage: no. $dB = 20 \log V_{out}/V_{in}$.

For power: no. $dB = 10 \log P_{out}/P_{in}$.

Therefore, it is important to consider *all the decibel references in the text as referring to voltage ratios, unless a power ratio is specifically mentioned.*

MENTAL CALCULATION OF GAIN

For any statement of voltage gain given in decibels, the numerical ratio can easily be estimated *mentally by remembering just two rules:*

Decibels		*Numerical Gain Ratio*
1. No. dB/20	=	power of 10
2. Adding dB as follows:	=	multiplying the ratio as follows:
+6 dB	=	2.0 ×
+3 dB	=	1.4 ×
+2 dB	=	1.25 ×
+1 dB	=	1.12 ×

EXAMPLE FOR 42 dB

1. Taking the nearest multiple of 20:
   40 dB/20 $= 10^2$
2. Adding 2 dB $= 1.25\times$
   Then 42 dB $= 100(1.25)$, or 125:1 as the gain ratio.

[1] Those equivalents that are printed in boldface are emphasized for convenience for mentally estimating the approximate numerical gain ratio that corresponds to a given decibel value.

$db = 20 \log \frac{V_1}{V_2}$

$54 = 20 \log X$

$\log X = \frac{54}{20}$

$\log X = 2.7$

$X = 10^{2.7}$

$X = 501.187233$

$60 db = 20 \log X$

$\log X = \frac{60}{20}$

$\log X = 3$

$X = 1000$

APPENDIX II

QUICK REFERENCE SELECTION GUIDE FOR OP AMPS

Note: The *cross-reference index* (in Appendix III) is more complete and lists alphanumerically manufacturers' identifications and the section numbers in the text for individual types of OP AMPS and other LICs.

CLASSIFICATION OF LIC GROUPS

Arrays (Chapter 2)

Audio power amps (Chapter 7)

Comparators (Chapter 10)

Consumer circuits (Chapter 8)

Digital interfacing (Chapter 10)

Instrumentation amplifiers (Section 11-9)

Multipliers (Section 12-12)

OP AMPS (See following selection guide)

Phase-locked loops (Section 8-8)

Regulators (Chapter 9)

Specialized systems (Chapter 12)

OP-AMP SELECTION GUIDE

1. General-Purpose Types[1]
   A. Internally compensated: LM107, MC1556, μA741
   B. Extended bandwidth: LM101/A, μA748, SN52660
   C. FET input: AD503, LH0042, μA740
   D. Dual types: MC1558, μA747
   E. Quad types: CA3060 (triple), HA2400/2405, MC3301/3401, LM3900
2. High-Performance Types[2]
   A. Wide band (fast slew rate): AD505, LM118, CA3100, SE/NE531, 1321(T/P)
   B. Low-bias (I_B), (FET input or super beta): AD523, BB3521, LM108/A, MC1556, SE/NE537
   C. High-output current: AD512, μA791, CA3047, MC1538
   D. Low power (micropower and operational transconductance amplifier (OTA): CA3080, μA776, 1404(T/P)
   E. Precision (low drift and high accuracy): AD508, μA777, CA6078, SN52770/771
   F. Instrumental amplifiers: See Section 11-9.

[1] Table 1-1 (Sec. 1-7) identifies "second-source" types; Table 4-1 (Sec. 4-6) gives comparative data on general-purpose OP-AMP types.

[2] Table 11-1 (Sec. 11-3) gives comparative data on high-performance OP-AMP types.

APPENDIX III

CROSS-REFERENCE LIST OF LINEAR INTEGRATED CIRCUITS

Key to Identification: gives manufacturers' names (see addresses in Appendix IV).

Order of Listing: first, alphabetically by manufacturers' letter prefixes: e.g., MC1303, MC1304 ... MFC ... μA (μA = muA)..., followed by purely numerical designations.

Italic Numbers: represent much-used type numbers also appearing in "second-sourced" types; e.g., SN52*741* is equivalent to 741 type.
Section Numbers: refer to discussion in text.

KEY TO IDENTIFICATION OF MANUFACTURERS' TYPES (SEE APPENDIX IV FOR MANUFACTURERS' ADDRESSES)

A = Hybrid Systems
A = Intech
AD = Analog Devices
AIM = Precision Monolithics
Am = Advanced Micro Devices
AM = Datel Systems
AMD = Amperex
BB = Burr–Brown
CA, CD = RCA

(DAC) = Hybrid Systems Corp.
EAA = Electronic Associates
HA = Harris
HC = RCA
HEP = Motorola (Hobby/Experimental Program)
ICL, IH = Intersil

L = Siliconix Inc.
LA = Nucleonic Products
LH, LM = National
LP, LS = Lithic Systems
MC, MFC = Motorola
MIC = ITT
μA = Fairchild
Mono = Precision Monolithics
NE = see SE/NE
N/S = Signetics (see also S/N)
PA = General Electric
QC = Qualidyne
RC, RM = Raytheon

S = see N/S
S/N = Signetics (see also N/S)
SE/NE = Signetics
SG = Silicon General
SN = Texas Instruments
SSS = Precision Monolithics
TOA = Transitron
UC = Solitron Devices

ULN, ULS, ULX = Sprague
WB = Data Devices
XR = Exar

Numerical for the following:
Analog Devices (modular only) Beckman/Helipot
Teledyne/Philbrick (both IC and modular)

Manufacturers' Type Number	*Description*	*Section in Text*
A-132	FET-input op amp (Intech)	11-7
A956 A/B	Op amp (high speed, internally compensated) Hybrid Systems	
AD*101/A*	Op amp (extended bandwidth)	See LM*101*/A
AD503	Op amp (FET input, high accuracy)	11-4, 11-5
AD504	Op amp (precision, ultrastable)	
AD505	Op amp (FET input, high speed)	12-9
AD506	Op amp (FET input, low drift)	
AD507	Op amp (high slew rate)	11-7
AD508	Op amp (chopperless, low drift)	11-5
AD512	Op amp (high output current)	7-2
AD516	Op amp (FET input, wide band)	12-9
AD518	Op amp (wide band) (12MHz)	See LM*118*/318
AD520	Op amp (instrumentation)	11-10
AD523	Op amp (micropower, electrometer)	12-8
AD530	Multiplier (self-contained)	12-12
AD*741*	Op amp (internally compensated)	See μA*741*
AIM/DAC100	Digital-to-analog converter	10-6
Am118	Op amp (wide bandwidth)	11-7
Am*1488*	Quad line driver	See MC1488
Am*1489/A*	Quad line receiver	See MC1489/A
Am9614	Dual differential line driver	
Am9615	Dual differential line receiver	
Am150031	Voltage comparator	
AM200	Instrumentation amplifier (hybrid)	
AM532	Multiplier/divider	
AM*715*	Op amp (high speed)	See μA715
AMD*112*	Op amp (super beta)	See LM112
AMD*318*	Op amp (wide band)	See LM*118*/318
AMD*739*	Op amp (dual, low noise)	See μA739
BB3329/03	Power booster	7-3
BB3500	Op amp (low drift)	
BB3505/7	Op amp (fast slew rate)	
BB3506/8	Op amp (wide band)	
BB3521	Op amp (combined low drift and FET input, electrometer)	11-5, 12-8

Manufacturers' Type Number	*Description*	*Section in Text*
BB3522	Op amp (low offset voltage)	
BB3524	Op amp (FET input)	
BB3542	Op amp (FET input)	
BB3625/A	Instrumentation amplifier (variable-gain module)	
CA2111/A	FM–IF amplifier; limiter/detector	
CA3005	RF/IF amplifier (similar to μA703)	
CA3008/16	Op amp (wide band)	
CA3010/15	Op amp (wide band)	
CA3014	TV sound IF amplifier and detector	
CA3018/A	Four-transistor array	2-11
CA3019	Quad diode array	
CA3020/A	Audio/video power amplifier (1 W)	
CA3026	Dual RF/IF differential amplifier array	2-11
CA3028/A/B	Differential cascode amplifier (to 120 MHz)	
CA3029/30	Op amp (wide band)	
CA3030	Op amp (wide band)	
CA3033/A	Op amp (high output current)	
CA3035	Triple wide-band amplifier	
CA3036	Dual-darlington array	
CA3037/38	Op amp (wide band)	
CA3039	Six-diode array	
CA3043	FM–IF amplifier limiter	8-3
CA3045/6	Five-transistor array	2-11
CA3047/A	Op amp (high output current)	
CA3048	Quad amplifier array	2-15
CA3049	Dual differential amplifier array	
CA3050	Dual-darlington differential amplifier array	
CA3053	Differential cascode amplifier (to 10.7 MHz)	
CA3054	Dual differential amplifier array	
CA3059	Zero-voltage switch	9-9
CA3060	Triple operation transconductance amp (OTA)	12-10
CA3062	Photodetector and power amplifier	
CA3064	TV (automatic fine-tuning control)	8-6
CA3065	TV/FM (sound IF system)	

Manufacturers' Type Number	*Description*	*Section in Text*
CA3066	TV (chroma amplifier)	8-6
CA3067	TV (chroma-demodulator system)	8-6
CA3068	TV (Video IF, also AM–IF)	8-6
CA3070	Chroma signal processor (PLL)	
CA3071	Chroma amplifier	
CA3072	Chroma demodulator (see also μA746)	
CA3075	FM radio system	
CA3078	Op amp (micropower-programmable OTA)	11-8
CA3080/A	Op amp (micropower-programmable OTA)	11-8, 12-12
CA3081	Seven-transistor common-emitter driver array	2-11
CA3082	Common-collector array	2-11
CA3083	High-current (five-transistor) array	2-11
CA3084	Transistor array (PNP)	2-11
CA3085	Regulator (*similar to LM105*)	
CA3086	Five-transistor array	See CA3045/6
CA3088	AM-radio (subsystem)	
CA3089	FM-radio (IF subsystem)	
CA3091	Analog multiplier	
CA3093	Transistor-Zener diode array	
CA3094/A	High-current output power switch (programmable)	7-2, 9-8, 12-2
CA3095/A	Array (super beta)	
CA3096/A	NPN-PNP array	
CA3100	Wide-band op amp	11-7
CA3102	Dual high-frequency differential amplifier	
CA3118/A	Four-transistor high-voltage array	2-11
CA3120	TV signal processor	
CA3123/E	AM-radio (subsystem)	
CA3146/A	Five-transistor high-voltage array	2-11
CA3183/A	High-current (five-transistor) array	2-11
CA3401/E	Quad op amp (single supply)	See MC3401/P
CA*3458*	Dual MC1*741* (dual type 741)	See MC*1558*/*1458*
CA3*541*	Similar to sense amplifier MC1*541*/*1441*	
CA3*558*	Dual type 741 (similar to MC1*558*)	
CA3600/E	Linear COS/MOS array	

Manufacturers' Type Number	*Description*	*Section in Text*
CA3*741*	Op amp (internally compensated)	See μA*741*
CA3*747*	Dual 741 general-purpose op amp	See μA*747*
CA3*748*	Type 748 (noncompensated, general purpose)	See μA*748*
CA6078	Low-noise version of CA3078 (micropower OTA)	
CA6*741*	Op amp (special 741)	
CD4007/A	COS/MOS for D/A conversion	10-7
DAC372-12	Hybrid digital-to-analog converter	10-6
EAA0015	Audio power amplifier (15 W) (Electronic Associates)	7-7
HA2000	Op amp (FET input, unity gain)	
HA2050	Op amp (high slew rate)	
HA2060	Op amp (FET input, wide band)	
HA2400/2405	Quad op amp (gate controlled)	12-10
HA2500	Op amp (high slew rate, internally compensated)	
HA2800/2820	Phase-locked loop (PLL) to 25 MHz	12-11
HA2825	Phase-locked loop (PLL) 0.1 Hz to 3 MHz	12-11
HA2900/4/5	Op amp (chopper-stabilized)	11-5
HC1000	High-power (100 W) audio power amplifier (RCA)	7-7
HC3000	Hybrid power dual darlington (RCA)	
(HEP)590	Op amp (high frequency)	Equiv. to MC1550
(HEP)C6002	See MFC8040 (preamplifier) (Motorola)	
(HEP)C6004	Audio power amplifier (1 W)	See MFC8010
(HEP)C6005	Audio power amplifier (2 W)	See MFC9020
(HEP)C6053	Op amp (high slew rate)	Equiv. to MC1539
(HEP)C6065	See MC1303 (dual stereo preamplifier)	
ICL*101*	Op amp (extended bandwidth)	See LM*101*/A
ICL*741* HS	Op amp (741, higher slew rate)	
ICL*748*	Op amp (extended bandwidth)	See μA*748*
ICL8001	Precision comparator	10-4
ICL8007	Op amp (FET input, precision)	12-7
ICL8021	Op amp (micropower, programmable)	
ICL8038/A	Waveform generator (VCO)	12-11
ICL8048	Log amplifier (monolithic)	12-7
ICL8*101*	Op amp (extended bandwidth)	See LM*101*/A

Manufacturers' Type Number	*Description*	*Section in Text*
ICL*8741*	Op amp (internally compensated)	See μA*741*
IH5009	Analog gate switch (15 volts)	
IH5010	Analog gate switch (5 volts)	
LA*709*	Op amp (external compensation)	See μA*709*
L144	Op amp (low power) (Siliconix)	
LH0001/A	Op amp (micropower)	11-7
LH0023	Sample and hold	12-9
LH0043	Sample and hold	12-9
LH0052	Op amp (FET input, electrometer)	12-8
LH101	Internally compensated, type 741	12-4
LH*740*/A	Op amp (FET input)	See μA*740*
LM100	Regulator (positive voltage)	
LM*101*/A	Op amp (extended bandwidth)	4-6, 5-8, 6-8
LM*102*	Op amp (voltage follower)	12-9
LM104	Regulator (negative voltage)	
LM105	Regulator (positive voltage)	
LM*106*	Voltage comparator (high speed)	10-4
LM*107*	Op amp (similar to type 741; internally compensated)	12-9, 12-11, 12-12
LM*108*/A	Op amp (super beta)	1-7, 4-6, 6-4, 11-2
LM*109*	Regulator (5 V)	9-8
LM*110*	Voltage follower (high slew rate)	6-9, 12-5
LM111	Voltage comparator	
LM112	Op amp (micropower)	
LM*118*	Op amp (high slew rate)	11-7
LM119	Dual comparator (precision high speed)	
LM121/A	Op amp (low-drift preamp)	
LM170	AGC/squelch amplifier	
LM171	RF/IF amplifier	
LM172/372	AM–IF strip	8-5
LM2*XX*	See corresponding LM1*XX* nos.	
LM3*XX*	See corresponding LM1*XX* nos.	
LM311	Voltage comparator (precision)	
LM339	Quad comparator	10-4
LM373	AM/FM SSB-IF strip	
LM376	Regulator (positive voltage)	9-8
LM377	Dual audio power amplifier (2 W/channel)	7-5
LM378	Dual audio power amplifier	7-6

Manufacturers' Type Number	*Description*	*Section in Text*
	(4 W/channel)	
LM380	Audio amplifier (2 W)	7-5
LM381/382	Dual stereo preamplifier	7-6
LM383	Audio power amplifier (5 W)	7-6
LM*565*	Phase-locked loop (PLL)	See SE/NE565
LM*709*	Op amp	See μA*709*
LM*710*	Voltage comparator	See μA*710*
LM*711*	Dual voltage comparator	See μA*711*
LM*723*	Voltage regulator	See μA*723*
LM*741*	Op amp (internally compensated)	See μA*741*
LM*747*	Dual op amp (dual 741 type)	See μA*747*
LM*748*	Op amp (extended bandwidth)	See μA*748*
LM*1303*	Dual stereo preamplifier	See MC*1303*
LM*1558*	Dual op amp	See MC*1558*
LM*2111*	FM–IF amplifier (limiter/detector)	See CA2111/A
LM*3028*/A/B	Differential cascode amplifier	See CA3028/A/B
LM*3064*	TV AFT subsystem	See CA3064
LM*3900*	Quad op amp	12-10
LM*4250*	Op amp (programmable)	
LM*7524*	Dual sense amplifier	See SN*7524*
LM*7525*	Dual sense amplifier	See SN*7525*
LP1000	"Miser" LED flasher	
LP2000	Microtransmitter subsystem	
LS170	Audio VOX/squelch amplifier	See LM170
LS*171*	RF/IF amplifier	See LM171
MC*1303*	Dual stereo preamplifier	7-7
MC1304	FM radio (stereo demodulator)	
MC1305	Stereo FM multiplex decoder	8-4
MC1306	Audio power amplifier ($\frac{1}{2}$ W)	7-5
MC1310	FM radio (stereo decoder)	
MC1312/1313	Quadraphonic decoder (SQ matrix)	
MC1326/1328	TV (chroma demodulator)	See CA3072
MC1345	TV (sync-separator, etc., "Jungle")	
MC1351	TV sound system	
MC1357	Radio and TV (video IF, also AM-IF)	See CA2111/A
MC1358	TV (sound IF)	See CA3065
MC1364	FM radio (AFT)	
MC1398	TV (chroma-signal processor)	
MC14*XX*	For all MC14*XX*, except those listed below, see MC15*XX*	

Manufacturers' Type Number	*Description*	*Section in Text*
MC1445/1545	Differential output, two-channel gated op amp	12-10
MC1488	Quad line driver	
MC1489	Quad line receiver	
MC1514	Dual differential comparator	10-4
MC1530	Op amp (general puspose)	
MC1536	Op amp (internally compensated)	
MC1537	Op amp (dual MC1709)	
MC1538	Power booster	7-3
MC1539	Op amp (fast slew rate)	
MC1540	Core memory sense amplifier	
MC1541	Sense amp (dual channel)	
MC1545	Wideband amplifier. Gate control	12-10
MC1550	AM-IF amplifier	8-5
MC1553	Video amplifier	
MC1554	Power amplifier (1 W)	7-5
MC*1556*	Op amp (internally compensated)	11-2, 11-4
MC*1558*	Dual MC1*741*	See μA*747*
MC1563	Voltage regulator (negative)	
MC1566	Regulator (voltage + current)	9-8
MC1568	Voltage regulator (dual polarity)	9-8
MC1569	Voltage regulator (positive)	
MC*1595*	Four-quadrant analog multiplier	8-7
MC*1596*	Double-balanced mixer/modulator	8-7, 12-12
MC1648	Voltage-controlled oscillator (200 MHz)	
MC1658	Voltage-controlled oscillator (125 MHz)	
MC*1709*	Op amp (general purpose)	See μA*709*
MC1*710*	Differential voltage comparator	See μA*710*
MC1*711*	Dual voltage comparator	See μA*711*
MC1*723*	Voltage regulator	See μA*723*
MC1*733*	Differential video amplifier	See μA*733*
MC1*741*	Op amp (internally compensated)	See μA*741*
MC1*748*	Op amp (extended bandwidth)	See μA*748*
MC3301/3401	Quad op amp	12-10
MC3302/P	Quad comparator	
MC4024/4324	Voltage-controlled multi-vibrator	12-11
MC4044/4344	Phase-frequency detector	12-11
MC4344	Phase-locked loop (PLL)	12-11

Manufacturers' Type Number	*Description*	*Section in Text*
MFC4050	Audio amplifier (4 W)	
MFC6010	FM/IF amplifier	
MFC8010	Audio amplifier (1 W, including preamplifier)	7-5
MFC8020/A	Audio class B driver	
MFC8040	Preamplifier (low noise)	
MFC9020	Audio power amplifier (2 W)	7-5
MIC*730*	Differential amplifier	See μA730
μA*702*	Op amp (wide band)	
μA*703*	Op amp (FM–RF/IF)	8-2
μA*706*	Audio power amplifier (monolithic, 5 W)	7-6
μA*709*	Op amp (general purpose)	4-6, 11-7
μA*710*	Differential voltage comparator (general purpose)	10-3
μA*711*	Dual voltage comparator (general purpose)	10-4, 10-5
μA715	Op amp (high slew rate)	11-7
μA720	AM radio (subsystem)	
μA*723*	Precision voltage regulator	9-5
μA*725*	Op amp (precision)	6-7
μA727	Temperature controlled preamplifier	
μA729	FM stereo MPX decoder	
μA*730*	Differential amplifier	
μA732	FM stereo MPX decoder	
μA*733*	Differential video amplifier	
μA734	Voltage comparator (precision)	10-4, 10-6
μA735	Op amp (micropower)	
μA739	Op amp (dual, low noise)	
μA*740*	FET-input op amp	4-6, 11-2, 11-4
μA*741*	Op amp (internally compensated)	1-7, 4-6
μA742	Zero-crossing ac trigger	
μA745	Dual ac amplifier	
μA746	Color TV chroma demodulator (see also CA3072)	
μA*747*	Dual μA*741*	4-6, 12-6
μA*748*	Op amp (extended bandwidth)	4-6, 5-5
μA749	Dual preamplifier	
μA750	Dual comparator (subsystem)	10-4
μA753	FM gain block	

Manufacturers' Type Number	*Description*	*Section in Text*
μA754	TV (sound IF) TV/FM sound system	
μA757	AGC-IF amplifier	
μA758	Stereo decoder (PLL)	
μA*760*	Differential comparator (high speed)	4-6, 5-8, 6-8, 11-4
μA761	Two-channel sense amplifier	
μA767	Stereo decoder	
μA768	Stereo decoder	
μA769	Stereo decoder	
μA771	Instrumentation Amplifier	11-10
μA772	Op amp (high slew-rate)	11-7
μA*776*	Op amp (programmable, micropower)	11-8
μA*777*	Op amp (precision)	11-3
μA780	Chroma signal processor (PLL) (see also CA3070)	
μA781	TV chroma amplifier (see also CA3071)	
μA782	TV (chroma processing)	
μA*791*	Op amp (high current output, 1 A	7-2
μA795	Four-quadrant multiplier	
μA796	Doubly balanced modulator/ demodulator	
μA78*XX*	Regulator, positive (Example: 7805 = 5 V up to 7828 = 28 V, preset output)	
Mono CMP-02	Precision comparator (Precision Monolithics)	10-4
N/S5*101*	Op amp (extended bandwidth)	See LM*101*/A
N/S5*108*	Op amp (super beta)	See LM*108*/A
N/S5*556*	Op amp (internally compensated)	See MC*1556*
N/S5*558*	Dual op amp	See MC*1558*
N/S5*595*	Linear four-quadrant multiplier	See MC*1595*
N/S5*596*	Balanced modulator/demodulator	See MC*1596*
N/S5723	Precision voltage regulator	See μA723
N/S5*733*	Differential video amplifier	See μA*733*
N/S5*741*	Op amp (internally compensated)	See μA*741*
N/S5*747*	Dual 741 op amp	See μA*747*
N/S5*748*	Op amp (extended bandwidth)	See μA*748*
N/S*7524*	Dual sense amplifier	See SN*7524*
N/S*7525*	Dual sense amplifier	See SN*7525*
	NE/SE	(See SE/NE)

Manufacturers' Type Number	*Description*	*Section in Text*
PA234	Audio amplifier (1 W) (discontinued by G.E.)	
PA*237*	Audio amplifier (2 W) (discontinued by G.E.)	7-5
PA*239*	Dual stereo preamplifier (discontinued by G.E.)	7-7
PA263	Audio amplifier (3.5 W) (discontinued by G.E.)	
QC*741*	Op amp (internally compensated)	See μA*741*
RC*105/A*	Voltage regulator (precision positive)	See LM105/A
RC*108/A*	Op amp (super beta)	See LM*108*/A
RC*109*	Voltage regulator (+5 V)	See LM*109*
RC*111*	Voltage comparator (precision)	See LM111
RC*112*	Op amp (super beta, internally compensated)	See LM112
RC*118*	Op amp (high slew rate)	See LM*118*
RC*723*	Voltage regulator (precision)	See μA*723*
RC*748*	Op amp (extended bandwidth)	See μA*748*
RC*1556/A*	Op amp (internally compensated, super beta)	See MC*1556*
RC*4136*	Quad op amp (quad 741)	See LM*3900*
RC4739	Dual-channel preamplifier (stereo, low noise)	
RC*7524*	Dual sense amplifier	See SN*7524*
RM*101/A*	Op amp (extended bandwidth)	See LM*101*/A
RM4131	Op amp (precision)	
RM4132	Op amp (micropower)	
RM4*531*	Op amp (high slew rate)	See SE/NE531
RM4*558*	Dual MC1741	See MC1558/1458
SXXXX	See N/S for Signetics SXXXX	
SE/NE259	Comparator (precision)	10-4
SE/NE516	Op amp (differential input and output)	
SE/NE526	Comparator (analog voltage)	
SE/NE528	Four-channel plated-memory sense amplifier	
SE/NE533	Op amp (micropower)	
SE/NE536	Op amp (FET input)	
SE/NE537	Op amp (precision)	
SE/NE540	Power driver (1 W)	

Manufacturers' Type Number	*Description*	*Section in Text*
SE/NE550	Voltage regulator (precision)	
SE/NE555	Timer	
SE/NE556/A	Dual timer	Dual *555*
SE/NE562	Phase-locked loop (PLL)	12-11
SE/NE565	Phase-locked loop (PLL)	8-8
SE/NE566	Voltage controlled Oscillator (VCO)	12-11
SE/NE567	Tone decoder (PLL)	8-8
SG*101*	Op amp (extended bandwidth)	See LM*101*/A
SG109	Regulator (5 V)	See LM*109*
SG*748*	Op amp (extended bandwidth)	See μA*748*
SG3818	Four-transistor array	See CA3018/A
SN*7524*	Dual sense amplifier	10-5
SN*7525*	Dual sense amplifier	10-5
SN14097	16-diode array	
SN21886	Array, quad (100 mA) NPN transistor	
SN52*101/A*	Op amp (extended bandwidth)	See LM*101*/A
SN52*108/A*	Op amp (super beta)	See LM *108*/A
SN52*558*	Dual MC1*741*	See MC*1558*
SN52*660*	Op amp (precision)	4-6, 6-8
SN52*709/A*	Op amp (general purpose)	See μA709
SN52*733*	Differential video amplifier	See μA*733*
SN52*741*	Op amp (internally compensated, 741 type)	1-7
SN52*747*	Dual μA*741*	See μA*747*
SN52*748*	Op amp (extended bandwidth)	See μA*748*
SN52*770*	Op amp (super beta)	4-6, 6-8, 11-3
SN52*771*	Op amp (super beta, internally compensated)	11-3
SN52*777*	Op amp (precision)	See μA*777*
SN52800	Op amp (high performance)	
(For SN72*XXX*, see also SN52*XXX*.)		
SN72*301*	Op amp (extended bandwidth)	See LM*101*/A
SN72*308*	Op amp (super beta)	See LM*108*/A
SN72*711*	Dual voltage comparator	See μA*711*
SN76011	Audio power amplifier (1 W)	7-5
SN76024	Audio power amplifier (4 W)	7-6
SN76242	Chroma signal processor (PLL)	See CA3070
SN76243	Chroma amplifier	See CA3071
SN76266	TV (chroma amplifier)	See CA3066

Manufacturers' Type Number	*Description*	*Section in Text*
SN76267	TV chroma-demodulator system	See CA3067
SN76564	TV (automatic) fine-tuning control	See CA3064
SN76665	TV/FM (sound IF system)	See CA3065
SSS*741*	Op amp (internally compensated)	See μA*741*
TOA17/2 *709*	Op amp (general purpose)	See μA*709*
UC4250	Op amp (micropower)	
UC4*741*	Op amp (internally compensated)	See μA*741*
ULN2026	Dual RF/IF differential amplifier array	See CA3026
ULN2031/A	Seven darlington-pair array	
ULN2046	Five-transistor array	See CA3046
ULN2081	Seven-transistor driver array	See CA3081
ULN2111/A	Quadrature FM detector	8-7
ULN2114/A	Chroma demodulator	See CA3072
ULN2124/A	Chroma signal processor (PLL)	See CA3070
ULN2127/A	Chroma amplifier	See CA3071
ULN2275	Dual 1-W amplifier	
ULN2276	Dual 4-W amplifier	
ULN2277	Dual 2-W amplifier	
ULN/S2175	Op amp (super beta)	See LM*108*/A
ULN/S2*741*	Op amp (internally compensated)	See μA*741*
ULS2045	Five-transistor array	See CA3045
ULS2171	Op amp (internally compensated)	See SN52*770*/*771*
ULS2173/77	Op amp (internally compensated)	
ULX2285	Audio amplifier (5 W)	7-6
WB-23	Op amp (high slew rate, modular) (Data Devices)	11-7
XR205	Wave-form generator	12-11
XR215	Phase-locked loop (PLL) system	
XR2207	Voltage-controlled oscillator (VCO)	12-11
XR2208/2308	Operational multiplier	12-12
141/142	Op amp (externally compensated) (Teledyne/Philbrick)	See LM*101*/A
234	Op amp, chopper-stabilized modulator (Analog Devices)	11-5
881	Universal active filter, hybrid (Beckman/Helipot)	12-6
1028	Op amp (general purpose) module (Teledyne/Philbrick)	
1029	Op amp (low bias current FET)	

Manufacturers' Type Number	*Description*	*Section in Text*
	(Teledyne/Philbrick)	
1319	Op amp (low drift) (Teledyne/Philbrick)	
1321	Op amp (wide band) (Teledyne/Philbrick)	
1322	Op amp (high slew rate) (Teledyne/Philbrick)	
1323	Op amp (micropower) (Teledyne/Philbrick)	
1324	Op amp (fast settling) (Teledyne/Philbrick)	
1404	Op amp (micropower) (Teledyne/Philbrick)	
1412	Op amp ("minichopper") module (Teledyne/Philbrick)	
1426	Op amp, FET-input (Teledyne/Philbrick)	11-4
1702	Parametric amplifier (full differential) (Teledyne/Philbrick)	
1703	Op amp (chopper) module (Teledyne/Philbrick)	
4253	Instrumentation amplifier (modular FET input) (Teledyne/ Philbrick)	11-11
4450	Four-quadrant multiplier (module) (Teledyne/Philbrick)	12-12
4452	Analog multiplier (module) (Teledyne/Philbrick)	12-12

APPENDIX IV

MANUFACTURERS' ADDRESSES

Advanced Micro Devices, Inc., 901 Thompson Place, Sunnyvale, Calif. 94086.

Analog Devices, Inc., Route 1, Industrial Park, P.O. Box 280, Norwood, Mass. 02062.

Beckman/Helipot, 2500 Harbor Blvd., Fullerton, Calif. 92634.

Burr–Brown Research Corp., P.O. Box 11400, International Airport, Industrial Park, Tucson, Ariz. 85706.

Data Devices Corp., Dept. 17, 100 Tec St., Hicksville, N.Y. 11801.

Datel Systems, 1020 Turnpike St., Canton, Mass. 02021.

Electronic Associates, Inc., 185 Monmouth Pkwy., West Long Branch, N.J. 07764.

Exar Integrated Systems, 750 Palomar Ave., Sunnyvale, Calif. 94086.

Fairchild Semiconductor, 464 Ellis St., Mountain View, Calif. 94040.

Ferranti Electric Co., East Beth Page Rd., Plainview, L.I., N.Y. 11803.

General Electric (no longer manufactures ICs).

Harris Semiconductor, P.O. Box 883, Melbourne, Fl. 32901.

Hybrid Systems Corp., 95 Terrace Hall Ave., Burlington, Mass. 01803.

Intech, Inc., 1220 Coleman Ave., Santa Clara, Calif. 95050.

Intersil, Inc., 10900 N. Tantau Ave., Cupertino, Calif. 95014.

ITT Semiconductor, West Palm Beach, Fla. 33047.

Lithic Systems, Inc., 10010 Imperial Ave., P.O. Box 869, Cupertino, Calif. 95014.

Motorola Semiconductor Products, Inc., P.O. Box 20924, Phoenix, Ariz. 85036.

National Semiconductor, Microcircuits Div., 2900 Semiconductor Drive, Santa Clara, Calif. 95051.

Precision Monolithics, Inc., 1500 Space Park Drive, Santa Clara, Calif. 95050.

Raytheon Company, 350 Ellis St., Mountain View, Calif. 94040.

RCA Corporation, Solid State Division, Route 202, Somerville, N.J. 08876.

Signetics Corporation, 811 E. Arques Ave., Sunnyvale, Calif. 94086.

Silicon General, Inc., 7382 Bolsa Ave., Westminster, Calif. 92683.

Siliconix, Inc., 2201 Laurelwood Road, Santa Clara, Calif. 95054.

Solitron Devices, Inc., 256 Oak Tree Road, Tappan, N.Y. 10983.

Sprague Electric Company, North Adams, Mass. 01247.

Teledyne/Philbrick, Allied Drive at Route 128, Dedham, Mass. 02026.

Texas Instruments, Inc., Semiconductor Components Group, P.O. Box 5012, Dallas, Tex. 75222.

INDEX